Analysis and Experiment for Electric Power Quality

Analysis and Experiment for Electric Power Quality

Guest Editor

Gabriel Nicolae Popa

MDPI

Basel • Beijing • Wuhan • Barcelona • Belgrade • Novi Sad • Cluj • Manchester

Guest Editor
Gabriel Nicolae Popa
Electrical Engineering and
Industrial Informatics
Politehnica University
Timisoara
Hunedoara
Romania

Editorial Office
MDPI AG
Grosspeteranlage 5
4052 Basel, Switzerland

This is a reprint of the Special Issue, published open access by the journal *Energies* (ISSN 1996-1073), freely accessible at: www.mdpi.com/journal/energies/special_issues/A_EEPQ.

For citation purposes, cite each article independently as indicated on the article page online and as indicated below:

Lastname, A.A.; Lastname, B.B. Article Title. *Journal Name* **Year**, *Volume Number*, Page Range.

ISBN 978-3-7258-3912-4 (Hbk)
ISBN 978-3-7258-3911-7 (PDF)
https://doi.org/10.3390/books978-3-7258-3911-7

Contents

energies

MDPI

Editorial

Issues in Power Quality

Gabriel Nicolae Popa

Department of Electrical Engineering and Industrial Informatics, Politehnica University of Timișoara, 5 Revolution Street, 331128 Hunedoara, Romania; gabriel.popa@fih.upt.ro; Tel.: +40-254-207-541

Power quality generally refers to a series of boundary conditions that allow electrical systems connected to the network to operate in the expected way without causing significant performance or life losses. Thus, the operation of power systems outside these borders has a direct impact on the overall economic performance of the whole system. The disturbances responsible for this degradation are classified as conducted low-frequency, radiated low-frequency, conducted high-frequency, radiated high-frequency, electrostatic discharge, and nuclear electromagnetic phenomena [1].

Power quality deterioration can cause problems with or the shutdown of processes and equipment. The consequences range from excessive energy costs to complete work stoppages. The interdependence of different systems increases the complexity of power quality issues. Some of the problems lie inside the facility: installation—inadequate grounding, inadvertent routing, or undersized distribution; operation—equipment operated outside the design parameters; mitigation—inadequate protection or lack of power factor correction; maintenance—deteriorating cable insulation or ground connection [2]. Even the equipment that is perfectly installed and maintained in a perfectly designed facility can cause problems in terms of power quality with age.

The effects of power quality problems can consist of supply voltage waveform distortion, deviation from its nominal value, or a complete rupture. The problems of power quality can last from milliseconds to hours [3]. Various power electronics, such as domestic, industrial, and office equipment, connected to power supplies may have non-linear load characteristics that lead to poor power quality. Equipment such as photocopiers, computers, printers, etc., can cause electrical disturbances that can destroy certain sensitive equipment. When connected to the same source of supply, in some cases, they may cause malfunction. Industrial motors powered by electronic converters produce electrical disturbances. When disturbances occur, the quality of electricity is poor and production losses occur, with resulting financial losses. The main effects of voltage failure include early equipment failures, cost-effectiveness loss in rotating machines, equipment failures in information technology, loss of data or stability, process interruptions, failures of measuring and control devices, etc.

Electromagnetic disturbances that directly affect the electrical network, and, thus, the power quality of consumers, are of low frequency and include the following: frequency and voltage variations in the electrical voltage; voltage dips and interruptions; harmonic distortions; flickers; asymmetries; transients; DC components on AC voltage; signal voltage disturbances; low-frequency voltages [4–6].

There are many engineering solutions used to reduce the impact of power quality problems, and this is a very active area of innovation and development. The articles presented in this Special Issue, "Power Quality Analysis and Experiments", briefly discuss issues of power quality and their solutions in low- and medium-voltage applications [7].

Received: 7 February 2025
Accepted: 18 March 2025
Published: 8 April 2025

Citation: Popa, G.N. Issues in Power Quality. *Energies* **2025**, *18*, 1874. https://doi.org/10.3390/en18081874

Most electrical consumers are single-phase and deforming (contain switch mode power supplies), and they are found especially in domestic, residential, and industrial fields. It is a challenge to balance single-phase consumers in three-phase circuits with the possibility of obtaining a high-power factor at the level of low-voltage power substations [8].

Power quality can be evaluated through characteristic indices. In [9], a new methodology is proposed for determining the indices of voltage variation in a short period of time; to implement the new method, real data from more distribution system companies are used. This analysis is useful for introducing short-term voltage variation indices.

High-power industrial applications include offshore oil and gas platforms, where current distortion is high and power factors are low. To improve these power quality indices, the use of active power filters (made with silicon and silicon carbide) was proposed in [10]; compensation strategies were presented, at two different voltage levels, to identify the best solutions for improving the deformation regime and increasing the power factor.

Also, active power filters have medium-voltage applications for improving power quality. The decision tree method (as an optimal solution) for choosing and dimensioning active power filters was proposed to minimize energy losses and investment costs in [11].

Residual current devices from low-voltage networks have, as their main application, preventing electric shocks, with direct contact considered an additional protection. Power electronics that use the pulse width modulation technique can determine ground fault currents that are not detected via residual current devices. The limitations of the standards in force are identified and proposals for their improvement are made in [12].

Capacitor banks can be used in low-voltage power substation networks, in addition for improving the power factor and creating passive power filters to reduce the deforming regime. Because the capacitors are very demanding in such applications, special attention must be paid to the dielectric to increase their lifetime [13]. Also, it is important to choose the proper configuration and the values of passive filter components to obtain the optimal performance of the passive filter [14].

Non-linear autoregressive-type recurrent artificial neural networks with external input control can estimate harmonic behavior in photovoltaic systems [15]. Dynamic control and harmonic distortion reduction can, thus, be evaluated.

The continuous insurance of electricity supply is a field of wide interest in power quality. It can be used for uninterruptible power sources powered by vanadium redox flow batteries to provide long-term electricity in critical installations [16].

Matrix converters have applications in hybrid power filters because they have low total harmonic distortion for the fundamental voltage. They can be controlled via a fuzzy system or PI regulator in terms of compensation speed, accuracy, total harmonic distortion of supply current, and overall integrity [17].

The power factor regulators installed in low-voltage power substations that measure the current on one phase and the voltage on the other phases have important limitations for the conditions of distorting currents and imbalances in three-phase networks [18]. An analysis of a low-cost system for the automatic regulation of the power factor, with a reduction in transients and an increase in the lives of contacts, can be used in low-voltage power substations with capacitor banks controlled by one three-phase solid-state relay; such as system is presented in [19].

Voltage dips and short-term voltage interruptions have major implications both in low- and medium-voltage networks. In [20], an analysis of these indices was carried out for several companies from different industries to evaluate the understanding of these power quality issues.

Short-term voltage dips are usually caused by failure, high-speed electrical large loads, or intermittently loose electrical connections in the power line. Voltage fluctuations are usually related to system problems, but they also occur when heavy loads are switched on or a large motor is initiated [1].

Voltage fluctuations are frequent and widespread, and they can be seen as the first power quality phenomenon affecting the industry. Interruptions are the result of electrical system failures, equipment failures, and control errors. The duration of interruption due to equipment failures or disconnected connections may be irregular. The most common problem associated with short-duration RMS variations is equipment shutdown. Short-term interruptions can lead to process shutdowns that require hours of restarting. In many facilities, if the equipment travels, the effects on the process are the same for short-term variations and long-term phenomena [21]. Short-term dips cause many process interruptions. In addition, many control and emergency switch circuits use relays and contactors that are very sensitive to voltage dips. The common solution to this problem is to provide a constant voltage. Momentary and temporary interruptions almost always cause the equipment to stop operating and may lead to the failure of the inductive motors.

Voltage problems and the production of a balanced current are the two main areas in which power quality problems occur. Dips and swells, voltage transients, power interruptions and voltage imbalances can be monitored, analyzed, and compared with the device operation history to determine the cause and severity of the problem of power quality. The same can be undertaken with different harmonic currents of a system. It is also important to note that the power quality problems are often inter-related. Power quality problems must be addressed through a whole-plant approach, without losing focus on how they affect individual loads; sometimes fixing one energy quality problem can exacerbate another problem. By using three-phase power quality analyzers, we can identify the root causes of power quality problems and correct both symptoms and overall problems [22].

Voltage spikes and swells cause damage to electronic components, the flammability of insulation materials, excessive screen brightness, the damage or interruption of sensitive equipment, data processing errors or data losses, and electromagnetic interference [8]. Harmonics causes power consumption and leads to inefficient electricity use and unintentional equipment malfunctions. It affects the smooth operation of industrial machines and causes production interruption. In hospitals, it can lead to loss of life. It affects the data processing activities of information technology equipment, such as losing transactions in real time, etc.

The overheating of wiring (in particular, neutral conductors in three-phase systems) and equipment can be caused by low power quality. When communications cables are in parallel with power cables, the harmonic frequency interferes with communication signals, resulting in incorrect signals. Harmonics can cause the protective relay to be incorrectly operated.

The economic costs of power quality problems are high, especially in industry. Costs include production losses, damaged equipment, salaries, and restart costs. These costs can be quantified as additional money that the customer wishes to pay to avoid this inconvenience [23].

The business risks posed by power quality issues are real and even low-tech industries are exposed to serious financial losses. On the other hand, prevention is relatively cheap, with solutions ranging from simple good practice design techniques to installing widely available support equipment. Energy quality problems cost the about EUR 10 billion per year, whereas preventive measures cost less than 5% of that figure. Understanding the nature of a problem and assessing how it affects the business, as well as the risk associated with it, is vital [24]. The reliability and coherence of electricity supply is important for many industrial and service activities. If power quality is low, the company suffers. It is surprising and disturbing that companies often do not recognize the causes of low reliability, even though cost-effective solutions are available to them [25].

In the future, the main approaches used in power quality analysis should be related to development of new types of batteries for use with uninterruptible power supplies, the use of new types of controls for hybrid and active power filters, studies on limiting passive filters operations, the realization of switched-mode power supplies that reduce current distortion, the use of new statistical methods for power quality analysis, the judicious design of electrical installations in deforming regimes (especially for neutral conductor), the analysis of deforming regime effects on switching and protection equipment and on electrical insulation, methods for reducing flicker and transient regimes for high-power consumers, and the realization of load balancing methods in three-phase systems.

Conflicts of Interest: The author declares no conflict of interest.

References

1. Rodríguez, A.; Bueno, E.J.; Mayor, A.; Sanchez, F.J.R.; García-Cerrada, A. Voltage support provided by STATCOM in unbalanced power systems. *Energies* **2014**, *7*, 1003–1026. [CrossRef]
2. *The Costs of Poor Power Quality*; Application Note; Fluke: Everett, WA, USA, 2012. Available online: https://docs.rs-online.com/f015/0900766b815556ba.pdf (accessed on 6 February 2025).
3. Pakere, I.; Lauka, D.; Blumberga, D. Does the balance exist between cost efficiency of different energy efficiency measures? DH systems case. *Energies* **2020**, *13*, 5151. [CrossRef]
4. *Power Quality Application Guide*; Copper Development Association: Hemel Hempstead, UK, 2001. Available online: https://pdfcoffee.com/power-quality-application-guide-pdf-free.html (accessed on 6 February 2025).
5. Shin, Y.J.; Powers, E.J.; Grady, M.; Arapostathis, A. Power quality indices for transient disturbances. *IEEE Trans. Power Deliv.* **2006**, *21*, 253–261. [CrossRef]
6. Iagar, A.; Popa, G.N.; Dinis, C.M. *Electric Power Quality—From Theory to Experiments*; Politehnica Publishing House: Timisoara, Romania, 2017. (In Romanian)
7. Popa, G.N. Electric power quality through analysis and experiment. *Energies* **2022**, *15*, 7947. [CrossRef]
8. Popa, G.N.; Iagăr, A.; Diniș, C.M. Considerations on current and voltage unbalance of nonlinear loads in residential and educational sectors. *Energies* **2021**, *14*, 102. [CrossRef]
9. Grillo de Sauza, P.V.; Maria de Carvalho Filho, J.; Ferreira, D.F.; Filho, J.M.; Filho, H.K.R.; Pereira, N.B. Cluster-based method to determine base values for short-term voltage variation indices. *Energies* **2021**, *14*, 149. [CrossRef]
10. Vitoi, L.A.; Brandao, D.; Tedeschi, E. Active power filter pre-selection tool to enhance the power quality in oil and gas platforms. *Energies* **2021**, *14*, 1024. [CrossRef]
11. Klimas, M.; Grabowski, D.; Bula, D. Application of decision trees for optimal allocation of harmonic filters in medium-voltage networks. *Energies* **2021**, *14*, 1173. [CrossRef]
12. Czapp, S.; Tariq, H. Behavior od residual current devices at frequencies up to 50 kHz. *Energies* **2021**, *14*, 1785. [CrossRef]
13. Park, B.; Lee, J.; Jang, G. Harmonic mitigation using passive harmonic filters: Case study in a steel mill power systems. *Energies* **2021**, *14*, 2278. [CrossRef]
14. Diniș, C.M.; Popa, G.N.; Cunțan, C.D.; Iagăr, A. Aspects regarding of passive filters sustainability for non-linear single-phase consumers. *Sustainability* **2024**, *16*, 2776. [CrossRef]
15. Jumilla-Corral, A.A.; Perez-Tello, C.; Campbell-Ramirez, H.E.; Medrano-Hurtado, Z.Y.; Mayorga-Ortiz, P.; Avitia, R.L. Modeling of harmonic current in electrical grids with photovoltaic power integration using a nonlinear autoregressive with external input neural network. *Energies* **2021**, *14*, 4015. [CrossRef]
16. Kuzmin, I.; Loskutov, A.; Osetrov, E.; Kurkin, A. Source for autonomous power supply system based on flow battery. *Energies* **2022**, *15*, 3027. [CrossRef]
17. Koduah, A.; Effah, F.B. Fuzzy logic controlled hybrid active filter for matrix converter input current harmonics. *Enegies* **2022**, *15*, 7640. [CrossRef]
18. Diniș, C.M.; Popa, G.N. An experimental analysis of three-phase low-voltage power factor controllers used in a deforming regime. *Energies* **2024**, *17*, 1647. [CrossRef]
19. Popa, G.N.; Diniș, C.M. Low-cost system with transient reduction for automatic power factor controller in three-phase low-voltage installations. *Energies* **2024**, *17*, 1363. [CrossRef]
20. Motoki, É.M.; Maria de Carvalho Filho, J.; Márcio da Silveira, P.; Pereira, N.B.; Grillo de Sauza, P.V. Cost of industrial process shutdowns duet o voltage sag and short interruption. *Energies* **2021**, *14*, 2874. [CrossRef]

21. Coll-Mayor, D.; Pardo, J.; P'erez-Donsi'on, M. Power Quality. Measurements and Analysis of Electromagnetic Perturbations; ENE2007-68032-C04-01/CON, Spain. *Economics of Power Quality, Technical Report*. 2010. Available online: https://www.donsion.org/investigacion/informacion-PI/economic-of-power-quality.pdf (accessed on 6 February 2025).
22. *Detecting Power Quality Issues*; Application Note. Fluke: Everett, WA, USA, 2022. Available online: https://www.fluke.com/en-ph/learn/blog/power-quality/a-fresh-look-at-power-quality-basics (accessed on 6 February 2025).
23. Johnson, D.O.; Hassan, K.A. Issues of power quality in electrical systems. *Int. J. Energy Power Eng.* **2016**, *5*, 148–154. [CrossRef]
24. Chapmann, D. *The Cost of Poor Power Quality*; Copper Development Association: McLean, VA, USA, 2001.
25. Poor power quality costs european business more than Euro 150 billion a year. In *European Power Quality Surve, Leonardo Power Quality Initiative*; Leonardo-ENERGY: Napoli, Italy, 2008. Available online: https://leonardo-energy.pl/wp-content/uploads/2017/08/Poor-power-quality-costs-european-business-more-than-150-billion-a-year.pdf (accessed on 6 February 2025).

energies

MDPI

Review

Electric Power Quality through Analysis and Experiment

Gabriel Nicolae Popa

Department of Electrical Engineering and Industrial Informatics, Politehnica University of Timișoara, 5 Revolution Street, 331128 Hunedoara, Romania; gabriel.popa@fih.upt.ro; Tel.: +40-254207541

Abstract: The quality of electrical energy is of particular importance for power engineering. This study presents an analysis of articles made in the Special Issue "Analysis and Experiments for Electric Power Quality". As techniques and technology advance, electrical consumers and equipment become more sensitive to disturbances in the electrical network (in particular, low- and medium-voltage). It can lead to costly outages and lost production, which affect productivity. The analyzed articles present interesting technical studies made on industrial and nonindustrial consumers, of low- and medium-voltage, from the point of view of the quality of electricity. Voltage and current harmonics, voltage sags and swells, interruptions, unbalance, and low power factor will lead to higher electricity bills, overloading, and rapid aging of electrical networks and electric equipment. The power quality depends not only on the supplier but also on all consumers connected to the same power network; some can cause disruptive influences in the supply network, affecting the operation of other consumers. Ensuring the power quality of industrial and nonindustrial applications is an objective difficult to achieve.

Keywords: electric consumers; industrial applications; power quality

1. Introduction

Electrical energy is used by a wide variety of consumers, from the industrial ones, which are fewer, but of high power, to the domestic ones, characterized by low power, but very numerous, used both in the urban and rural environment. Electromagnetic disturbance is any electromagnetic phenomenon that can degrade the performance of an electrical, electronic, or radio device, and a consumer, equipment, or system can adversely affect life or inert matter [1].

Electromagnetic disturbances can be classified according to several criteria [2–4]:

By frequency:

- Low-frequency disturbances (refer to signals with a frequency below 1 MHz);
- High-frequency disturbances (signals with a frequency above 1 MHz);

According to the mode of propagation:

- Disturbances conducted through the network conductors (including currents and voltage differences);
- Radiated disturbances (in the air), through electric and magnetic fields;

By duration:

- Permanent or sustained disturbances (affecting analog electronic circuits);
- Transient, random, or periodic disturbances (affecting digital electronic circuits).

The origin of electromagnetic disturbances can be both in the electrical network (e.g., incidents or wrong maneuvers, and defects) and in the consumer's electrical network. Currently, ensuring the power quality of electricity has become an increasingly complex task of major interest to both electricity suppliers and consumers [3,5,6]. The power quality concerns have been guiding the following major issues:

Citation: Popa, G.N. Electric Power Quality through Analysis and Experiment. *Energies* **2022**, *15*, 7947. https://doi.org/10.3390/en15217947

Academic Editor: Abu-Siada Ahmed

Received: 9 October 2022
Accepted: 23 October 2022
Published: 26 October 2022

- To increase the yield in the production, transport, and use of electricity, power electronics have been introduced on a large scale to control the energy conversion processes, and electronic equipment have been introduced to control the power factor;
- The complexity of energy systems and the mutual influences between them and users, as well as between consumers connected to the same power system, are constantly growing;
- The amount of nonlinear electrical equipment, generating electromagnetic disturbances, has grown at an impressive rate in recent decades;
- Modern electrical equipment is more sensitive to the decrease in the power quality of electrical energy because they include sensible electronic devices and control systems based on microprocessors/microcontrollers, whose operating characteristics are affected by electromagnetic disturbances from the electrical network;
- Consumers have become more aware and better informed about the impact that different electromagnetic disturbances have on electrical equipment and technological processes and, as a result, they ask suppliers to provide them with electricity at the contracted electrical quality parameters.

Two essential aspects must be followed when supplying electricity. On the one hand is the quality of the product (refers to the technical parameters, such as voltage amplitude, frequency, harmonics content, and symmetry of three-phase systems), and on the other hand is the quality of the service (refers to the continuity of the supply and refers to short/long interruptions and safety in supplying) [3]. The most important types of electrical consumers that cause electromagnetic disturbances are [7–12]:

- Nonlinear consumers, for example, modern household appliances, electric tools, electric arc furnaces, and electromagnetic induction furnaces, which absorb a nonsinusoidal current, whose harmonics, passing through the harmonic impedances of the electrical supply, lead to harmonic voltages on the bars;
- Unbalanced consumers, for example, electric arc welding equipment, public lighting, and interurban electric traction, which absorb currents of different amplitudes on the three phases and, passing through the upstream impedances of the electrical network, cause voltage asymmetry on the power bars;
- Consumers with variable loads produce voltage fluctuations on the power bars (for example, power mill, mechanical-processing-equipment-driven electrical, and starting large power motors; electric arc furnaces and spot welders cause flickering).

The electromagnetic disturbances that directly influence the electrical network and, therefore, the power quality of the electricity supplied to consumers, are the low-frequency conducted ones (with frequencies up to 9 kHz at the most). This category includes [3,4,6]:

- Variations in the frequency of the supply voltage;
- Variations in the supply voltage;
- Gaps and interruptions (short and long) of voltage;
- Harmonic distortion (harmonics and inter-harmonics);
- Voltage fluctuations/flicker;
- Asymmetries;
- Temporary over-voltages and transient phenomena;
- The continuous component in the applied voltage curve;
- Signaling voltages;
- Voltages induced by low-frequency.

2. A Short Review of the Contributions in Special Issue "Analysis and Experiments for Electric Power Quality"

This volume contains the successful invited and peer-reviewed submissions [13–22] to a Special Issue of Energy "Analysis and Experiments for Electric Power Quality" on the subject area of power quality.

Power quality is covered on two main subjects: the development of power quality indices and the detection, analysis, and correction of electrical disturbances.

This Special Issue of Energies, "Analysis and Experiments for Electric Power Quality", published outstanding contributions on electric power quality, in low- and medium-voltage applications:

- Harmonics;
- Blackouts;
- Under- and over-voltage;
- Sags and swells;
- Unbalance;
- Flickers.

A variety of engineering solutions are available to eliminate or reduce the effects of electric power quality problems and it is a very active area of innovation and development. The articles presented different power quality problems in power systems and had brief ideas about their solutions with comparative studies.

The articles made for the Special Issue "Analysis and Experiments for Electric Power Quality" had the following topics: unbalance [13]; short-term voltage variation indices [14]; passive filters [18]; active power filters [15,16]; hybrid active filters [22]; residual current devices at high frequency [17]; voltage sags and short interruptions [19]; photovoltaic integration using neural networks [20]; autonomous power supply [21].

Article [13] presented a study of the balancing of single-phase electrical consumers in a three-phase system, and improving the power factor in low-voltage power substations in residential and educational areas. Industrial electrical consumers are usually three-phase (with three or four wires) and high-power, with voltage and current unbalance being at a low level; consumers in the residential and educational sectors are usually single-phase, in large numbers; unbalanced voltages and, especially, currents are important. To perform the study, experiments were carried out in the laboratory and the low-voltage electrical power substation, before and after balancing the single-phase electrical consumers per phase, on workweek and weekend days. It was found that after balancing the electrical consumers by phases, the current unbalance in the three-phase system was reduced and the power factor was improved by using single-phase capacitive electrical consumers (for example, personal computers, which are in large numbers in such sectors) distributed equally on all phases.

Following a uniform distribution and balancing of the electrical consumers (e.g., single-phase consumers from classrooms, offices, and libraries) among the three phases, the measurement data were taken in the power substation of the residential and educational sectors. All of the nonlinear consumers (particularly single-phase) connected at various locations throughout the low-voltage network have the effect of deforming the voltage and current waveforms from point of common coupling (PCC).

As a result, the balanced consumers are connected to a network of unbalanced consumers, turn into active power unbalance consumers, report a higher amount of power consumption than is necessary, and consequently perform worse overall. Additionally, supply conductor losses (personal technological usage) rise. Unbalanced consumers are to blame for these losses, but the power systems support the growth.

Balance was achieved by using capacitive electrical consumers (e.g., PCs) and uniformly distributing electrical consumers among the three phases (within the technical possibilities). Additionally, we improved the power factor without using fixed capacitor banks or a power factor controller with capacitor banks connected to the PCC of the power substations, and we were able to achieve the relative balance of the current and voltage, respectively. Additionally, the unbalance of the voltage was only impacted little by the kind of electrical consumers and the amplitude of the supply voltage, although the unbalance of the current, PF, and DPF was significantly altered [13].

Determination and knowledge of short-term voltage variation indices are important for the power quality of electricity. In paper [14], a new methodology was proposed for determining indices of voltage variation over a short period. The variables that best describe the short-term voltage disturbance were established, as well as the clusters that allow more adequate definition of the basic values for the indices. To implement the new method, real data from 19 distribution systems of a national energy company were used. The study (including the proposed flowchart—Figure 1) can be useful as a basis for making regulations regarding short-term voltage variation indices and establishing clusters of electricity distribution systems.

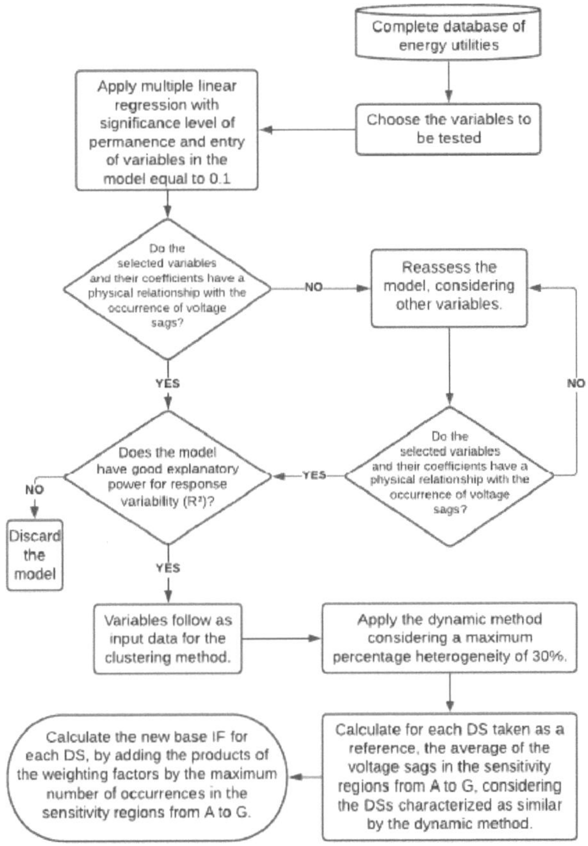

Figure 1. Flowchart of the proposed methodology [14].

Figure 1 provides an overview of the suggested methodology. Industrial users with sensitive loads suffer significant financial losses because of voltage sags. The criteria for recommending limits will vary in the future. Accordingly, it is thought that the most appropriate technique to use is to construct a unique base impact factor for each distribution system by the system performance that it most closely matches. This work, which presented a methodology for the establishment of the base impact factor that was utilized in the computation of the index that regulates voltage sags, is in line with the goals of the electrical industry in this context.

Offshore oil and gas platforms are large consumers of electricity and represent important industrial applications of power quality (Figure 2), in which power quality indices are low (especially current distortion and low power factor). In the design of article [15], a selection and evaluation tool for active power filters (power electronic devices made with silicon and, in particular, with silicon carbide, taking into account the reduced number of components, the power losses, and filter size) was proposed. For active power filters used in these industries, size and weight are critical constraints in offshore applications. At the same time, compensation strategies are presented, at two different voltage levels, to identify the best solutions for improving the deformation regime and increasing the power factor.

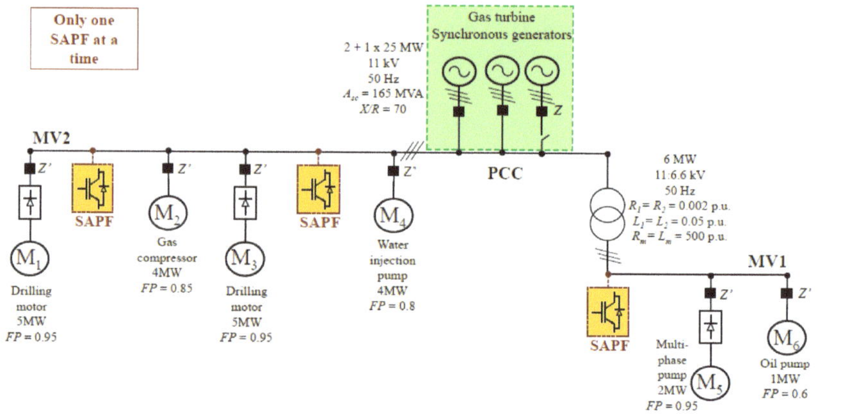

Figure 2. Diagram of a typical power grid of an offshore oil and gas platform (SAPF—Shunt Active Power Filters) [15].

In isolated power networks, such as those found on oil and gas platforms, where poor power quality necessitates reactive and harmonic compensation, this paper demonstrated that SAPF can be a workable option. Based on several factors, including the SAPF connection point, losses, passive components, power quality, and semiconductor type, a SAPF pre-selection tool was created.

The power quality of electrical energy is also important in medium-voltage electrical networks, in conditions where nonlinear electrical loads are connected to the network. Article [16] applied the decision tree method for choosing and dimensioning active power filters, which represents a method of improving the power quality. An analysis was made of the number and location of active power filters so that energy losses and investment costs are minimal, under the conditions of permanent monitoring of the total harmonic distortion of voltage in the nodes of the medium-voltage network. In these applications, the decision tree method allows the selection of the optimal solution (Figure 3).

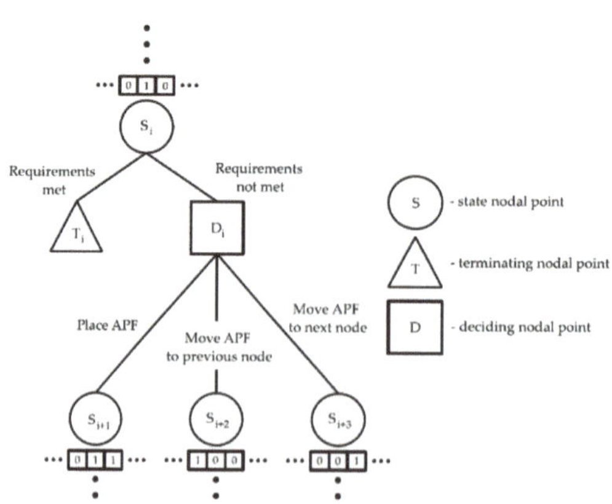

Figure 3. Schematic of recurring part of decision tree used for the optimization of active power filter (APF) placement [16].

This study presented the issue of harmonic filter allocation optimization in terms of lowering power losses and APF expenses. It was highlighted that the decision tree approach is well-known in many research fields but has not yet been used for power filter allocation in medium-voltage networks.

Residual current devices are protective devices found in almost all low-voltage networks, being common nowadays [17]. Their importance in preventing electric shocks in the case of indirect contact is known and, in the case of direct contact, can be considered as additional protection. Pulse-width-modulation-controlled power electronic converters (Figure 4) produce ground fault currents made up of high-frequency components (can be tens of kHz). The usual residual current devices are not designed to detect high-frequency currents, and they are ineffective. In article [17], an analysis of residual currents with frequencies up to 50 kHz was made on residual current devices. As expected, most residual current devices (especially F and B types) do not work under the conditions in which they are tested according to the standards. Limitations of the standards in force were identified and proposals for their improvement were made.

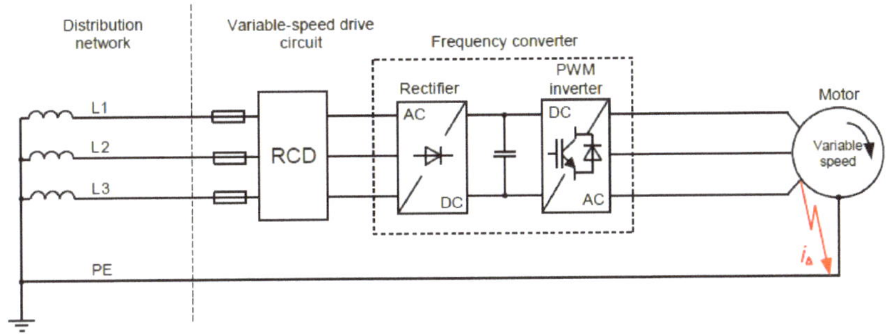

Figure 4. A variable-speed drive circuit producing residual currents of high-frequency components; RCD—residual current device; iD—residual current; PWM—pulse width modulation [17].

Utilization of residual current devices has become mandatory for industrial as well as modern domestic applications. This device's main goal is to protect users from electric shock in the event of direct or indirect contact. For such devices, exposure to residual currents with high-frequency components poses the greatest challenge. In certain situations, it is possible that the device will not trip at the anticipated level, meaning that electric shock protection may not be guaranteed.

Power electronics have a special impact on low- and high-power electric drives [18]. In many of today's electric drives, the load is variable, and the static frequency converter-motor assembly is a strongly nonlinear element with direct implications on reducing the power quality of electricity, e.g., flicker and distortion of voltage waveforms. Power capacitor batteries can be used to reduce the voltage drop, and well-sized passive filters can be used to reduce the current deforming regime. Using capacitor batteries and passive filters (Figure 5), there is the possibility that with a small mechanical load of the motor, over-voltages and electrical resonance phenomena may occur between the passive filter and the power transformer used to supply the drive. Thus, the capacitors in the application must have a performing dielectric to increase their lifetime. In article [18], an analysis and design were made regarding the harmonic filters used in such applications.

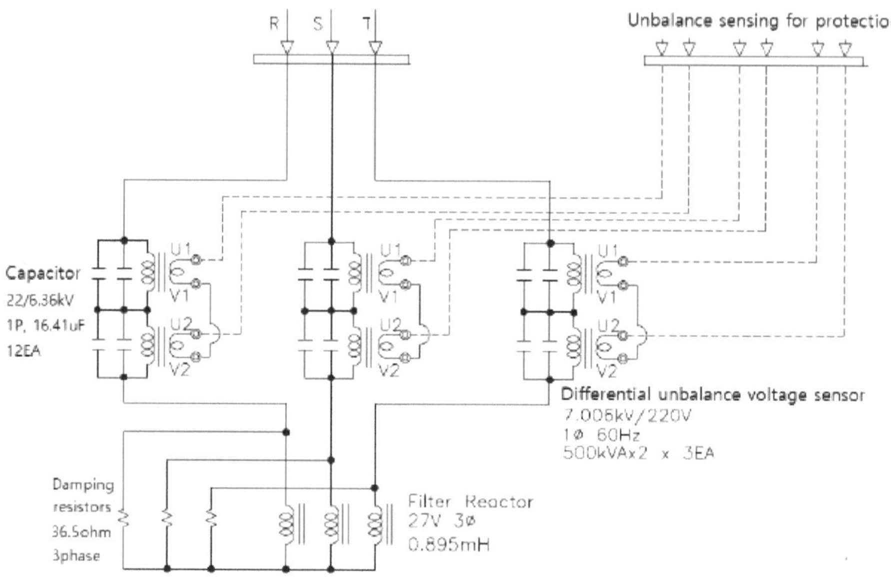

Figure 5. Single line drawing of passive filter [18].

In this study, it was proven that a passive harmonic filter system in the mill motor drive system with an ideal capacity could offer a cost-effective solution that simultaneously compensated for reactive power and absorbed harmonics. The following topics were also covered: attenuation of harmonic voltage by current divider Hid; key components of the filter capacitor and SR design; harmonic filter bank setting parameters. Voltage fluctuation characteristics and voltage harmonics were measured to confirm the harmonic filter's performance.

Voltage gaps (voltage drops below a certain value) and short voltage interruptions, from industrial applications in both low- and medium-voltage networks, are important components of the power quality of electricity that deserve to be studied more deeply. In article [19], a specific questionnaire was made for the field study, for industrial consumers (an analysis was made of 33 companies from 12 distinct types of industrial activities) connected to medium-voltage networks. The study carried out led to an important contribution to the analysis of voltage gaps and short interruptions of medium-voltage in industrial applications, which completes the knowledge in this area of power quality.

The survey results allowed for the quantification of the losses experienced by 33 small- and medium-sized businesses with an average of 349 employees, distributed across 12 different business sectors, and all connected to medium-voltage networks (11.9 kV and 13.8 kV). The average cost per incident was USD 7364.75 and the average cost per interrupted kW was USD 6.72. The objectives of this study were achieved, and it significantly benefited the electricity industry in particular. As a recommendation to continue this line of investigation, it is proposed to include more activity segments and increase the sample size of the segments now being studied (food industry, furniture, mining, stones and granites, oil). The success rate of responses to the survey form was 47.1% when considering the initial sample of 70 firms and 33 responses; however, when accounting for the sample of 59 companies due to the withdrawal of 11 companies, the rate improved to 55.9% [19].

Currently, the emphasis is on the production of green electricity using photovoltaic systems, wind turbines, etc. When using photovoltaic systems, the output voltage is dc, and to transform it into ac (used most often by consumers), power inverters are used, which have an impact on the power quality. In work [20], an analysis was made of the neural control (recurrent artificial neural networks of nonlinear autoregressive type with external input) applied to power inverters with the aim of estimating the harmonic behavior in photovoltaic systems. Following the acquired and measured data, it was found that the neural network (NARX networks, Figure 6) captures the dynamics of the system to control and reduce harmonic distortion.

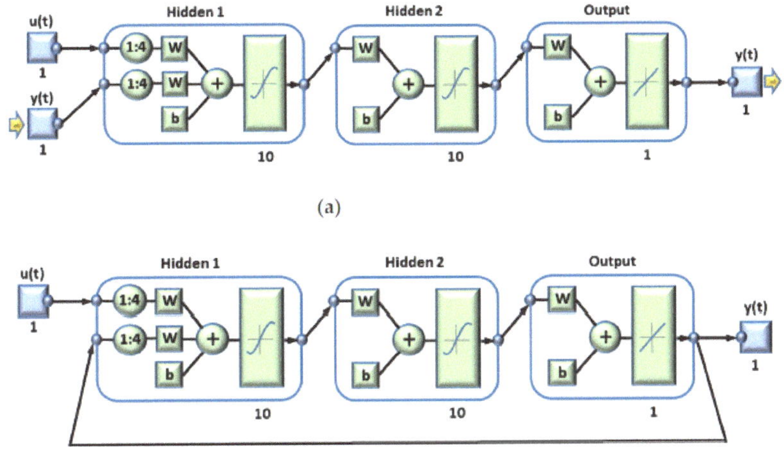

(a)

(b)

Figure 6. Architecture of configured NARX networks: (**a**) serial–parallel architecture (open loop); (**b**) parallel architecture (closed loop) [20].

It was found that integrating the PV system power obtained through electronic inverters into the PCC has an impact on the sinusoidal waveform of current in the electrical supply grid. The establishment of a highly efficient pattern in terms of execution times and computational resources as a result of modeling the dynamic and nonlinear behavior of that signal using NARX networks produced an MSE of 0.0067 with respect to the actual behavior of the signal, demonstrating the high performance of the neural network. When employing the closed-loop NARX to anticipate the results, an MSE of 0.0094 was obtained, demonstrating the model's viability and demonstrating a significant correlation between inputs and error values.

In terms of the type and volume of data that can be managed, the resultant model exhibits remarkable flexibility, enabling representation and prediction of the behavior of the system under investigation over extended time periods and under diverse operating situations. The resulting algorithm can be used to create real-world or virtual systems for reducing or controlling harmonic disturbances that impact electrical grids.

Ensuring the continuity of electricity supply is an important area of power quality. An extremely important field of research has been represented by electrical energy storage systems that can be used together with uninterruptible power sources to ensure the continuity of the electrical energy supply. In article [21], an analysis was made of flow batteries (vanadium redox flow batteries) that can provide long-term electricity in critical installations. The advantage of flow batteries is that they can be designed from independent blocks. The study presented a detailed experimental analysis of a vanadium redox flow battery (VRFB, Figure 7), especially from the point of view of the electrolyte used (electrolyte with the addition of hydrochloric acid).

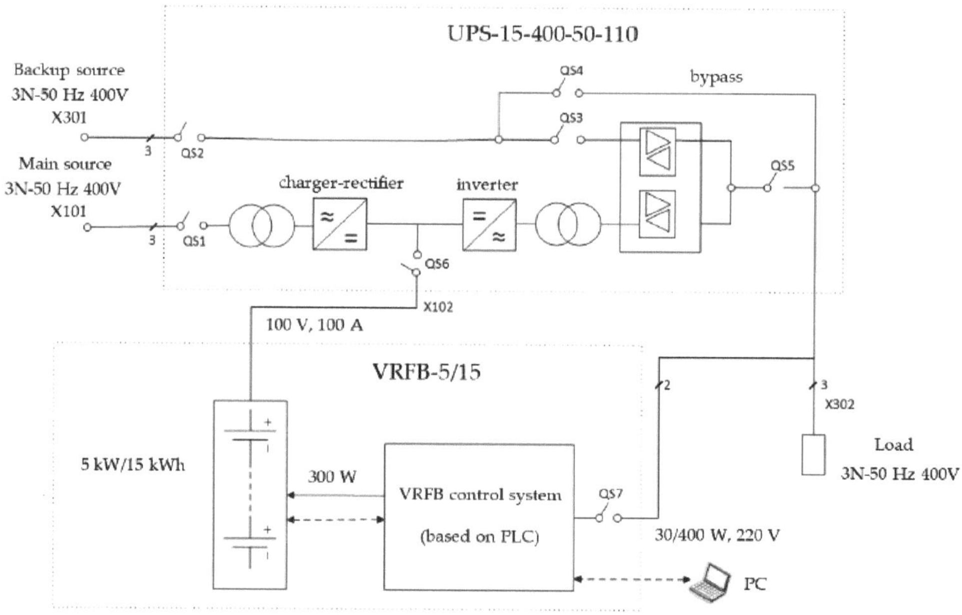

Figure 7. Block diagram of the UPS system based on 10 kW/30 kWh vanadium redox flow batteries [21].

A study of the characteristics of the cells with a change in the electrolyte pumping rate was conducted, and a VRFB hydraulic system was developed. It was demonstrated that a change in the electrolyte pumping rate had little impact on the power and efficiency of the stack (10%), while the stack was running in one of its operational modes. A VRFB-based UPS electrical circuit and an associated control algorithm were suggested. After researching the dynamic characteristics of the UPS in the VRFB charge/discharge modes, diagrams were offered. The demonstrated VRFB could function with a 1.5-fold overload without efficiency degradation and with a rise in efficiency without significantly reducing capacity, demonstrating the electrolyte's strong compatibility with the device.

Matrix converters are used more and more often. If they are connected to ordinary electrical loads, they produce harmonics of the order of kHz and even tens of kHz. Article [22] presented a study (through simulations and experiments) on a hybrid power filter (Figure 8) that is controlled by a fuzzy system to reduce very-high-frequency signals. The hybrid power filter had a low total harmonic distortion for the fundamental voltage (50 Hz). A comparative analysis was performed with a PI-controlled hybrid filter that demonstrated the superior performance of the fuzzy-system-controlled hybrid filter in terms of compensation speed, accuracy, total harmonic distortion of supply current, and overall integrity of the matrix converter.

High-frequency harmonics in supply lines heat transformers and motors and interfere with metering and telecommunications equipment, as well as protective relays. Interference with hospital and laboratory settings and measurement tools is the worst-case scenario. There is a need to safeguard delicate loads and equipment that is not designed to handle high frequencies, particularly harmonic frequencies, as high-frequency transmission applications are expanding globally and the issue of using the power lines as communication lines is increasing. After active filter activation, the suggested HAPF only received a response for less than half a cycle. The right control approach for producing the compensational currents is the basis of the HAPF. The hysteresis control approach, which was used in this study, places restrictions on the ability to adjust the switching frequency of the HAPFs [22].

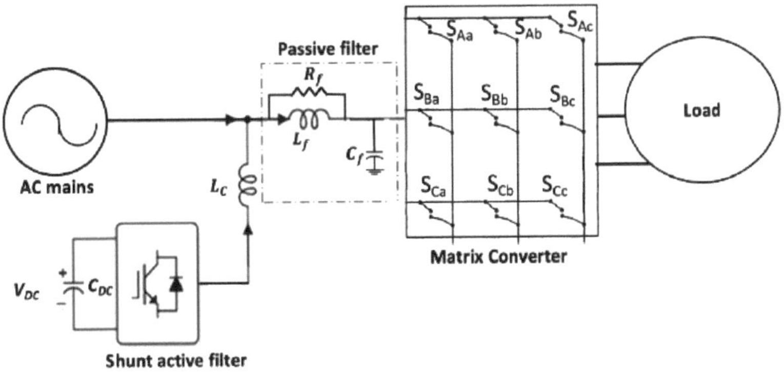

Figure 8. Proposed hybrid active power filter (HPAF) architecture.

3. Challenges on Power Quality

As electromagnetic disturbances affect both the economic and functional parameters of the electricity supplier and the consumers, appropriate power quality of electricity requires their joint actions [1,5,6]. In this regard:

- The electricity supplier must monitor the level of electromagnetic pollution of the electrical network and establish acceptable levels for different types of disruptive emissions of consumers, so that all equipment connected to the electrical network may have normal operating conditions;
- The electricity user is responsible for keeping the emissions they generate at the common connection point limited, below the limits specified by the electricity supplier. It is also responsible for drawing up studies and choosing methods to limit the emission of electromagnetic disturbances.

The most obvious defects determined by the power quality of electricity at consumers are interruptions and voltage gaps, in which the voltage increases or decreases for a short time. The transport and distribution systems of electricity can cause electromagnetic disturbances, lightning, wind, ice deposits, etc. The negative impact of voltage gaps (typically 0.3 s at most) on the power quality of electricity supplied to consumers is particularly important and depends on the type of gaps (accounting for the percentage in which the voltage drops, but also the duration of the gaps), as well as on the acceptability curves of different classes of electrical energy equipment. Security in the supply of electricity is a very important aspect of the power quality for large consumers because interruptions in the supply of electricity cause great damage to users [1,2]. At present, the securing of the electricity supply to consumers is achieved with the help of classic devices of automatic activation of the reserve, usually powered by two independent energy sources. Power-outage-sensitive and critical consumers must be equipped with uninterruptible power supplies to ensure power continuity in the event of a power outage.

Due to the size, weight, and cost advantages, switch mode power supplies have been used almost exclusively for all types of low- and medium-voltage power consumers. Practically, switching sources are present in almost all household and industrial appliances, such as computers, monitors, laptops, electronic ballasts for fluorescent lamps, and induction furnaces. These types of consumers show the highest harmonic distortion of the current absorbed from the network.

Harmonic pollution also causes negative effects on equipment in electrical networks [8]. Thus, additional losses occur, which reduce the efficiency of the electrical energy transformation, the lifetime of the equipment, and the functioning of the equipment, and their operating regimes are negatively influenced (this effect is manifested in power transformers and electronic equipment). A high power factor reduces reactive power, reduces electrical energy losses, and increases electrical-energy-carrying capacity. If the waveform of the current is nonsinusoidal, then the power factor is lower relative to the power factor of the fundamental, with a higher value of the total harmonic distortion.

The methods of limiting the deforming regime can be divided into three groups: passive filters, active filters, and hybrid filters [3,9]. Each option presents its advantages and disadvantages, so the choice of a certain type of filter requires a careful analysis of the efficiency of all types of filters for the specific situation, under the conditions of the respective electrical network configuration. Electrical filters can be used in medium- and low-voltage electrical networks, but must be used with caution in order not to create unwanted resonances, and possibly additional losses, in the electrical network. In practice, the combined solution of power factor improvement and electrical filtering of current harmonics can be used for deforming electrical consumers. The solution of filtering at each electrical consumer separately, although it is a more expensive solution, is a more effective solution in the long term than filtering, with high power filters, in the point of common coupling.

In three-phase networks, when using nonlinear consumers, the load capacity of three-phase transformers is reduced. Electric motors fed by static frequency converters experience additional thermal stresses, inadequate ventilation, and strong mechanical stresses (with direct implications on the life of the motor). The load capacity of electrical cables is reduced when using nonlinear and phase-unbalanced electrical consumers. The most requested conductor may be the neutral conductor if it is designed improperly.

When choosing and adjusting protection and switching devices, the deformation regime must be taken into account [5]. An incorrect adjustment of the protections causes their untimely actuation, and the de-energization of electrical consumers, even during their normal operation. In the deforming mode, when measuring voltages and currents, measuring devices (e.g., multi-meters) of the true RMS type must be used to ensure the correct measurement of these quantities. Sometimes, in practice, it is possible to reach a measurement even 40% lower than these quantities, if inappropriate measuring devices are used.

The operation in a distorting mode and the irrational consumption of reactive power lead both to penalizing the consumer, due to noncompliance with the technical norms in force, and to the ageing of the consumer's electrical equipment.

In the case of sinusoidal regimes, the solution adopted for reactive power compensation consists of the use of capacitor batteries [6]. Knowing the active daily load curves and the reactive daily load curves allows the modification of some technological parameters, but also the appropriate connection of the capacitor batteries and the regulation of the reactive power according to the inductive reactive power requirement of the electrical consumers.

In electrical installations, the symmetry of electrical consumers on phases is difficult to achieve, especially in the case of single-phase ones, different in power and sometimes fed by switching power sources.

The negative effects of the asymmetry of the supply voltages consist of [3]:

- Heating in the three-phase rotating electric machines, due to the additional losses introduced by the negative and zero sequence currents, which pass through the windings of the machines;
- High-frequency pulsating torques in rotating electric machines, which represent parasitic inverse braking torques; thanks to these torques, vibrations appear, which increase (in the case of fluctuating asymmetries);
- Negative influences on telecommunications lines;
- Reducing the reactive power provided by the capacitor banks, and, implicitly, the power factor.

Currently, the power systems do not have systems for measuring the additional circulation of electricity and, therefore, the related losses, conditioned by the presence of asymmetries [11]. To be able to highlight the energetic effects of the circulation of asymmetric powers, it would be sufficient that, in parallel with the current means of measuring powers and energies, the means of measuring the circulation of asymmetric powers and energies should also be provided, because they are kept separately. For this purpose, some counters equipped with filters of symmetrical components (negative and zero) would be sufficient. As unbalanced receivers can give, but also receive, the energy of asymmetry, the meters should be designed with the double-way operation, having one dial for one direction and another for the other direction.

Another direction of research refers to the different strategy control of equipment (e.g., for dynamic voltage restorer [23]) used in power quality.

Due to topographical and/or utility investment constraints, there may be several isolated microgrids in remote locations. The configuration, energy sources, and types of loads present problems for the isolated microgrids that affect power quality. Organizing these microgrids into many clusters depending on their relative locations and connecting them may also have other advantages including increased dependability, stability, and cost-effective operation [24].

Accumulating low-power sources in a microgrid with solar and wind power plants is a realistic way to improve the efficiency of distributed generation and renewable energy sources [25]. These systems have a lot of characteristics, including a lot of semiconductor equipment and bi-directional power flows. Therefore, ensuring the necessary power quality indicators is a crucial responsibility.

Stand-alone microgrids are those that can run independently from the bulk power supply or the national grid. Microgrids are typically connected to the national grid or a bulk power supplier. Power quality is the main concern for both standalone and microgrids connected to the national grid. For many years, research has been conducted to find a solution and raise the standard of power in microgrids, as renewable energy sources (including solar, wind, and fuel cells) are frequently connected to microgrids to meet local consumer demands and to lower operational costs. The microgrid will have a reactive voltage problem as a result of the renewable energy sources' failure to supply the microgrid with reactive power instead of real power [26–28].

4. Conclusions

Ensuring the power quality in household and industrial applications is a complex and difficult objective to achieve. It is found that the disturbances that occur in the operation of energy systems affect practically all the characteristics of the voltage and current: shape, frequency, amplitude, interruptions, voltage gaps, flicker, symmetry (in the case of three-phase systems), and continuity of electricity supply.

The power quality, unlike other sectors of activity, therefore depends not only on the supplier, but also on all consumers connected to the same power network; some of them can cause disruptive influences in the supply network, affecting the operation of other consumers connected to the same network.

Solving electricity quality problems requires the assessment of the quality of the supplied energy, by monitoring the voltages and currents in the point of common coupling, but also by monitoring the electromagnetic disturbances introduced by the consumers. The distorting and nonsymmetrical regime is a difficult process that requires a detailed analysis, both at the level of suppliers and consumers of electricity.

Short and, especially, long power interruption affects electrical consumers (in particular, industrial ones). Ensuring the continuity of electricity supply is perhaps one of the most important research directions in power quality. Currently, there is a tendency toward the use of microgrids that have renewable energy sources (e.g., solar, wind, and fuel cells). The use of renewable energy sources will determine new challenges.

Funding: This research received no external funding.

Acknowledgments: The author is grateful to the Energies MDPI Publishing House for the invitation to act as guest editor of Special Issue "Analysis and Experiment for Electric Power Quality"—Energies and thanks the "Energies" editorial office for their kind cooperation, patience, and committed engagement. I would like to give special thanks to my colleagues Eng. Angela Iagăr and Eng. Corina Maria Diniș for interesting discussion on the main topics and, also, for analysis of the submitted articles at this Special Issue "Analysis and Experiment for Electric Power Quality", Energies MDPI.

Conflicts of Interest: The author declares no conflict of interest.

References

1. Balasubramaniam, P.M.; Prabha, S.U. Power quality issues, solutions and standards: A technology review. *J. Appl. Sci. Eng.* **2015**, *18*, 371–380.
2. Mehebub, A.; Mandela, G. Power quality problems and solutions: An overview. *Int. J. Sci. Res.* **2014**, *3*, 1024–1030.
3. Copper Development Association. *Power Quality Application Guide*; Copper Development Association: Hemel Hempstead, UK, 2001.
4. Shin, Y.J.; Powers, E.J.; Grady, M.; Arapostathis, A. Power quality indices for transient disturbances. *IEEE Trans. Power Deliv.* **2005**, *21*, 253–261. [CrossRef]
5. Reid, W.E. Power quality issues-standards and guidelines. *IEEE Trans. Ind. Appl.* **1996**, *32*, 625–632. [CrossRef]
6. Iagar, A.; Popa, G.N.; Dinis, C.M. *Electric Power Quality—From Theory to Experiments*; Politehnica Publishing House: Timisoara, Romania, 2017. (In Romanian)
7. Rodríguez, A.; Bueno, E.J.; Mayor, A.; Sanchez, F.J.R.; García-Cerrada, A. Voltage support provided by statcom in unbalanced power systems. *Energies* **2014**, *7*, 1003–1026. [CrossRef]
8. Seguí-Chilet, S.; Gimeno-Sales, F.; Orts, S.; Garcera, G.; Figueres, E.; Fillol, M.A.; Masot, R. Approach to unbalance power active compensation under linear load unbalances and fundamental voltage asymmetries. *Int. J. Electr. Power Energy Syst.* **2007**, *29*, 526–539. [CrossRef]
9. Pomilio, J.A.; Deckmann, S.M. Characterization and compensation of harmonics and reactive power of residential and commercial loads. *IEEE Trans. Power Deliv.* **2007**, *22*, 1049–1055. [CrossRef]
10. Czarnecki, L.S.; Haley, P.M. Unbalanced power in four-wire systems and its reactive compensation. *IEEE Trans. Power Deliv.* **2015**, *30*, 53–63. [CrossRef]
11. Pakere, I.; Lauka, D.; Blumberga, D. Does the balance exist between cost efficiency of different energy efficiency measures? DH systems case. *Energies* **2020**, *13*, 5151. [CrossRef]
12. Soni, J.S.; Jangalwa, N.K.; Gupta, R.; Palwalia, D.K. Various power quality challenges and solution techniques using FACTS technology for power system. *AIP Conf. Proc.* **2016**, *1715*, 020024.
13. Popa, G.N.; Iagăr, A.; Diniș, C.M. Considerations on current and voltage unbalance of nonlinear loads in residential and educational sectors. *Energies* **2021**, *14*, 102. [CrossRef]
14. Grillo de Sauza, P.V.; Maria de Carvalho Filho, J.; Ferreira, D.F.; Filho, J.M.; Filho, H.K.R.; Pereira, N.B. Cluster-based method to determine base values for short-term voltage variation indices. *Energies* **2021**, *14*, 149. [CrossRef]
15. Vitoi, L.A.; Brandao, D.; Tedeschi, E. Active power filter pre-selection tool to enhance the power quality in oil and gas platforms. *Energies* **2021**, *14*, 1024. [CrossRef]
16. Klimas, M.; Grabowski, D.; Bula, D. Application of decision trees for optimal allocation of harmonic filters in medium-voltage networks. *Energies* **2021**, *14*, 1173. [CrossRef]
17. Czapp, S.; Tariq, H. Behavior od residual current devices at frequencies up to 50 kHz. *Energies* **2021**, *14*, 1785. [CrossRef]
18. Park, B.; Lee, J.; Jang, G. Harmonic mitigation using passive harmonic filters: Case study in a steel mill power systems. *Energies* **2021**, *14*, 2278. [CrossRef]
19. Motoki, É.M.; Maria de Carvalho Filho, J.; Márcio da Silveira, P.; Pereira, N.B.; Grillo de Sauza, P.V. Cost of industrial process shutdowns due to voltage sag and short interruption. *Energies* **2021**, *14*, 2874. [CrossRef]
20. Jumilla-Corral, A.A.; Perez-Tello, C.; Campbell-Ramirez, H.E.; Medrano-Hurtado, Z.Y.; Mayorga-Ortiz, P.; Avitia, R.L. Modeling of harmonic current in electrical grids with photovoltaic power integration using a nonlinear autoregressive with external input neural network. *Energies* **2021**, *14*, 4015. [CrossRef]
21. Kuzmin, I.; Loskutov, A.; Osetrov, E.; Kurkin, A. Source for autonomous power supply system based on flow battery. *Energies* **2022**, *15*, 3027. [CrossRef]
22. Koduah, A.; Effah, F.B. Fuzzy logic controlled hybrid active filter for matrix converter input current harmonics. *Energies* **2022**, *15*, 7640. [CrossRef]
23. Omar, A.I.; Abdel Aleem, S.H.E.; El-Zahab, E.E.A.; Algablawy, M.; Ali, Z.M. An improved approach for robust control of dynamic voltage restorer and power quality enhancement using grasshopper optimization algorithm. *ISA Trans.* **2019**, *95*, 110–129. [CrossRef] [PubMed]
24. Elmetwaly, A.H.; ElDesouky, A.A.; Omar, A.I.; Saad, M.A. Operation control, energy management, and power quality enhancement for a cluster of isolated microgrids. *Ain Shams Eng. J.* **2022**, *13*, 101737. [CrossRef]

25. Shalukho, A.V.; Lipuzhin, I.A.; Voroshilov, A.A. Power Quality in Microgrids with Distributed Generation. In Proceedings of the 2019 International Ural Conference on Electrical Power Engineering (UralCon), Chelyabinsk, Russia, 1–3 October 2019.
26. Zahira, R.; Lakshmi, D.; Ravi, C.N. *Power Quality Issues in Microgrid and its Solutions*; Microgrid Technologies, Wiley online Library: Hoboken, NJ, USA, 2021.
27. Tascikaraoglu, A.; Uzunoglu, M.; Erdinc, O. Power quality assessment of wind turbines and comparison with conventional legal regulations: A case study in Turkey. *Appl. Energy* **2011**, *88*, 1864–1872. [CrossRef]
28. Igual, R.; Medrano, C. Research challenges in real-time classification of power quality disturbances applicable to microgrids: A systematic review. *Renew. Sustain. Energy Rev.* **2020**, *132*, 110050. [CrossRef]

energies

MDPI

Article

Fuzzy-Logic-Controlled Hybrid Active Filter for Matrix Converter Input Current Harmonics

Asare Koduah [1,*] and Francis Boafo Effah [2]

[1] Department of Electrical Power System, Kaunas University of Technology, 51394 Kaunas, Lithuania
[2] Department of Electrical Engineering, Kwame Nkrumah University of Science and Technology, Kumasi 00233, Ghana
* Correspondence: asakod@ktu.lt

Abstract: The proliferation of matrix converter interfaces coupled with traditional loads produces nonstandard and high-frequency harmonics in the range of (2 to 150) kHz in the power system. Although several research works have been conducted on passive and active filter solutions, most of these are low-frequency (below 2 kHz) solutions and are not effective under supraharmonic frequencies. An experimental study of a fuzzy-inference-system-controlled hybrid active power filter (HAPF) for the attenuation of higher frequency harmonics (above 8 kHz) is proposed. The compensational approach introduced is different from traditional approaches and the use of the fuzzy logic controller eliminates complexities involved in active filter designs. The proposed filter obtained a total harmonic distortion (THD) of 1.16% of the fundamental 50 Hz supply frequency. The performance of the proposed hybrid filter was compared with that of the proportional and integral (PI) controlled topology. The results obtained indicated superior performance of the fuzzy logic controller over the PI in terms of compensational speed, accuracy, the THD of the supply current and the overall integrity of the matrix converter. Illustrative design blocks and simulation in MATLAB/Simulink environment are provided to buttress these findings.

Keywords: active harmonic filter; hybrid harmonic filter; matrix converter; passive filter

Citation: Koduah, A.; Effah, F.B. Fuzzy-Logic-Controlled Hybrid Active Filter for Matrix Converter Input Current Harmonics. *Energies* **2022**, *15*, 7640. https://doi.org/10.3390/en15207640

Academic Editor: Gabriel Nicolae Popa

Received: 25 September 2022
Accepted: 14 October 2022
Published: 16 October 2022

1. Introduction

Since their introduction in the 1970s, matrix converters (MC) have gained popularity due to their many applications [1]. However, they produce significantly higher frequency harmonic pollution, around their switching frequency, in the supply. These high frequencies are due to their large number of switching states and switching frequency. The switches of the MC are required to block voltages and conduct currents bidirectionally. This is because the MC is mostly fed by a voltage source, hence its input terminals must not be short circuited. Connected loads, to the MC, are mostly inductive in nature, and hence the output terminals must also not be open circuited. With the above conditions, the total switching states required to implement the MC are reduced from 512 to 27. Irrespective of this, its protection scheme is that of a simple, relatively lower cost hardware architecture. Several research works have been conducted on the MC due to its wide application [1,2], particularly in applications where weight and size are of much concern such as in electric vehicles, aircraft, ships and submarines. Other research works are focused on improving its voltage transfer ratio of 0.867, regarding which several research works have successfully obtained more than unity voltage transfer ratio [3–7].

In a typical power system, domestic and industrial nonlinear loads result in harmonic pollution of the source voltage and current. The presence of these harmonic voltages and currents in the power system may result in copper, iron and dielectric losses. Transformers and rotating machines may experience heating, dielectric stress, hysteresis and eddy current losses due to the presence of voltage harmonics [8]. Capacitor banks may experience overloads. Protective relays' time delay characteristics are deteriorated in the presence of harmonics [9]. Although higher frequency switching of power converters results in relatively smaller sizes for design components, they tend to produce higher switching losses and frequency harmonics for which passive filters may not be adequate to minimize [10]. Relatively higher frequency harmonics from cycloconverters are also challenging to compensate by the shunt active power filters (SAPF) due to non-availability of required switches and economic considerations [6,11]. The presence of these higher frequency harmonics poses threats to the nonlinear loads themselves, as well as to sensitive electrical loads such as industrial controllers, hospital equipment and laboratory and experimental setups [12–14]. Also, high losses may be recorded in utility transformers, transmission lines and motors. Relay settings are also susceptible to high frequency ac line harmonics.

The objectives and contributions of this paper are:

1. To propose a HAPF consisting of a SAPF and a series passive R-L-C filter.
2. To obtain a THD within the recommended IEEE 519-2014 requirements with faster dynamic response time using fuzzy inferencing system.
3. Separation of the high frequency currents was accomplished in the $\alpha - \beta$ domain using a second order analogue filter with cutoff frequency of 30 Hz.

The remainder of this paper is arranged as follows: Section 2 is a review of selected literature on the topic. The methodology used is introduced and explained in Section 3, which involves four subsections. Section 3.1 explains the mathematical modelling and analysis of the passive section of the proposed HAPF, while Section 3.2 introduces the design of the active section. Sections 3.3 and 3.4 address the reference current extraction and the control strategies used in the design. Section 4 summarizes the experimental simulations and obtained results while Section 5 discusses the obtained results. Section 6 concludes the study with future research directions.

2. Review of Selected Literature

Periodic voltages and currents with frequencies that are integer multiples of the fundamental supply frequency are termed as harmonic voltage or current, respectively. Supply voltage harmonics are mostly a result of supply impedance and load current distortions [15]. The supply current harmonics are mostly due to the nonlinearity of connected load, as experienced in AC/DC converters and controls of industrial and domestic appliances. Cell phone chargers, laptop chargers and all DC loads which require rectifier circuits and power semiconductor switches are the primary sources of harmonic currents in the power system [16,17]. Due to the fact that DC supply is not readily available, various international bodies have proposed standards for minimum harmonic production of various manufactured electric and electronic products. The IEEE-519-1992 recommends a maximum 5% harmonic content for voltages below 65 kV [18]. The European harmonic standard, IEC-555, also recommends an absolute harmonic limit for all consumer manufactured loads [8].

The harmonic disposition of the voltage and current and their characteristics can be identified at the point of common coupling (PCC) and corrective measures implemented. Traditionally, passive filters were used in mitigating power quality issues, specifically, harmonics [6]. The authors of [19] performed a numerical investigation on the stability of an electric drive system coupled with a passive LC filter topology with a dumping resistor parallel to the inductor, to form an RLC filter topology. The RLC filter was put in series with a MC feeding a three-phase symmetrical R-L load. They concluded that their proposed model maintained the stability of the MC during harmonic compensation. However, because the damping resistance was limited to a minimum value of 4 Ω, there was a significant reduction of the output power limit of the MC. Also, the authors of [20] supplemented the traditional passive RLC filter with a relatively small capacitor in series to the damping resistor in the RLC topology to form a CLCR filter. The authors concluded that the frequency response characteristics of the CLCR topology were relatively lower compared with the traditional RLC topology. Hence, their topology had a harmonic damping rate with a decrease of THD level of 13–16%. Although the increase of the overall capacitance may result in further deterioration of the input power factor, these were skipped and very little was said about it by the authors.

Furthermore, in [21,22] the authors presented Lyapunov-controlled-matrix-based unified power flow controllers coupled at the input side with a passive RLC filter. Their model improved the steady state errors. Their obtained THD for the line current and voltage were 4.86% and 4.53% of the fundamental frequency, respectively. However, the disadvantages of passive filters are well known and have been explored extensively by [23], and they range from their bulky size and non-reliability to the risk of resonance with the line impedance. Their fixed compensational property diminishes their advantages in an ever growing and dynamic load system. The problem of consistent component (capacitor) replacement with load changes also affects their use. To overcome these challenges, active power filters (APF) have been proposed by researchers.

The authors of [1,24] modelled an SAPF for supply current harmonic reduction in matrix converters. They employed the instantaneous reactive power control theory for the generation of compensation currents. Although the authors did not specifically state the numerical THD of their obtained results, analysis of their input current waveform suggests a THD of not less than 25% of their fundamental frequency. The authors did no analysis on the effects of their filter on the voltage transfer ratio or on the dynamic performance of the MC.

The authors of [2,7] focused on HAPF and heuristic approaches in harmonic mitigations in order to harness the advantages of both active and passive filters. Ref. [2] presented a topology without the use of input or output filters, rather being based on the modulation of the matrix converter by feedforward and fuzzy logic control (FLC) system feedback method. In their model, the input and output voltage and current parameters of the MC were measured and compared with referent values to modulate the matrix converter appropriately. Although this model worked correctly to maintain the voltage transfer ratio of the MC and reduce harmonics, the stability and output power limit are questionable. Also, the response time of the MC needs to be investigated as well as the overall power factor since there was no analysis done to verify the effects on these parameters.

On the other hand, Ref. [7] modelled a HAPF comprising two active power filters separately controlled (shunt and series active filters). The SAPF was modelled for the input side harmonics to compensate the supply current ripples and the series active filter for the output voltage ripples of the MC. The authors concluded that there were 18% input current harmonic distortions and 3% output voltage distortions of their fundamental 60 Hz supply frequency. The problem with their model was the use of two active filters with separate control strategies which are likely to generate undesired higher frequency components at their switching frequencies and tend to slow down the response of the coupled converter.

The authors of [25] proposed an advance common control method for a HAPF topology consisting of two active filters (shunt and series filters) for a six-pulse thyristor rectifier. The authors recommended an improved reference signal tracking (RST) control architecture to overcome phase lag effects often associated with this type of hybrid filter topologies. In all, their work showed promising results with 1.5% THD and it was within the accepted 5% recommendations per IEEE values. Irrespective of this, their solution may not be applicable to the MC. Table 1 shows similar works on the harmonic mitigations by some authors and their results.

Table 1. Summary of reviewed literature.

References	Filter Type	Topology	THD%	Limitations
[1,14]	Active	Shunt active	>20	Reduced power limit
[19]	Passive	R-L-C	<15	Reduced power limit at Rd < 4 Ω
[20]	Passive	CLCR	<16	Poor power factor, power limited below Rd < 4 Ω
[21,22]	Passive	R-L-C	<16	Reduced power limit at Rd < 4 Ω
[2]	Hybrid	FIS	4.16	Not stable

Table 1. Cont.

References	Filter Type	Topology	THD%	Limitations
[7]	Hybrid	Shunt and series active filters	18 and 3	Slow response
[25]	Hybrid	Shunt and series active filters	1.5	Good performance

3. Methodology

This research considers a 400 $V_{L\text{-}L}$ 50 Hz, three-phase balanced supply connected to a resistive load through a MC with power rating (S_{MC}) and switching frequency of 50 kW and 8 kHz, respectively. The input to the MC is connected to the passive filter section of the hybrid filter. This is to reduce the infinite rate of change of the MC input current to a finite rate of change to enable the SAPF to compensate the harmonics effectively. Figure 1 shows the proposed HAPF. The components of the supply current, due to the converter, are the DC component, active and reactive powers, as well as harmonics, as illustrated in (1).

$$i_S(t) = i_L(t) = i_0 + i_1 \sin(\omega t + \theta_1) + \sum_{n=5,7.9.11\ldots}^{\infty} i_n \sin(n\omega t + \theta_n) \tag{1}$$

where

$i_{S(t)}$: Supply current.
$i_{L(t)}$: Load current.
i_0: DC current component.
i_1: First harmonic component.
ω: Angular frequency.
θ: Phase shift.
n: Harmonic number.

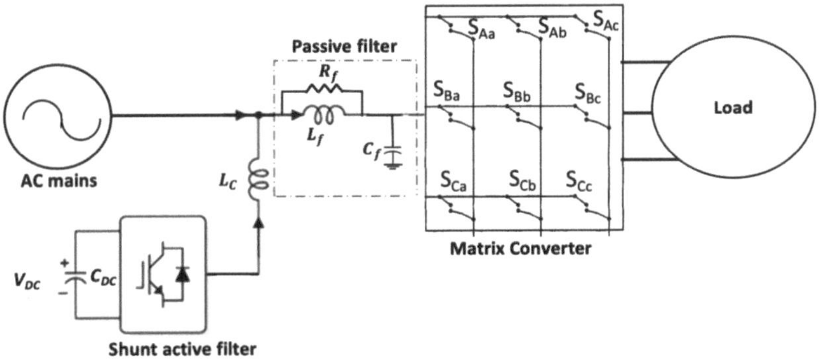

Figure 1. Proposed HAPF architecture.

3.1. Design of Passive LC Filter Section of the HAPF

For an effective implementation of the HAPF, the passive filter must be designed to reduce the rate of change of the input current and attenuate the higher order harmonics while the active filter compensates the lower order harmonics, particularly the 5th, 7th, 11th and 13th harmonic range [26]. The design therefore consists of the two sections: the passive filter section and the SAPF section. A power system consisting of an LC filter in series with the MC and a resistive load was considered. Figure 2 shows the series connection of the passive LC filter and its equivalent circuit diagram.

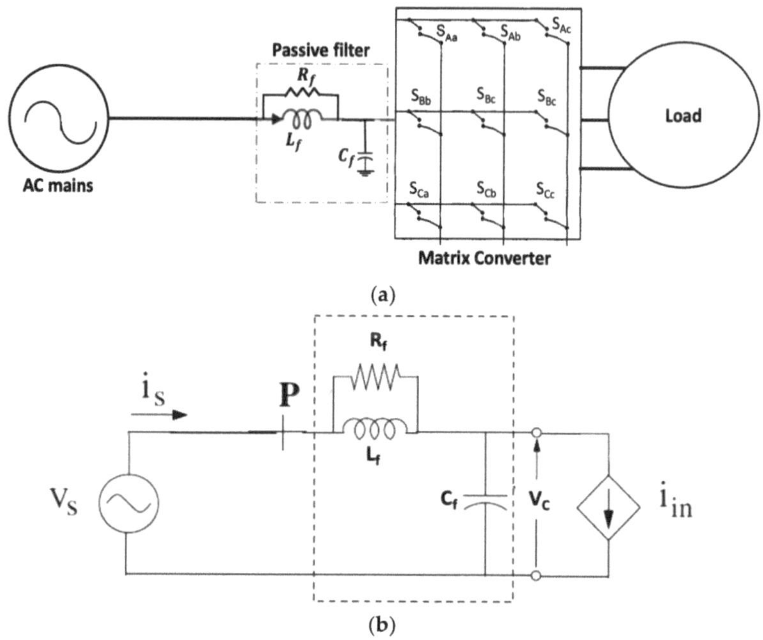

Figure 2. Passive LC filter for MC input harmonic mitigation: (**a**) single line diagram; (**b**) equivalent circuit diagram.

From Figure 2b, the transfer function of the passive filter can be represented under the following equation sequence:

$$G(j\omega) = \frac{j\omega\frac{L_f}{R_f} + 1}{1 - \omega^2 L_f C_f + j\omega\frac{L_f}{R_f}} \tag{2}$$

$$|G(j\omega)| = \sqrt{\frac{1 + \frac{r_\omega^2}{Q^2}}{(1 - r_\omega^2)^2 + \frac{r_\omega^2}{Q^2}}} \tag{3}$$

where

$$r_\omega = \frac{\omega}{\omega_C} \quad Q = R\sqrt{\frac{C_f}{L_f}} \tag{4}$$

with

V_S: Supply voltage.
i_S: Supply current.
P: Point of common coupling.
L_f: Filter inductance.
C_f: Filter capacitance.
R_f: Damping resistor.

The filter values were selected with respect to established constraints of the maximum attenuation of the filter to be less than 26 dB. Again, the criteria for selecting the filter inductance was based on 5% voltage drop over the filter inductance, and 10% of I_{MC} represents the reactive currents over the filter capacitor. A damping resistor of 0.5 Ω was selected. These constraints will restrict the high rate of change (di/dt) of the MC input current. The filter elements C_F and L_F, as well as the MC input current, were calculated as:

$$I_{MC} = \frac{S_{MC}}{\sqrt{3}V_S} = \frac{50e3}{400\sqrt{3}} = 72 \text{ A} \tag{5}$$

$$C_f \leq \frac{0.866K_C I_{MC}}{\omega_g V_S} = \frac{0.1 \times 72 \times 0.866}{2pi \times 50 \times 400} = 49.6 \text{ μF} \tag{6}$$

$$L_f \leq \frac{K_L V_S}{\omega_g \sqrt{(i_{in}^2 + i_c^2)}} = \frac{0.05 \times 400}{2\pi \times 50 \times \sqrt{(i_{in}^2 + i_c^2)}} = 1 \text{ mH} \tag{7}$$

3.2. Design of Active Section of HAPF

The design of the active section of the HAPF includes the power stage and the control strategy. The power stage consists of an IGBT with anti-parallel diodes voltage source converter (VSC). Current source converters (CSC) could be used but would require extra design modifications as IGBTs with series diodes are not readily available. The design of the VSC consists of three component selections:

1. The selection of the DC voltage, V_{dc}.
2. The selection of coupling inductance, L_c.
3. The selection of the DC side capacitance, C_{DC}.

3.2.1. Selection of V_{dc} and L_c

The selection of V_{dc} and L_c required the assumption that the supply ac voltage remained sinusoidal under all conditions. Also, the peak-to-peak ripple line current distortions were assumed to be 5% of the capacitor current. Finally, the PWM inverter was assumed to operate in the linear modulation mode.

From Figure 3, it can be seen that the HAPF must adjust the inverter current, i_{inv}, to compensate the reactive power in (1). An efficient compensation will result in the supply current, i_s, being in phase with the supply voltage, V_s, and the inverter current, i_{inv}, being orthogonal to the supply voltage, as seen in Figure 3b.

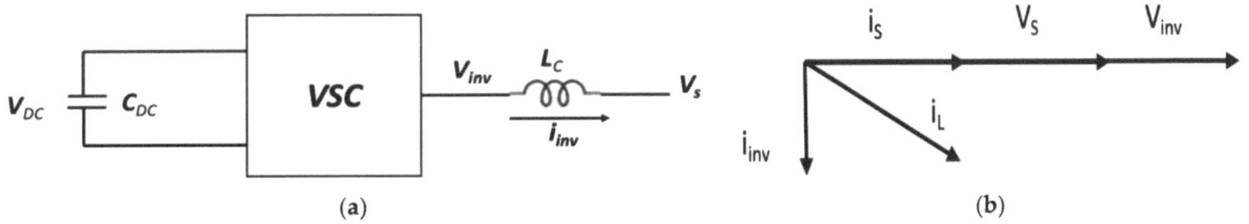

Figure 3. Active power filter (HAPF) architecture: (**a**) reactive power flow; (**b**) vector diagram.

Again, from Figure 3, the reactive power produced by the inverter can compensate the reactive power in the supply when $V_{inv} > V_s$ under linear modulation mode [27]. The capacitance voltage, V_{dc}, was obtained as:

$$m_a = \frac{2\sqrt{2}V_{inv}}{V_{dc}} \tag{8}$$

If $m_a = 1$, then:

$$V_{dc} = 2\sqrt{2}V_{inv} = 2\left(\sqrt{2}/\sqrt{3}\right)Vs = 653 \text{ V} \approx 700 \text{ V} \tag{9}$$

The coupling inductance, L_c, filters out the ripples in the inverter output current and limits the high rate of change of the inverter current. Hence, its selection was based on the peak-to-peak ripple current, i_{ripp}, which was calculated as (10). From Figure 4, and considering the period of the current change as 10% of the MC switching period (1/8 kHz = 1.25×10^{-4}), the harmonic currents are assumed to be 10% of the input current to the MC, and then the maximum di/dt of the inverter current can be calculated as:

$$max\left|\frac{di_L}{dt}\right| = \frac{72 \times 0.1}{1.25 \times 10^{-4}} = 0.576 \times 10^{6}\ \text{s} \tag{10}$$

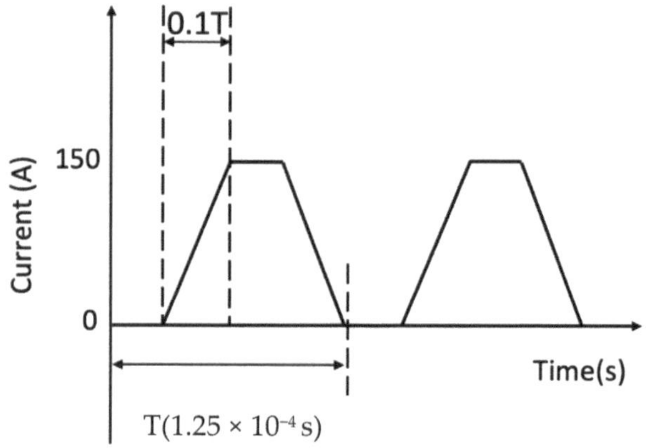

Figure 4. Theoretical input current rate of change after passive filter.

Hence, from (11), minimum Lc can be obtained as:

$$L_C \geq \frac{\frac{2}{3}V_{dc} - V_s}{max\left|\frac{di_L}{dt}\right|} = 115\ \mu\text{H} \tag{11}$$

3.2.2. Selection of the DC Capacitance

The selection of the DC capacitance value was based on the instantaneous power exchange between the inverter and the grid during transients. The peak-to-peak ripple voltage of the inverter was assumed to be 15% of the DC bus voltage. Hence, the lost energy in the inverter is compensated for as:

$$E = \frac{1}{2}C_{dc}\left(V_1^2 - V_2^2\right) \tag{12}$$

where $V_2 = V_1 - \Delta V_{ripp}$ and ΔV_{ripp} is the peak-to-peak ripple voltage. V_1 is the DC voltage. If the power rating of the inverter is P, then the energy stored in the capacitor for a period T can be estimated by:

$$E = P \times T \tag{13}$$

$$\frac{1}{2}C_{dc}\left(V_1^2 - V_2^2\right) = P \times T = K3V_S I_f aT = 5624.9\ \mu\text{F} \tag{14}$$

where

a: Overload factor (1.2).
Vs: Phase voltage.
I_f: Active filter ac side current.
T: Recovery period (30 ms).
K: Proportionality constant (0.1).

3.3. Control and Reference Current Extraction for the SAPF

The effectiveness of the SAPF is dependent on the appropriate control strategy used for the generation of compensational currents as well as gating signals for the voltage source inverter. Voltage and line current measurements were obtained to extract the reference currents. The instantaneous reactive power (P-Q) theory proposed by [28] was used to generate the compensating currents in the time domain. Finally, the gating signals for the voltage source inverter were generated using the hysteresis current loop control.

The instantaneous reactive power theory gives the flexibility of deciding the kind of compensation required. There are basically two kinds:

1. Total compensation.
2. Partial compensation.

Total compensation involves reactive power compensation, harmonic attenuation, power factor correction and three-phase power system balancing. Total compensation is mostly recommended for low power applications (Akagi, 2005). Implementing reactive power compensations in high power applications is not an economically viable option due to the current and voltage magnitudes that will be handled.

Partial compensation is mostly recommended for harmonic content compensation only. It includes either the voltage, current or both harmonic contents compensation. Most publications are about partial compensation since it is implemented in low power applications. The shape of the supply voltage is crucial in both the current and voltage harmonic compensations. A distorted supply voltage waveform increases the cost and difficulty in current harmonic compensations. In most cases, phase locked loop (PLL) circuits are recommended for unbalanced and distorted supply voltage applications. From (1), the magnitude of the instantaneous complex power (S) can be given as:

$$s = p + jq = 3/2v(t)i^*(t) \tag{15}$$

where

s: Instantaneous complex power.
$v(t)$: Instantaneous voltage vector.
$i(t)$: Instantaneous current vector.
p: Active power.
q: Reactive power.

With the instantaneous space vectors of the voltage and current represented as 'v' and 'i', respectively, then the current space vector can be expressed as

$$i(t) = {}^{2}/_{3}\frac{v}{|v|^2}S^* = {}^{2}/_{3}\frac{v}{|v|^2}(p - jq) \tag{16}$$

with

$$|v|^2 = v_\alpha^2 + v_\beta^2 \text{ in the } \alpha - \beta \text{ domain}$$

The active and reactive powers (p and q) can be decomposed into their direct and oscillatory components, where;

$$P = \widetilde{P} + \overline{P} \;\; and \;\; Q = \widetilde{q} + \overline{q} \tag{17}$$

with

S*: Desired apparent power.
$\overline{P}$ and $\overline{q}$: Direct component of the active and reactive power.
$\widetilde{P}$ and $\widetilde{q}$: Oscillatory component of the active and reactive power.

Equation (16) was used to calculate the reference current in the SAPF topology. By the instantaneous reactive power theory, the desired apparent power vector has two components: the direct component and the oscillatory component for both the instantaneous active and reactive powers as seen in (18):

$$S_{desired} = \begin{cases} \tilde{p} - j\tilde{q} & Partial\ compensation \\ \tilde{p} - jQ & Total\ compensation \end{cases} \tag{18}$$

From (18), the compensation of the oscillatory components of the instantaneous active and reactive powers of the desired complex power will result in partial compensation, while the compensation of the oscillatory component of the instantaneous active power and the instantaneous reactive power will result in total compensation by the SAPF. This research was based on total compensation of the desired apparent power.

The instantaneous P-Q theory involves the transformation, using the Clarke and its inverse transformations, of the three-phase a-b-c reference frame into a two-phase orthogonal stationary reference frame, α-β-0, as referred to in (19)–(21). Figure 5 depicts the transformation.

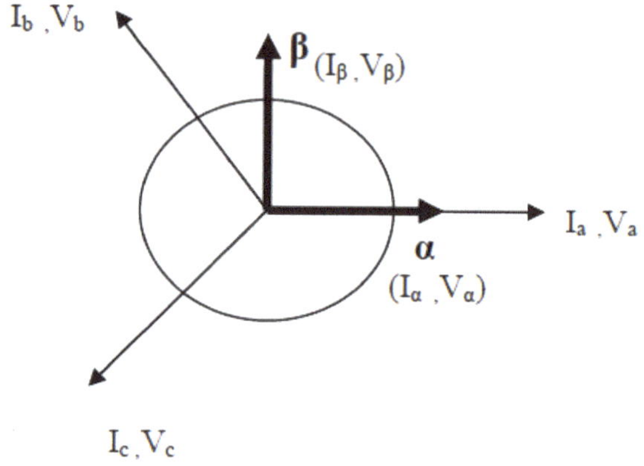

Figure 5. Three-phase a-b-c to $\alpha - \beta$ transformation.

$$\begin{bmatrix} i_\alpha \\ i_\beta \\ i_0 \end{bmatrix} = \sqrt{\frac{2}{3}} \begin{bmatrix} 1 & -1/2 & -1/2 \\ 0 & \sqrt{3}/2 & -\sqrt{3}/2 \\ 1/2 & 1/2 & 1/2 \end{bmatrix} \begin{bmatrix} i_{La} \\ i_{Lb} \\ i_{Lc} \end{bmatrix} \tag{19}$$

$$\begin{bmatrix} i_{Ca} \\ i_{Cb} \\ i_{Cc} \end{bmatrix} = \sqrt{\frac{2}{3}} \begin{bmatrix} 1 & 0 & 1/\sqrt{2} \\ -1/2 & \sqrt{3}/2 & 1/\sqrt{2} \\ -1/2 & -\sqrt{3}/2 & 1/\sqrt{2} \end{bmatrix} \begin{bmatrix} i_\alpha \\ i_\beta \\ i_0 \end{bmatrix} \tag{20}$$

and

$$\begin{bmatrix} p \\ q \end{bmatrix} = \begin{bmatrix} V_\alpha & V_\beta \\ V_\beta & V_\alpha \end{bmatrix} \begin{bmatrix} I_\alpha \\ I_\beta \end{bmatrix} \tag{21}$$

The harmonic contents of the apparent power were extracted using low pass filters [29]. Reference currents were then derived from the extracted harmonic contents by means of the inverse Clarke transformation. The inverter gating signals were generated by comparing the reference currents with the inverter currents by means of the hysteresis current loop, as shown in Figure 6. The reference current falls within the hysteresis band (HB), bounded above and below by setpoints.

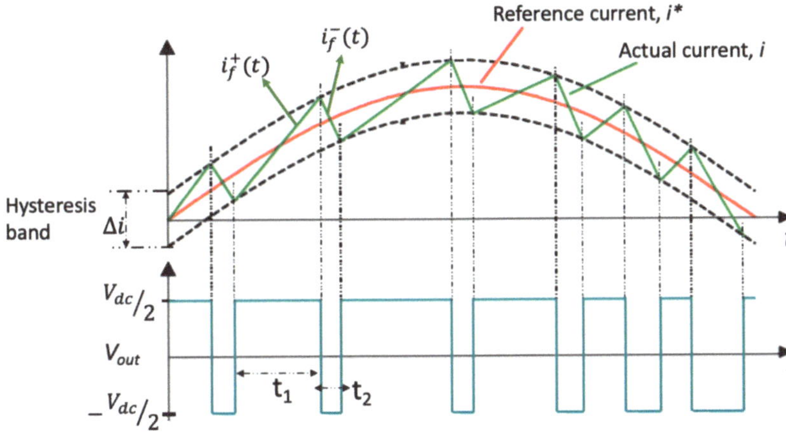

Figure 6. Hysteresis current loop control.

The generated compensating currents are 180° out of phase with the load current and contain the harmonics required by the load currents. Hence, the supply current is rendered free from the load harmonics, as depicted theoretically in (23).

$$i_S + i_{SH} = i_L + i_{LH} - i_C \tag{22}$$

with

$$i_S = i_L + i_{LH} - (i_{LH} - i_{SH}) - i_{SH} \tag{23}$$

where

i_{SH}: Supply current harmonics.
i_{LH}: Load current harmonics.
i_C: Compensational current.

3.4. Fuzzy Inferencing Control System (FIS)

One of the most important factors for classification of active harmonic filters is the regulation of the DC capacitor voltage. There is the need to maintain constant capacitor voltage so as to control the dynamic performance as well as the overall performance of the active filter. Traditionally, PI controllers are used for this function. The complexity in linearizing the system to tune the PI controller and its integral delay diminishes the usefulness of the PI controller as the control algorithm for the DC capacitor voltage. This research employed the FIS control model to regulate the DC capacitor voltage and compared its performance with that of the PI-controlled version. The selection of FIS over PI controllers was based on the avoidance of the complex tuning processes associated with PI controllers. Also, the simplicity and faster response time during transients, as well as better voltage tracking, encouraged the use of FIS controllers over the PI controller architecture.

Fuzzy logic enables computers to mimic human reasoning in quantifying incomplete and imprecise data to achieve definite conclusions. The authors of [30] presented extensive work on the implementation and architecture of fuzzy logic controls and further explanation of the topic can be obtained in their work. Figure 7a summarizes the FIS in a synoptic diagram, and Figure 7b shows its architecture for controlling the DC capacitor voltage in the proposed HAPF using MATLAB/Simulink building blocks.

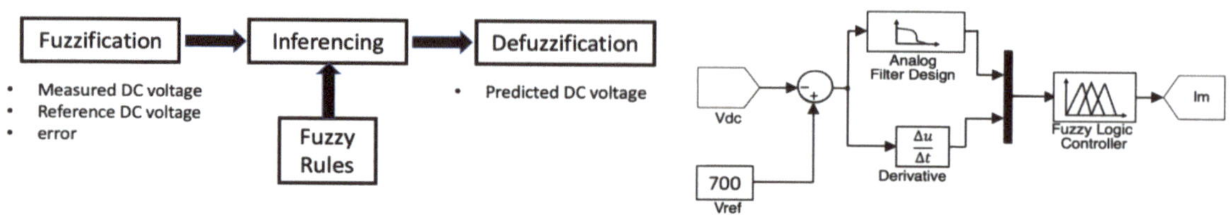

(a) (b)

Figure 7. (**a**) Synoptic fuzzy inference architecture; (**b**) DC capacitor voltage control block.

The DC capacitance voltage, V_{dc}, is compared with its reference, V_{ref}, and the error, e_{rr} ($V_{ref} - V_{dc}$), and its rate of change were used as the inputs to the fuzzy controller. The output of the fuzzy controller is added to the active power compensation block in the instantaneous power calculation. This increases the active power drawn from the supply just enough to compensate for the losses in the inverter operation.

The fuzzification process converted the input and output variables, in this case the DC capacitor voltage error and its rate of change, into linguistic fuzzy sets. These sets are assigned membership values based on the Mamdani style of referencing which indicates the level of belongingness. Figure 8a shows the adopted triangular membership functions and seven sets or levels of belongingness: more negative (MN), negative (N), partially negative (PN), zero error (Z), partially positive (PP), positive (P) and more positive (MP). Figure 8b shows the output of the fuzzy process in a surface plot view. The inferencing process involves a set of rules to control the FIS input to output coordination through modus ponens means. Table 2 shows the rules generated for the implementation. The degree of truth that a particular input has with the rules is measured and contributes to a specific output behavior [27,31,32]. The output of the FIS is finally converted into crisp values to be interpreted. The process of converting the FIS output into crisp quantities is the so-called defuzzification process.

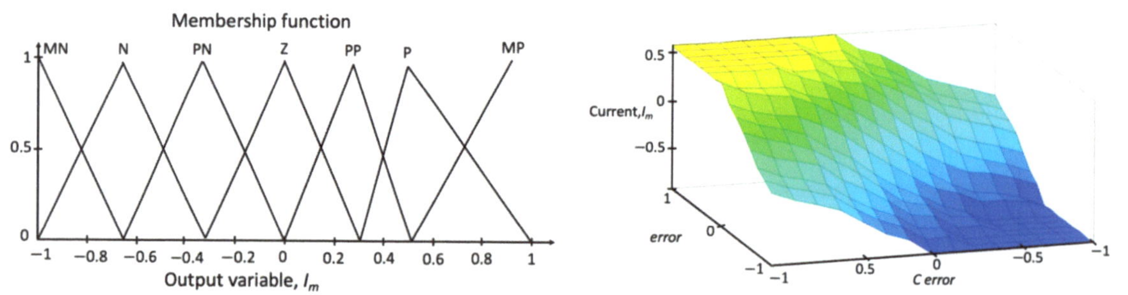

(a) (b)

Figure 8. (**a**) Level of membership function; (**b**) filter output as shown in surface view.

Table 2. FIS rule table.

	MN	N	PN	Z	PP	P	MP
MN	MN	MN	MN	MN	N	PN	Z
N	MN	MN	MN	N	PN	Z	PP
PN	MN	MN	N	PN	Z	PP	P
Z	MN	N	PN	Z	PP	P	MP
PP	N	PN	Z	PP	P	MP	MP
P	PN	Z	PP	P	MP	MP	MP
MP	Z	PP	P	MP	MP	MP	MP

4. Simulation and Results

The proposed HAPF was modelled in MATLAB/Simulink environment as shown in Figure 9. It consists of the parallel combination of a SAPF and an RLC filter in series to the input side of the MC. The model was simulated for 150 ms. The SAPF is only activated after 20 ms of simulation time by means of a circuit breaker action. Table 3 shows the parameters for the simulation.

Table 3. Simulation parameters.

Parameter	Value
Supply voltage ($V_{L\text{-}L}$)	400 V, 50 Hz
Matrix converter	50 kW, 8 kHz
LC low pass filter values, L_F, C_F	1 mH, 50 µF
DC reference voltage, V_{dc}	677.69 V min, 700 V max
Coupling inductance, L_{inv}	115 µH
DC capacitance, C_{DC}	600 µF

Equations (18)–(21) were used to separate the higher frequency current harmonics from the load current to generate reference currents. These generated reference currents are opposite and 180° out of phase with the load current. Hence, the algebraic sum of the load current and the reference current signals result in the fundamental supply current. This proves the effectiveness of the alpha–beta transformation. Figure 10 illustrates the simulated results obtained under the principle in MATLAB/Simulink, with the compensation current, load current and the sum as I_C, I_L and their sum, respectively.

An R-L load of similar rating was coupled to the load by means of a three-phase breaker which activated after 50 ms. It was observed that the proposed HAPF topology obtained harmonic compensation within half a cycle of the activation of breaker 2. All results obtained showed no phase lag between the voltage and the supply current. However, with the introduction of distorted supply voltages, a minimal lag was observed. This was corrected with the introduction of a phase locked loop (PLL) in the control design. The passive input filter impedance was carefully designed to attenuate the high-frequency harmonics enough to maintain the stability of the MC by reducing the passive filter capacitance.

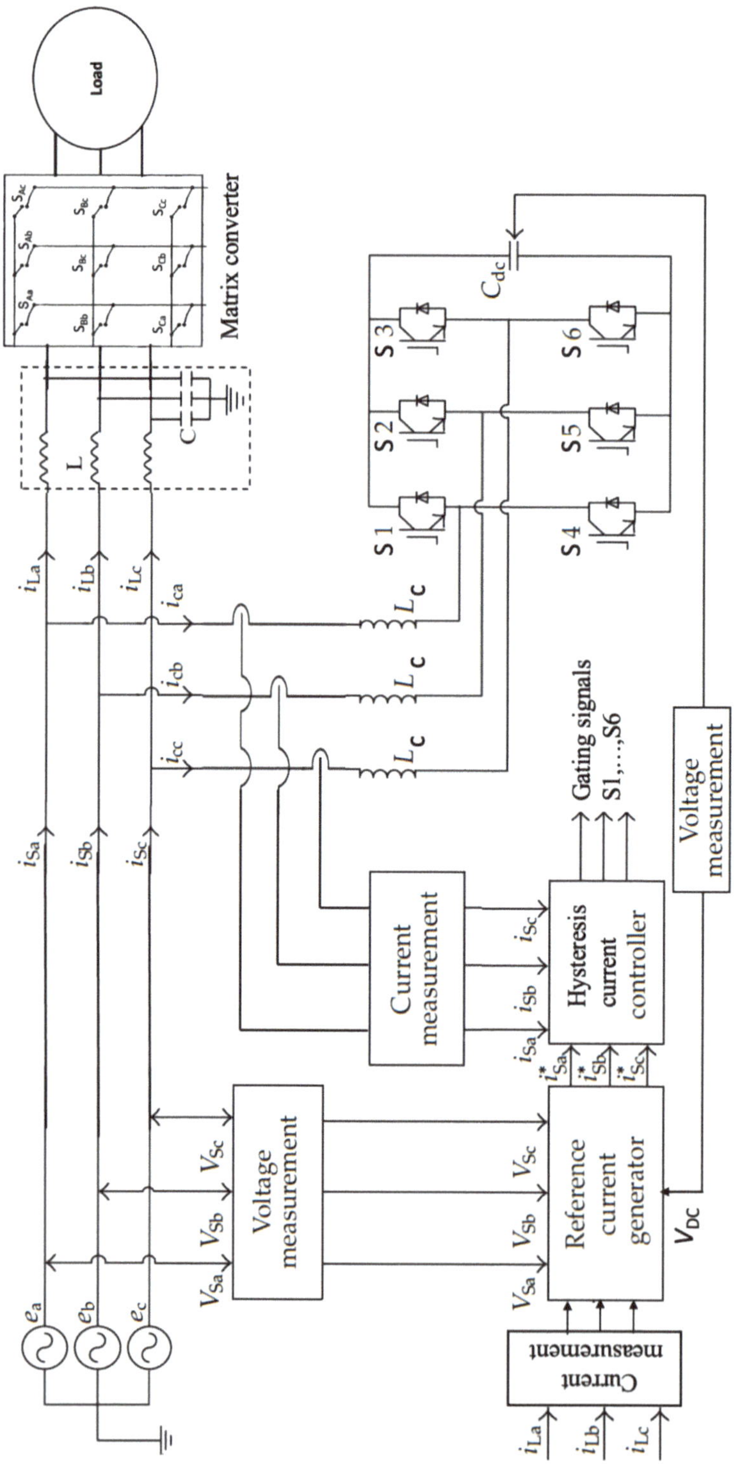

Figure 9. FIS-controlled HAPF architecture in MATLAB/Simulink environment.

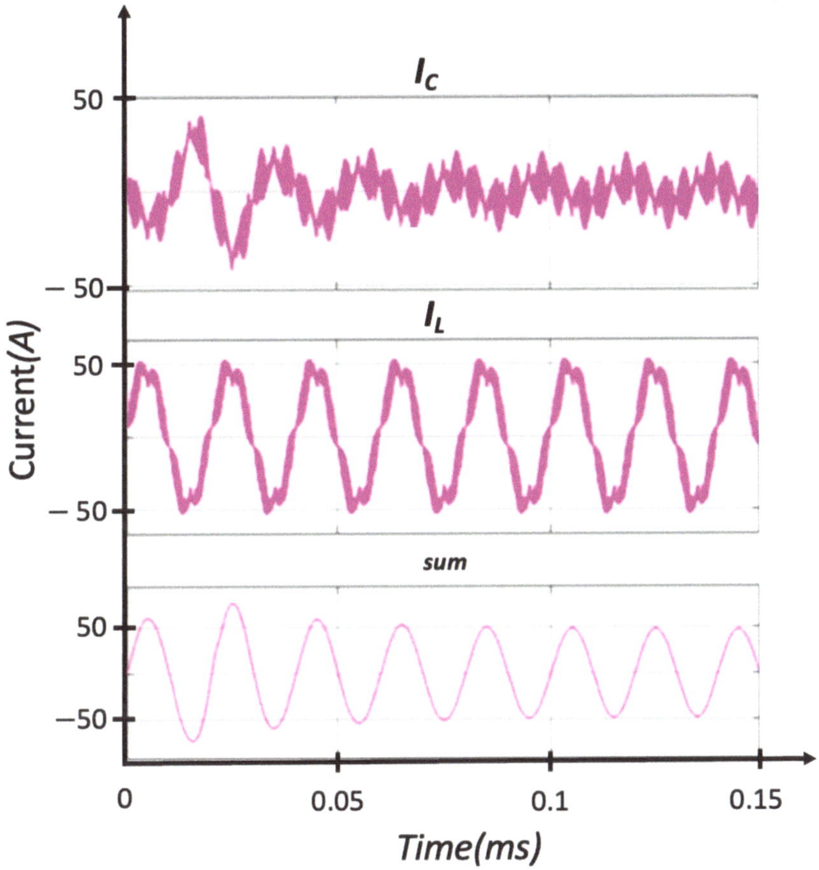

Figure 10. High-frequency error separation.

5. Discussion

Figure 11a shows the three-phase supply voltage (V_S) and current (I_S), as well as the load current (I_L) of the setup without any filter. It can be observed that the supply current is rich in high-frequency harmonics. These harmonics are obviously from the load current. FFT analysis of the supply current as well as the load current shows a 56.88% THD. High-frequency harmonics of a magnitude of 19.56% of the fundamental were observed at the switching frequency of the MC, as can be seen from Figure 12a. The simulation was performed with three different switching frequencies of the MC (6, 8 and 10) kHz. In all control experiments, the high frequencies were observed around the switching frequency of the MC. It was then concluded that the high frequencies propagated from the MC. Figure 11b shows the setup with the passive filter only. It was observed that the passive RLC filter successfully attenuated the high frequencies from 56.88% to 17.29% THD, and the higher frequencies from 19.56% to 7.64% THD, as could be observed from Figure 12b. The passive filter could not have further filtered without affecting the input power quality and the maximum power transfer of the MC. With these values as the new set point, the active filter now has a much lower rate of change of the input current.

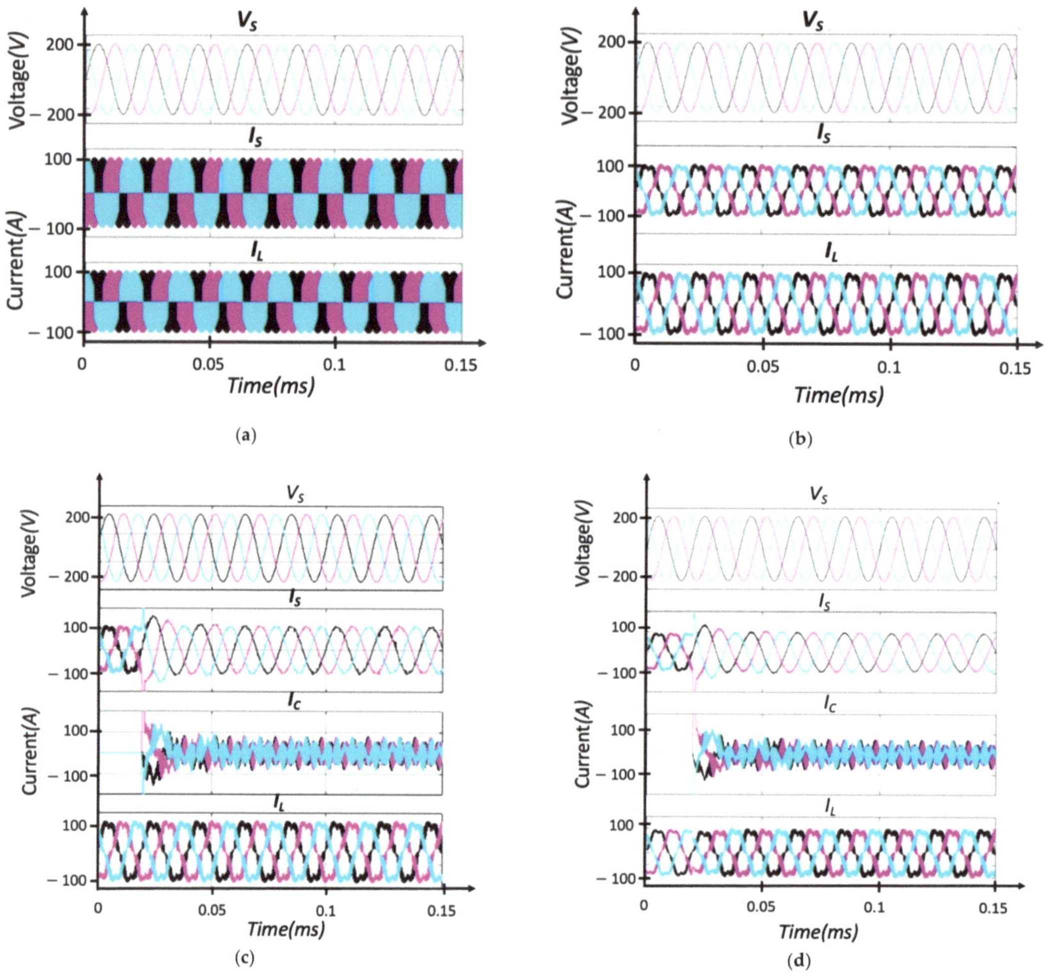

Figure 11. Setup with: (**a**) no filter; (**b**) passive R-L-C filter only; (**c**) HAPF with PI-controlled DC voltage; (**d**) HAPF with FIS-controlled DC voltage.

Figure 11c shows the shape of the three-phase supply voltage (*Vs*) and current (*Is*) as well as the compensating and load currents (I_C and I_L) of the setup with the PI-controlled SAPF. It could be observed that the supply voltage remained sinusoidal throughout the simulation time. In addition, after 20 ms of simulation time the supply current became sinusoidal and the load current remained distorted. The compensating current was highly distorted and 180° out of phase with the load current, which is most expected. FFT analysis of the supply current from Figure 12c indicated a 1.25% THD of the fundamental 50 Hz supply frequency. The higher ripple frequencies from the matrix converter were also recorded at the switching frequency with a reduced magnitude of the dominant harmonic, h155, as 0.15% of the fundamental. This shows a drastic improvement of the harmonic content in the supply current.

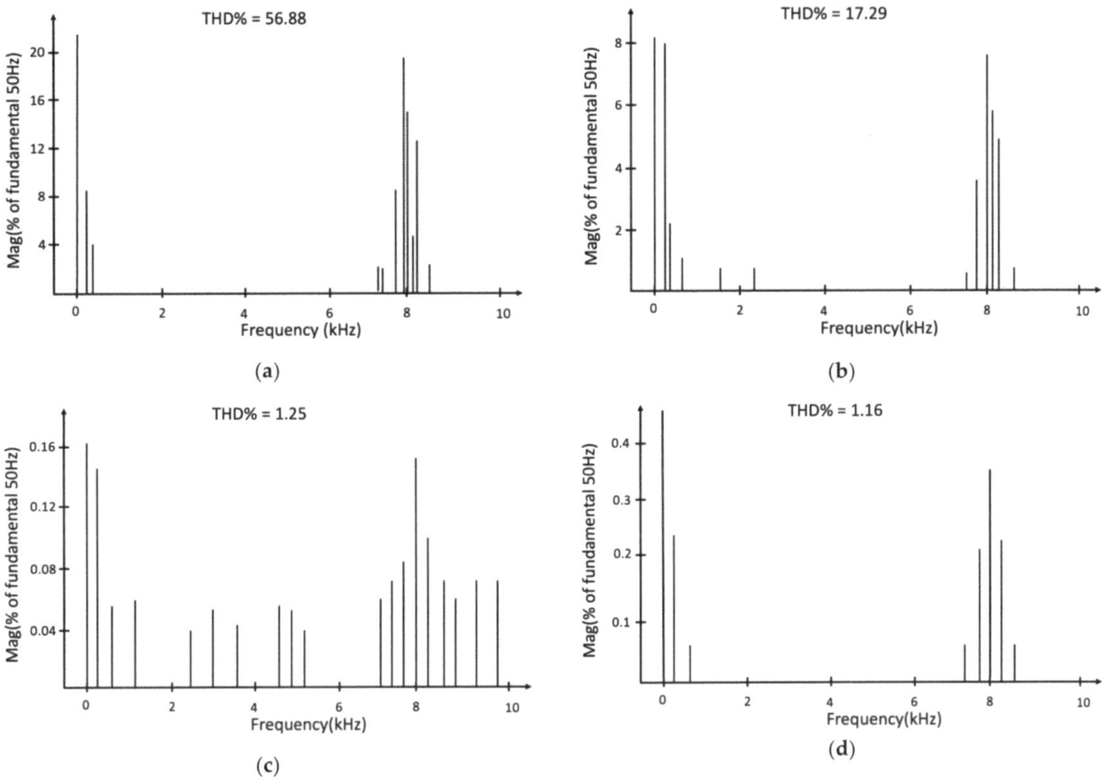

Figure 12. THD of the supply current: (**a**) without any filter action; (**b**) passive R-L-C filter only; (**c**) HAPF with PI-controlled DC capacitor voltage; (**d**) HAPF with FIS-controlled DC capacitor voltage.

Figure 11d shows the three-phase supply voltage (Vs) and its current (Is) shape, that of the compensating current (I_C), as well as the load current (I_L) of the setup with the FIS-controlled SAPF. While the supply voltage remained sinusoidal, the supply current became sinusoidal and in phase with the voltage after 20 ms of simulation time within half a cycle of settling time. FFT analysis of the supply current indicated a THD magnitude of 1.16% of the fundamental, and that of the dominant h155 from the MC as 0.37%, as shown in Figure 12d. Table 4 shows the comparison of harmonic contents before and after simulation under steady state operation for all scenarios.

Table 4. Comparison of results.

Simulation	THD% before Compensation	THD% after Compensation	THD% of Dominant 155th Harmonic before	THD% of Dominant 155th Harmonic after
R-L-C filter only	56.88	17.29	19.56	7.64
PI-controlled HAPF	56.88	1.25	19.56	0.15
FIS-controlled HAPF	56.88	1.16	19.56	0.37

6. Conclusions

High-frequency harmonics present in supply lines cause heating of transformers and motors, and interference with protective relays, metering devices and telecommunications equipment. The worst-case scenario is interference with hospital and laboratory setups and measuring equipment. As applications of high-frequency transmission are increasing worldwide, and with the challenge of using the power lines as communication lines, there is a need to protect sensitive loads and equipment that are not equipped to handle high frequencies, particularly harmonic frequencies. The proposed fuzzy-controlled HAPF limits the high-frequency harmonics to the input side of the MC, thereby reducing the overall harmonic content of the supply current. The results obtained show a massive improvement in the harmonic content of the supply current in conformity with the recommended EMI and IEEE Std. 519-2014 requirements.

The proposed HAPF obtained less than half a cycle of response after active filter activation. The backbone of the HAPF is the appropriate control strategy for the generation of the compensational currents. This research employed the hysteresis control strategy which introduces limitations to the control of the switching frequency of the SAPF section of the HAPF. The solution would be to employ PWM-controlled SAPF, which will further introduce complexities to the design. This research is part of an ongoing investigation into supraharmonic frequencies generated by the matrix converter and their compensation using active filters. There is the possibility of further reducing the harmonic content of the line current, as well as the rate of change of the input current, with minimal use of passive elements and future works are aimed at that. Model predictive controllers exhibit promising results in inverter applications and future works are aimed at replicating their use in the active part of the proposed HAPF.

Author Contributions: Conceptualization, F.B.E.; methodology, F.B.E. and A.K.; software, A.K.; validation, F.B.E. and A.K.; formal analysis, F.B.E.; investigation, F.B.E. and A.K.; resources, F.B.E. and A.K.; writing—original draft preparation, A.K.; writing—review and editing, F.B.E. and A.K.; visualization, F.B.E. All authors have read and agreed to the published version of the manuscript.

Funding: This research received no external funding.

Conflicts of Interest: The authors declare no conflict of interest.

Nomenclature

C_{DC}	DC link capacitance
C_f	Passive filter capacitance
FIS	Fuzzy inferencing control system
HAPF	Hybrid active power filter
I_{MC}	Matrix converter current at grid frequency
I_S	Supply current
I_C	Compensating current
I_L	Load current
L_f	Passive filter inductance
L_C/L_{inv}	Active filter coupling inductance
MC	Matrix converter
SAPF	Shunt active power filter
S_{MC}	Matrix converter total power
V_S	Supply Voltage
VSC	Voltage source converter
ω	Grid frequency in radian
ω_C	Conner frequency in radian

References

1. Mutharasan, A.; Niranjan, K.; Rameshkumar, T.; Ajitha, A. Analysis of Power quality improvement in Matrix Converter for WEC System. *Int. J. Appl. Eng. Res.* **2014**, *9*, 11365–11372.
2. Karaca, H.; Akkaya, R. A Novel Hybrid Compensation Method Reducing the Effects of Distorted Input Voltages in Matrix Converters. *Elektron. Ir Elektrotech.* **2019**, *25*, 15–21. [CrossRef]
3. Chiang, G.T.; Itoh, J. Voltage Transfer Ratio Improvement of an Indirect Matrix Converter by Single Pulse Modulation. In Proceedings of the 2010 IEEE Energy Conversion Congress and Exposition, Atlanta, GA, USA, 12–16 September 2010; p. 9. [CrossRef]

4. Dabour, S.M.; Rashad, E.M. Improvement of Voltage Transfer Ratio of Space Vector Modulated Three-Phase Matrix Converter. In Proceedings of the 15th International Middle East Power Systems Conference, Alexandria, Egypt, 23–25 December 2012; pp. 581–586.
5. Goel, S.; Tiwari, D.; Pandey, M.K.; Bajpai, S. Power Quality Conditioners for Matrix Converter Using Shunt and Series Active Filters. *Int. J. Electron. Electr. Eng.* **2013**, *6*, 119–130.
6. Li, D.; Wang, T.; Pan, W.; Ding, X.; Gong, J. A comprehensive review of improving power quality using active power filters. *Electr. Power Syst. Res.* **2021**, *199*, 107389. [CrossRef]
7. Paul, P.J. Shunt Active and Series Active Filters-Based Power Quality Conditioner for Matrix Converter. *Adv. Power Electron.* **2011**, *2011*, 930196. [CrossRef]
8. Karaman, O.A.; Erken, F.; Cebeci, M. Decreasing Harmonics via Three Phase Parallel Active Power Filter Using Online Adaptive Harmonic Injection Algorithm. *Teh. Vjesn.—Tech. Gaz.* **2018**, *25*, 157–164. [CrossRef]
9. Wakileh, G.J. Effects of Harmonic Distortion on Power Systems. In *Power Systems Harmonics*; Springer: Berlin/Heidelberg, Germany, 2001; pp. 81–104. [CrossRef]
10. Shah, A.; Vaghela, N. Shunt Active Power Filter for Power Quality Improvement in Distribution Systems. *Int. J. Eng. Dev. Res.* **2005**, *13*, 22–26.
11. Koduah, A.; Svinkunas, G. Switching Harmonic Ripple Attenuation in a Matrix Converter-Based DFIG Application. In Proceedings of the 2022 IEEE 7th International Energy Conference (ENERGYCON), Riga, Latvia, 9–12 May 2022; pp. 1–7. [CrossRef]
12. Yanchenko, S.; Meyer, J. Impact of Network Conditions on the Harmonic Performance of PV Inverters. In Proceedings of the 2018 Power Systems Computation Conference (PSCC), Dublin, Ireland, 11–15 June 2018; pp. 1–7. [CrossRef]
13. Klatt, M.; Stiegler, R.; Meyer, J.; Schegner, P. Generic frequency-domain model for the emission of PWM-based power converters in the frequency range from 2 to 150 kHz. *IET Gener. Transm. Distrib.* **2019**, *13*, 5478–5486. [CrossRef]
14. Meyer, J.; Haehle, S.; Schegner, P. Impact of higher frequency emission above 2kHz on electronic mass-market equipment. In Proceedings of the 22nd International Conference and Exhibition on Electricity Distribution (CIRED 2013), Institution of Engineering and Technology, Stockholm, Sweden, 10–13 June 2013; p. 0999. [CrossRef]
15. Choudhary, J.; Singh, D.K.; Verma, S.N.; Ahmad, K. Artificial Intelligence Based Control of a Shunt Active Power Filter. *Procedia Comput. Sci.* **2016**, *92*, 273–281. [CrossRef]
16. Kaufhold, E.; Meyer, J.; Schegner, P. Impact of harmonic distortion on the supraharmonic emission of pulsewidth modulated single-phase power electronic devices. *Renew. Energy Power Qual. J.* **2021**, *19*, 577–582. [CrossRef]
17. Bollen, M.H.J.; Ribeiro, P.F.; Anders Larsson, E.O.; Lundmark, C.M. Limits for Voltage Distortion in the Frequency Range 2 to 9 kHz. *IEEE Trans. Power Deliv.* **2008**, *23*, 1481–1487. [CrossRef]
18. *IEEE Std 519-1992*; IEEE Recommended Practices and Requirements for Harmonic Control in Electric Power Systems. Institute of Electrical and Electronic Engineers: New York, NY, USA, 1992.
19. Casadei, D.; Serra, G.; Tani, A.; Zarri, L. Stability analysis of electrical drives fed by matrix converters. In Proceedings of the IEEE International Symposium on Industrial Electronics ISIE-02, L'Ayuila, Italy, 8–11 July 2002; Volume 4, pp. 1108–1113. [CrossRef]
20. Petrauskas, G.; Svinkunas, G. Innovative Filter Topology for Power Grid Protection from Switching Ripple Harmonics Produced by Matrix Converter. *Iran. J. Sci. Technol. Trans. Electr. Eng.* **2018**, *43*, 495–505. [CrossRef]
21. Monteiro, J.; Silva, J.F.; Pinto, S.F.; Palma, J. Matrix Converter-Based Unified Power-Flow Controllers: Advanced Direct Power Control Method. *IEEE Trans. Power Deliv.* **2011**, *26*, 420–430. [CrossRef]
22. Monteiro, J.; Pinto, S.; Delgado Martin, A.; Silva, J. A New Real Time Lyapunov Based Controller for Power Quality Improvement in Unified Power Flow Controllers Using Direct Matrix Converters. *Energies* **2017**, *10*, 779. [CrossRef]
23. Siva, B.V.; Babu, B.M.; Srinivas, L.R.; Tulasiram, S.S. Design of Shunt Active Power Filter for Improvement of Power Quality with Artificial Intelligence Techniques. *Int. J. Adv. Res. Electr. Electron. Instrum. Eng.* **2014**, *3*, 11304–11314. [CrossRef]
24. Koduah, A.; Svinkunas, G.; Ampofo, D.O. Design of a Shunt Active Power Filter for Direct Matrix Converter Application. In Proceedings of the 2021 IEEE PES/IAS PowerAfrica, Nairobi, Kenya, 23–27 August 2021; pp. 1–5. [CrossRef]
25. Alali, M.A.E.; Chapuis, Y.-A.; Saadate, S.; Braun, F. Advanced common control method for shunt and series active compensators used in power quality improvement. *IEE Proc.—Electr. Power Appl.* **2004**, *151*, 658. [CrossRef]
26. Janabi, A.; Wang, B. Hybrid matrix converter based on instantaneous reactive power theory. In Proceedings of the IECON 2015—41st Annual Conference of the IEEE Industrial Electronics Society, Yokohama, Japan, 9–12 November 2015; pp. 003910–003915. [CrossRef]
27. Jain, S.K.; Agrawal, P.; Gupta, H.O. Fuzzy logic controlled shunt active power filter for power quality improvement. *IEE Proc.—Electr. Power Appl.* **2002**, *149*, 317–328. [CrossRef]
28. Akagi, H.; Watanabe, E.H.; Aredes, M. *Instantaneous Power Theory and Applications to Power Conditioning*, 2nd ed.; IEEE Press/Wiley: Hoboken, NJ, USA, 2017.
29. Watanabe, E.H.; Monteiro, L.F.C.; Aredes, M.; Akagi, H. Instantaneous p-q Power Theory for Control of Compensators in Micro-Grids. In Proceedings of the 2010 International School on Nonsinusoidal Currents and Compensation, Lagow, Poland, 15–18 June 2010; p. 11.
30. Bissey, S.; Jacques, S.; Le Bunetel, J.-C. The Fuzzy Logic Method to Efficiently Optimize Electricity Consumption in Individual Housing. *Energies* **2017**, *10*, 1701. [CrossRef]

31. Roy, R.B.; Cros, J.; Basher, E.; Taslim, S.M.B. Fuzzy logic based matrix converter controlled indution motor drive. In Proceedings of the 2017 IEEE Region 10 Humanitarian Technology Conference (R10-HTC), Dhaka, Bangladesh, 21–23 December 2017; pp. 489–493. [CrossRef]
32. Tsengenes, G.; Adamidis, G. Shunt active power filter control using fuzzy logic controllers. In Proceedings of the 2011 IEEE International Symposium on Industrial Electronics, Gdansk, Poland, 27–30 June 2011; pp. 365–371. [CrossRef]

energies

MDPI

Article

Source for Autonomous Power Supply System Based on Flow Battery

Ivan Kuzmin [1], Alexey Loskutov [1], Evgeny Osetrov [2] and Andrey Kurkin [3,*]

[1] Department of Electric Power Engineering, Power Supply and Power Electronics, Nizhny Novgorod State Technical University n.a. R.E. Alekseev, 603950 Nizhny Novgorod, Russia; ink0981@mail.ru (I.K.); loskutov@nntu.ru (A.L.)
[2] Technocomplekt LLC, Moscow Region, 141981 Dubna, Russia; osetrov@techno-com.ru
[3] Department of Applied Mathematics, Nizhny Novgorod State Technical University n.a. R.E. Alekseev, 603950 Nizhny Novgorod, Russia
* Correspondence: aakurkin@nntu.ru

Abstract: The article deals with the urgent task of creating a technological and production basis for the development and serial production of energy storage systems with flow batteries and uninterruptible power systems based on them. Flow batteries are a highly efficient solution for long-term energy storage in critical and alternative energy facilities. The main advantage of the flow batteries is the ability to create a system with the required power and capacity without redundant parameters due to the fact that the characteristics of the system are regulated by independent blocks, as in a fuel cell. Among flow batteries, vanadium redox flow batteries (VRFB) are of particular interest, as they have a long service life. The main elements of a flow battery are the stack, which determines the power of the battery and its efficiency, and the electrolyte, which determines the energy capacity of the battery and its service life. A stand for testing the operating modes of the flow battery stack has been developed. A 5 kW flow battery operating on an electrolyte with the addition of hydrochloric acid, which is a stabilizer in new generation electrolytes, has been tested.

Keywords: flow battery; vanadium flow battery; vanadium redox flow battery; electrolyte synthesis; control system

Citation: Kuzmin, I.; Loskutov, A.; Osetrov, E.; Kurkin, A. Source for Autonomous Power Supply System Based on Flow Battery. *Energies* **2022**, *15*, 3027. https://doi.org/10.3390/en15093027

Academic Editor: Gabriel Nicolae Popa

Received: 17 March 2022
Accepted: 19 April 2022
Published: 21 April 2022

1. Introduction

As a rule, two independent power supply sources are used to ensure reliability at facilities requiring uninterrupted power supply. A backup source is installed additionally if there is a special group of consumers. An uninterrupted power supply system that ensures a constant quality of power supply may also be needed if power outages are too long. An uninterruptible power supply (UPS) or automatic transfer switch (ATS) to an alternative network is installed in the system, which allow a qualitative transition to the backup power line to be performed during the shutdown of the main source. To do this, energy storage devices are introduced into the UPS: batteries [1–4], supercapacitors [5], mechanical energy storage devices [6], or others.

Flow batteries (vanadium redox flow battery, VRFB) are one of the most efficient energy storage devices. They are suitable for long-term energy storage and have good cost characteristics with prospects for reducing the cost of stored kWh of energy [7].

The main technical characteristics of VRFB are as follows:

Energy storage density: 30–70 J/g.

Charge/discharge efficiency: 75–80%.

Service life: 10–20 years as part of an UPS system.

Number of charge/discharge cycles: 10,000.

Nominal cell voltage: 1.15–1.55 V.

The advantages of VRFB are durability, four-fold overload capability, low self-discharge, and low cost [8].

The disadvantages of VRFB are the complexity of the control system and the low energy storage density, which leads to an increase in weight and dimensions.

The energy of VRFB is stored in separated positive and negative electrolytes, which provide the driving force that initiates the oxidation–reduction reaction [8]. The use of energy stored in the form of chemical energy is a mode of storage that does not entail any geographical restrictions. This is expected to facilitate large-scale energy storage and has significant environmental and socio-economic advantages. VRFBs have been utilized as large-scale energy storage systems in the last decade [9–12], because they are one of the most promising technologies for mid-to-large-scale (kW to MW) energy storage and was first proposed by Skyllas-Kazacos in 1985 [9,13].

Currently, electric energy storage systems (ESS) (which also include VRFB-based uninterruptible power systems) are widely used in various sectors of the economy. There are already more than 100 facilities equipped with such systems with a total capacity of more than 300 MW, according to the US Department of Energy (DOE). The main producers are the USA and China. The share of the installed capacity of flow batteries in relation to other electrochemical energy storage devices is 10% and is increasing every year [5].

To optimize the flow battery design, it is necessary to understand the flow distribution, local current distribution, limits, and maximum current density. Understanding the shunt current and pressure distribution allows the flow battery stack to be designed with high power, large capacity, and high system efficiencies [9,10].

The purpose of the article is to develop and manufacture a prototype of an UPS system based on VRFB with a rated power of 5 kW and a capacity of 15 kWh.

A technology for manufacturing an electrolyte for a flow battery has been developed, and it has been synthesized at a volume sufficient for laboratory research. A 5 kW VRFB has been tested, and a layout of laboratory technological equipment for industrial electrolyte synthesis has been developed [9].

To achieve this goal, the following main tasks have been performed:

- Analysis of new-generation UPS systems based on flow batteries.
- Study of the influence of the characteristics and design of electric energy storage devices on the operation of UPS systems.
- The development of power converters for hybrid specialized sources.
- Mathematical description and control algorithms for power supply systems with specialized power supplies.
- Experimental studies of operating modes of specialized sources in power-supply systems with developed infrastructure.

The scientific novelty lies in the development and research of unique control algorithms for a UPS system based on batteries, depending on the consumer's load, which will increase the ESS efficiency. It is also necessary to conduct a study of the operability of the UPS system in emergency situations in order to optimize the operation of the control algorithm. Within the framework of this work, the stability of the electrolyte during cycling has been determined, and ways of restoring the resource of a vanadium redox flow battery without changing the electrolyte in the system have been investigated.

2. VRFB Cell Operation

The principle of a redox flow battery with vanadium as the active material is shown in Figure 1. A VRFB consists of flow-type cells, electrolyte tanks, pumps, and piping. The VRFB cell consists of two porous graphite electrodes through which electrolytes separated by an ion-exchange membrane are passed [1,5]. The electrolytic reactions take place in the cell, and each electrolyte tank stores a solution of the active material (electrolyte). The electrolytes (positive and negative) are pumped by displacement pumps and circulate between the tank and electrolytic cells. When an electric current is loaded to the cell, the battery reactions, which change the valence of the vanadium, occur in both the positive and negative electrodes, as shown in Figure 1 and expressed by the following equations. The valence change moves protons through the membrane, charging or discharging the battery according to the mechanism.

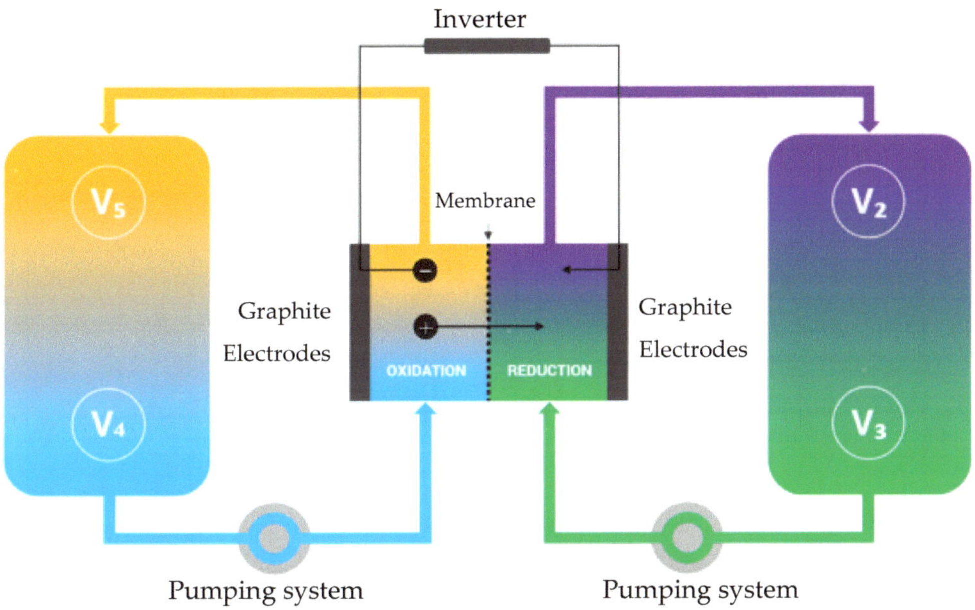

Figure 1. VRFB cell: V_2—discharged cathode electrolyte V^{2+}: V_3—charged cathode electrolyte V^{3+}; V_4—discharged anode electrolyte V^{3+}: V_5—charged anode electrolyte V^{5+}.

The reactions during the discharge process are follows:

- At the negative electrode:

$$V^{2+} + e^{-} = V^{3+};\ E^{0} = -0.255\ V; \quad (1)$$

- On the positive:

$$VO_2^{+} + e^{-} + 2H^{+} = VO^{2+} + H_2O;\ E^{0} = 1.000\ V; \quad (2)$$

- Total reaction:

$$VO_2^{+} + V^{2+} + 2H^{+} = VO^{2+} + V^{+3} + H_2O;\ E^{0} = 1.255\ V. \quad (3)$$

The difference between these electrode potentials results in a total cell voltage of 1.25 V in 1 mol acid solution, which increases with decreasing pH, since only the second reaction is proton-dependent and therefore has a higher potential at lower pH.

The ion exchange membrane (solid electrolyte) that separates the electrolyte streams from each other is the weakest element in the VRFB. The ion exchange membrane degrades during VRFB operation [1,5].

3. Materials and Methods

A laboratory stand was developed and manufactured for testing the VRFB. The hydrodynamic scheme of the VRFB consists of two tanks with positive and negative electrolytes, with separate pumps creating electrolyte flows (Figure 2).

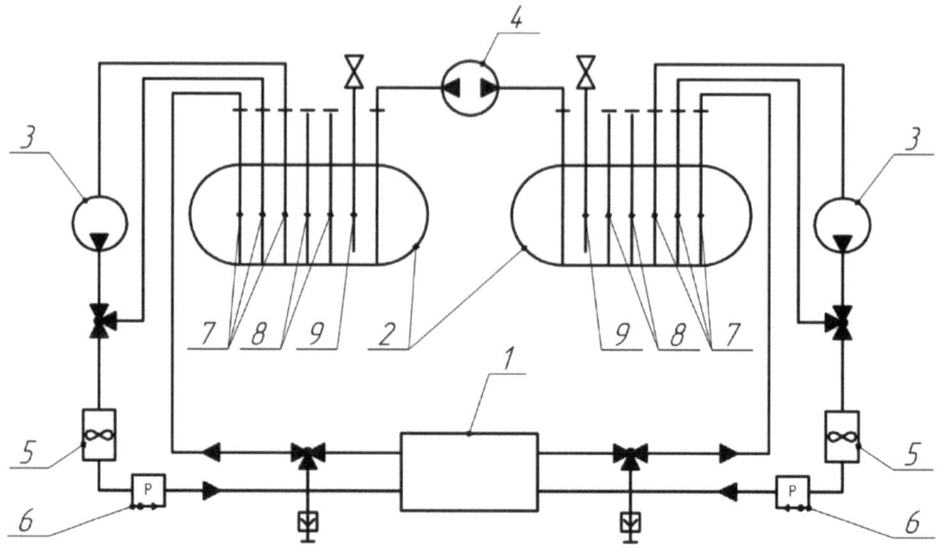

Figure 2. Hydrodynamic scheme of the stand: 1—flow battery stack; 2—tanks with electrolyte; 3—main electrolyte pumps; 4—balancing pump; 5—electrolyte flow sensors; 6—pressure sensors; 7—outlet pipes R-32; 8—branch pipes with an outlet in the form of a fitting; 9—branch pipes KD-10 with pressure relief valves.

The flow cell is the main element of the flow battery, namely the vanadium acid battery (VRFB). In the flow cell, some of the power is lost due to the need to pump the electrolyte through the porous carbon electrode and the hydraulic resistance of the materials. One of the approaches to reduce hydrodynamic resistance was to change the design of bipolar plates, which, according to the proposed concept, should include a serpentine channel, and the exchange of electrolyte with a porous carbon material is carried out due to diffusion of ions, osmotic pressure, and turbulent flows at the boundary, where the electrolyte flowing through the channel is in contact with the porous carbon electrode.

Mathematical modeling of hydrodynamics was carried out in the COMSOL Multiphysics. The purpose of this model is to compare 2 different flow cell designs [1]. Figure 3 shows models of half cells that simulate fluid flows through the channel. The size of the calculation area was 150 × 120 mm, the depth of the channels varies from 0.5 to 1.5 mm, the width of the channels varies from 3 to 8 mm, and the number of channels varies from 3 to 19. The material of the plate is graphite with a resistivity of 8 μΩ·m. The size of the porous carbon material in the case of a design with a serpentine channel is 150 × 120 × 1 mm, and the porosity is 92%. The most optimal design, i.e., Design 2, was then compared with Design 1.

The calculation of the pressure drop and the velocity field has been made by numerically solving the system of equations in the COMSOL. The electrolyte flow in the flow channel is described by the Navier–Stokes equations and the continuity equation:

$$\begin{cases} \rho(u\cdot\nabla)u = -\nabla p + \mu\nabla^2 u, \\ \rho\nabla\cdot u = 0, \end{cases} \tag{4}$$

where ρ is an electrolyte density, u is an electrolyte flow rate, p is an electrolyte pressure, and μ is an electrolyte viscosity.

The flow in carbon felt is described by the Brinkman model. Here, the modified Navier–Stokes [1] equations are written for the velocity field averaged over volumes much smaller than the size of the structure, but much larger than the size of the pores (according to the comparative analysis presented in [12,13], this model is the most suitable for Designs 1 and 2):

$$\begin{cases} \rho(u\cdot\nabla)u = -\nabla p - \frac{\mu}{\kappa}u + \mu\nabla^2 u, \\ \rho\nabla\cdot u = 0, \end{cases} \tag{5}$$

where κ is a permeability of carbon felt.

The electrolyte parameters are as follows: electrolyte density $\rho = 1.355\ \mathrm{g/cm^3}$, dynamic viscosity μ = 2.5 MPa·s. For various combinations of parameters, the Reynolds number varied from 15 to 43. A laminar flow model was chosen for the calculation.

Figure 4 shows the power curves of the cells and their efficiency $\eta(Q)$ depending on the electrolyte flow rate through the cell, taking into account electrical resistance losses (since the electrical contact in the case of a serpentine channel is worse than in the case of a solid plate).

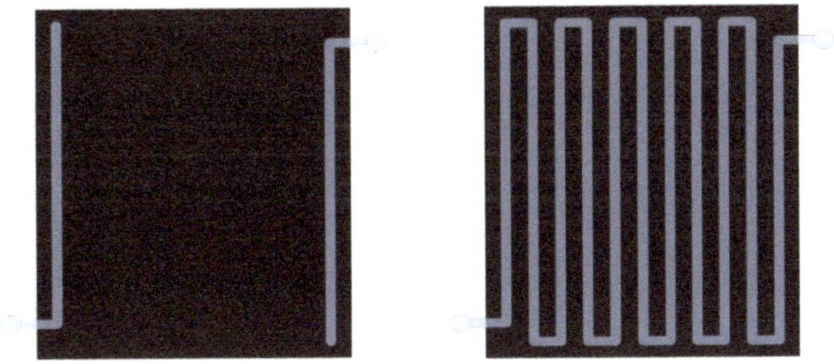

Figure 3. Type of flow half-cells of various designs.

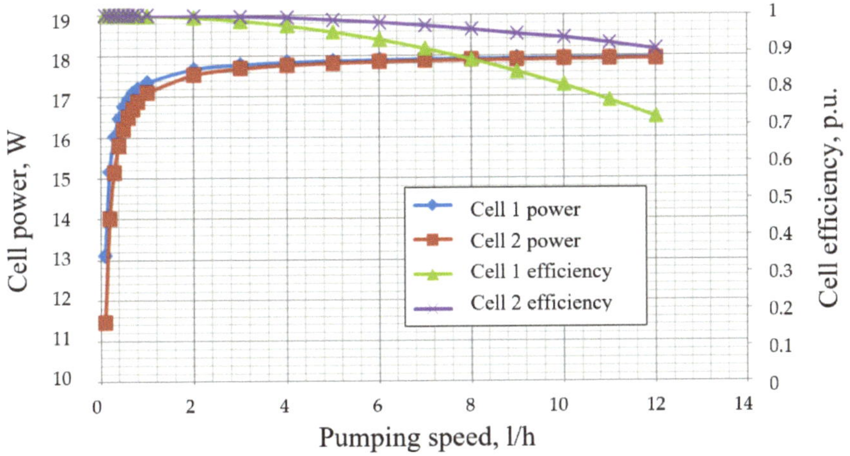

Figure 4. Dependences of the cells power and their efficiency on the rate of electrolyte flow through the cell.

It can be seen that the power depending on the electrolyte flow rate for Design 1 changes almost the same as for Design 2, but at the same time, the efficiency in the cells of Design 1 decreases with the increasing electrolyte flow rate more significantly than for the cell of Design 2, which is associated with a stronger increase in hydraulic resistance compared to a cell with a serpentine channel. Thus, the design of a flow cell with a serpentine channel is more acceptable in terms of reducing losses due to hydrodynamic resistance.

Based on the results, an algorithm was developed to determine the power of a flow cell depending on the electrolyte flow rate, and based on the results of the cell power, a comparison was made of the 2 designs of a flow cell. According to a comparative assessment, the most promising is the design of a flow cell with a serpentine channel for electrolyte flow, because in this case, the value of hydrodynamic losses is comparatively lower.

The thermal effect analysis results showed that the parameters of the charge/discharge test (flow rate and current density) affect the thermal effect of the VRFB. High current density causes a phenomenon of concentration overvoltage, and the thermal release is more serious. When the flow rate is too slow, this causes the concentration gradient and a higher ohmic resistance heat release, which reduce the efficiency.

The battery performance is significantly influenced by the flow rate of both electrolyte and coolant. Regulating the coolant flow rate is an effective approach to control the electrolyte temperature within the expected range. A comparison shows that the electrolyte flow rate has a negligible impact on electrolyte temperature within the considered range of flow rate but significantly influences the pressure drop and battery efficiencies. The increase in electrolyte flow rate improves the coulombic efficiency, voltage efficiency, and energy efficiency simultaneously. However, the system efficiency does not improve monotonically with the increase of electrolyte flow rate, because the associated pressure drop and pump power also increase. Instead, an optimal flow rate exists to maximize the system efficiency under a specific operating condition. The optimal flow rate of the electrolyte also differs when the applied current density changes [14].

Figure 5 shows the temperature curves. The temperature for the positive electrolyte rises faster than for the negative ones. This trend is observed for all modes. As is known, in addition to the release of heat due to electrical resistance in these batteries, there is an electrolyte crossover [8,10]. Due to the fact that V^{2+} and V^{3+} have a smaller diameter than the VO_2^+ and VO^{2+} ions, the flow of ions from the tank with a negative electrolyte to the tank with a positive electrolyte is stronger than in the opposite direction. This, most likely, causes additional overheating of the positive electrolyte. The difference in temperature between the positive and negative electrolytes ranges from 7 to 11%.

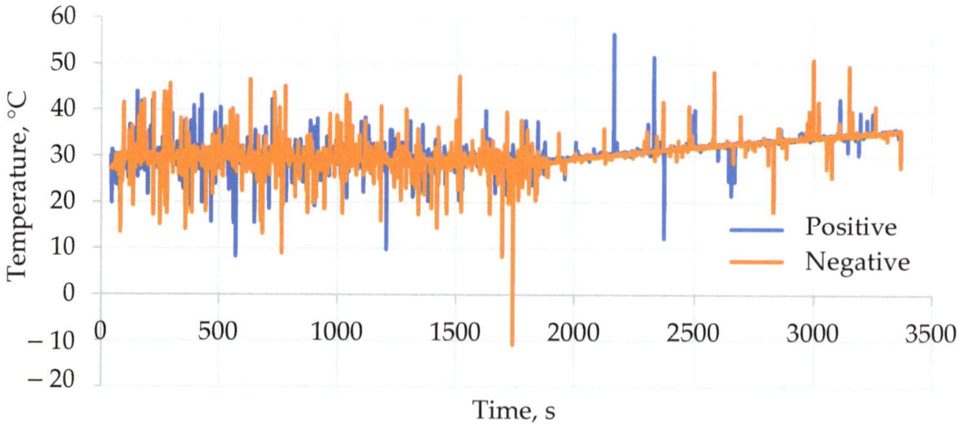

Figure 5. Positive and negative electrolyte temperature curves.

The anion exchange membrane FAP450 showed a similar order of magnitude of permeability rates for all four vanadium ions, while in the case of the cation exchange membranes F930 and VB2, the permeability of V^{2+} and VO_2^+ ions was almost an order of magnitude lower than that of V^{3+} and VO^{2+}. As expected, the thicker VRFB of the two cation exchange membranes showed the lowest permeability values. The experimental data were applied in simulation studies to investigate the effect of ion crossover on thermal behavior associated with self-discharge reactions caused by ion diffusion across the membrane. In order to reduce the rise in stack temperature when pumps were switched off during standby periods at high SOC, lower order of magnitudes of the permeability rates of V^{2+} and ${VO_2}^+$ ions are essential [11,13,15,16].

By comparing different electrolyte pumping speeds (Figure 4), it can be seen that the pumping speed has practically no effect on the stack efficiency (Table 1). This can be explained by the absence of an effect of the electrolyte pumping rate on the electrolyte crossover through the membrane and the overvoltage on the electrodes. However, increasing the pumping speed slightly increases the battery capacity, which is associated with a more complete use of the electrolyte. The optimal value of the stack pumping speed is 2 m^3/h.

Table 1. Test results.

No.	Discharge Current, A	Positive Electrolyte Temperature T+, °C	Negative Electrolyte Temperature T−, °C	Efficiency η(Q), %	Efficiency η(E), %	Discharge Capacity, kWh
1	99	6.1	5.5	91.2	68.4	2.16
2	101	5.9	5.3	89.8	68.9	2.25
3	98	5.0	4.6	82.3	64.0	1.71
4	100	5.9	5.5	92.5	68.3	2.09
5	122	5.7	5.3	98.1	69.9	2.05

As can be seen from the data (Figure 6), a decrease in the charge voltage leads to a sharp decrease (by 24%) in battery capacity and a decrease in stack efficiency by 4.9%. The decrease in efficiency is largely due to the drop in charge efficiency. The low coulomb efficiency can be associated with a high electrolyte crossover due to the fact that this charge mode took almost 2 times more time to charge, which led to large losses and waste of excess energy during charging, and this reduced the overall efficiency. Thus, a decrease in the operating voltage steadily leads to a decrease in efficiency.

The battery not only retains its high efficiency but also practically does not lose capacity compared to the nominal modes when overloaded by 1.5 times during charging. An increase in efficiency with an increase in the discharge current, as in the previous case, is associated with a decrease in time and, as a consequence, a decrease in Coulomb losses. The decrease in capacity, although not significant, is a consequence of the incomplete charge of the battery at the initial stage. Thus, it can be seen that a VRFB can operate well in overload mode without significant loss in efficiency. This also demonstrates the good compatibility of this stack with the developed and used electrolyte.

Side reactions such as hydrogen evolution due to water decomposition and CO_2 evolution due to the oxidation of carbon-based electrode may occur during operation [3,5]. The battery performance is generally evaluated with three efficiencies: coulombic efficiency (CE), voltage efficiency (VE), and energy efficiency (EE).

Figure 6 shows the current and voltage curves during charging and discharging. The energy efficiency was calculated from the graphs using (6), and the charge efficiency was calculated using (7)

$$\eta(E) = \frac{\int_0^{t_{Dch}} U_{Dch}(t) I_{Dch}(t) dt}{\int_0^{t_{Ch}} U_{Ch}(t) I_{Ch}(t) dt} \cdot 100\%; \quad (6)$$

$$\eta(Q) = \frac{\int_0^{t_{Dch}} I_{Dch}(t) dt}{\int_0^{t_{Ch}} U_{Ch}(t) I_{Ch}(t) dt} \cdot 100\%, \quad (7)$$

where $\eta(E)$ and $\eta(Q)$ are energy and charge efficiency, respectively, and I_{Dch}, U_{Dch}, t_{Dch}, and I_{Ch}, U_{Ch}, t_{Ch} are currents, voltages, and times in the discharge and charge modes, respectively.

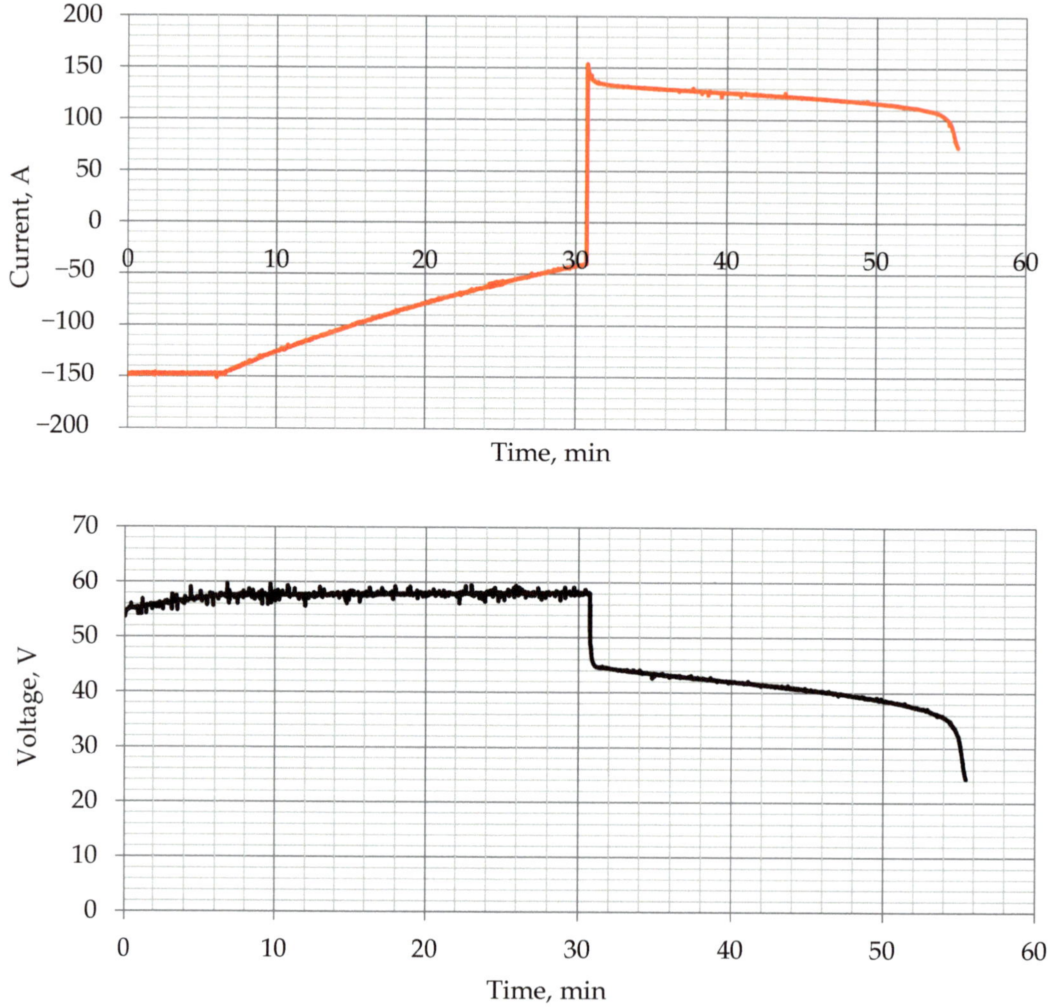

Figure 6. Current and voltage curves when charging and discharging.

As can be seen from Figure 6, at a discharge voltage of 33–34 V on the stack and a current of 100 A, an inflection begins, which indicates the end of the discharge. Thus, the spread between the charge and discharge voltage is 24–25 V, which is 43% of the charge voltage. Such a voltage spread requires special designs of inverters.

4. Prototype

The VRFB source electrical circuit consists of inverters that convert the VRFB voltage into load and mains voltage, as well as a PLC-based VRFB control system (Figures 7 and 8) [6,7]. A similar setup with VRFB was described in detail in [17].

The VRFB control system collects, analyzes, and stores data from sensors, as well as automatically and manually controls pumps in order to pump the electrolyte through the cells of the flow accumulator.

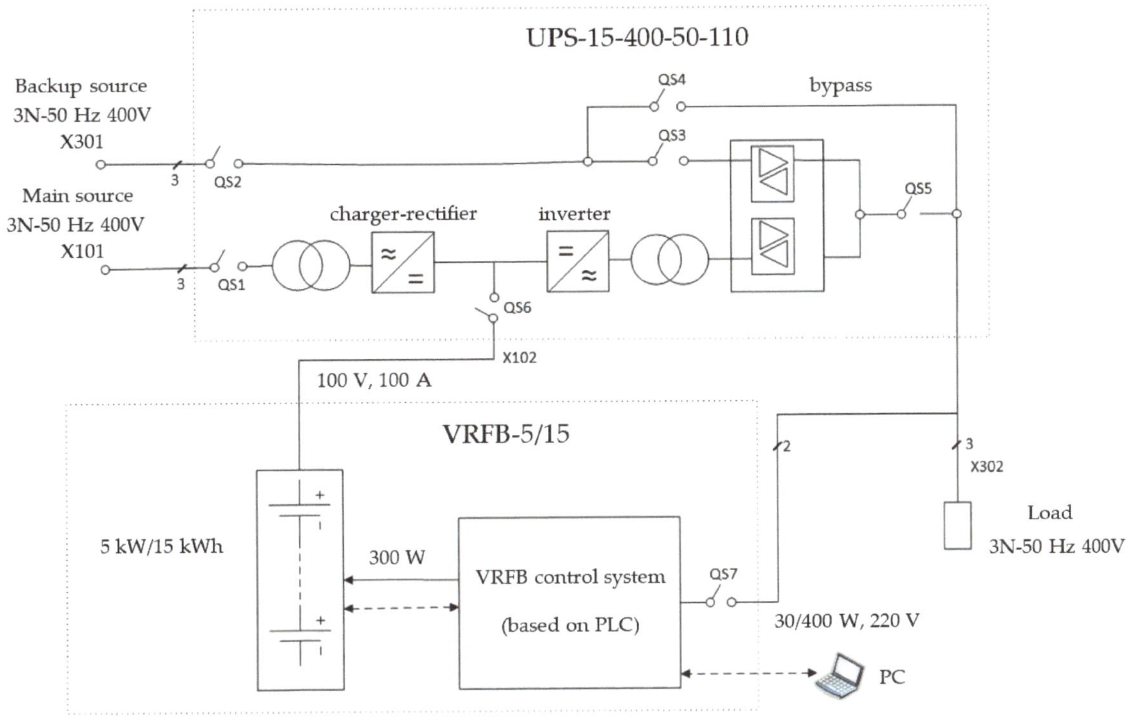

Figure 7. Block diagram of the UPS system based on 10 kW/30 kWh VRFB.

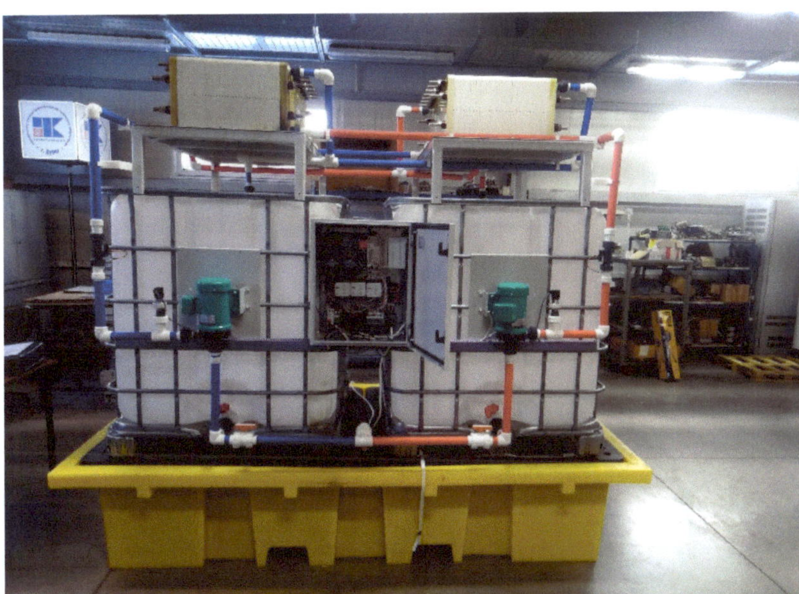

Figure 8. Prototype of the UPS system based on 5 kW/15 kWh VRFB.

A software for online remote monitoring of the VRFB operation allows the user to remotely receive all the necessary data from the PLC, as well as control all the functions provided (Figure 9).

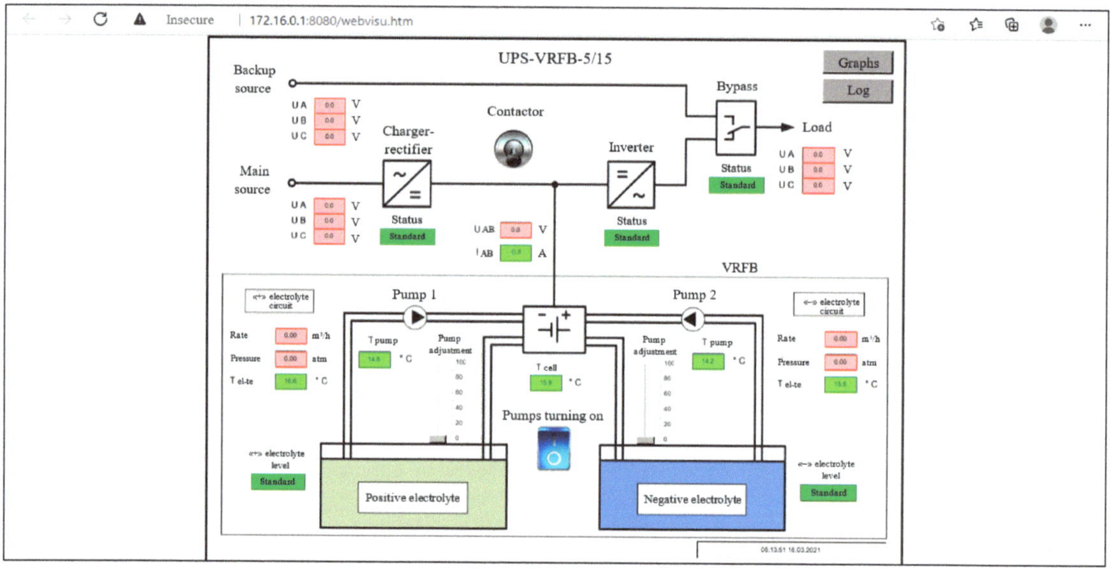

Figure 9. Main window of the computer application interface.

The program code of the UPS-VRFB-5/15 monitoring and control system is implemented in the Codesys V3.5 sp14 Patch3 in the ST (Structured Text) language. The program loaded into the PLC receives the following data:

- Pressure of positive and negative electrolyte;
- Consumption of positive and negative electrolyte;
- Temperature of positive and negative electrolyte;
- Temperature of pumps;
- Temperature of the cells of the flow battery;
- Voltage and current of the flow battery.

The above data are processed and transmitted to the web interface of the device connected to the PLC.

The program automatically controls the system, starting the pumps once every 24 h. To determine the pumping speed, the program monitors the battery mode—discharge or charge. The pumps are switched on at a specific constant speed during charging, and their speed can be regulated depending on the battery power during discharge.

5. Experimental Results

The VRFB charge/discharge tests have been carried out and the graphs of currents, voltages, and power in various modes have been obtained (Figure 10).

According to the test data, the total charge energy was 2.94 kWh, and the discharge energy was 2.05 kWh. Thus, the energy efficiency is 69.9%. The average power dissipated at the load (4.99 kW) almost coincided with the rated power of the cell. The specific capacity of the electrolyte was 17.11 kWh/m^3, which is at the level of the best results.

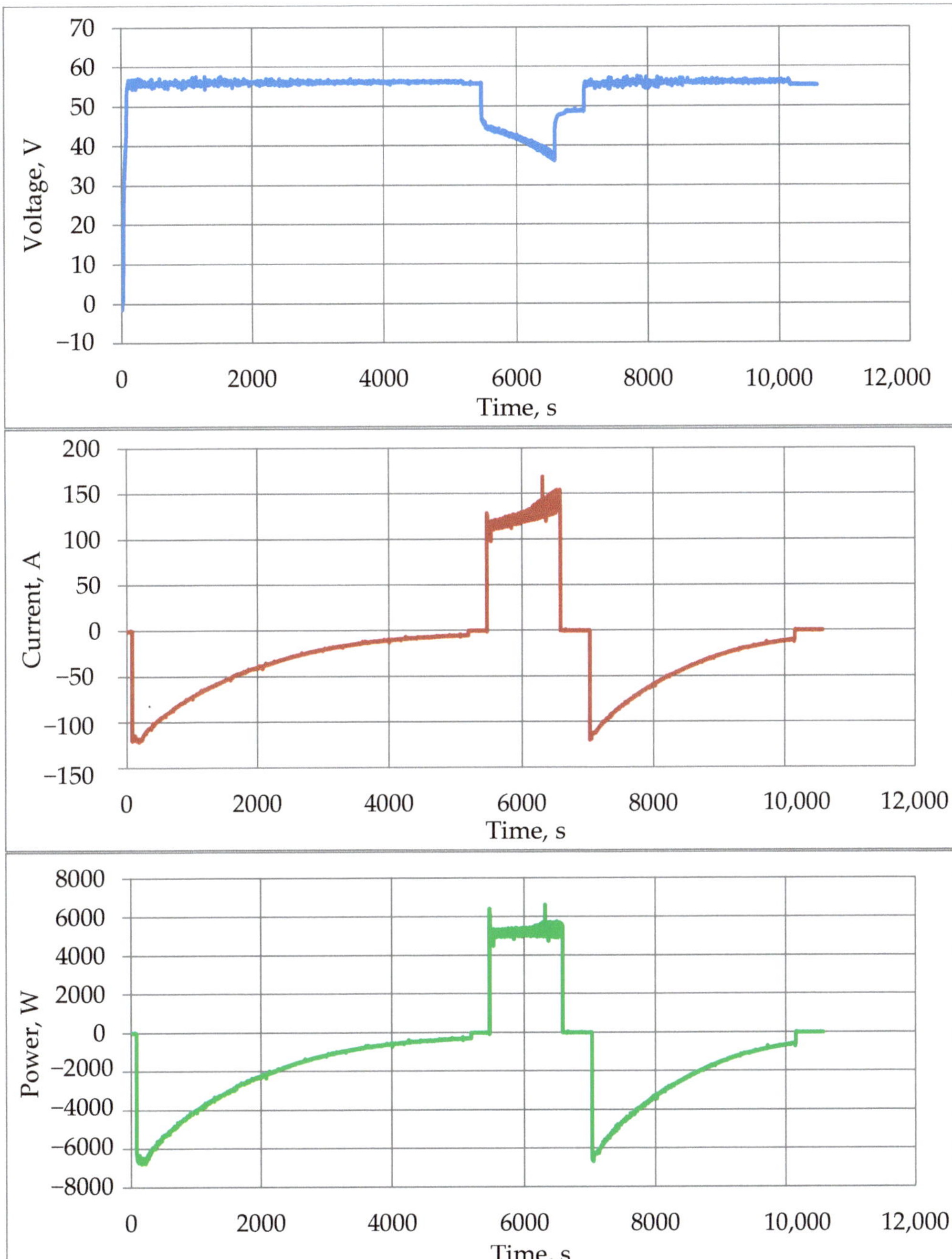

(**a**)

Figure 10. *Cont.*

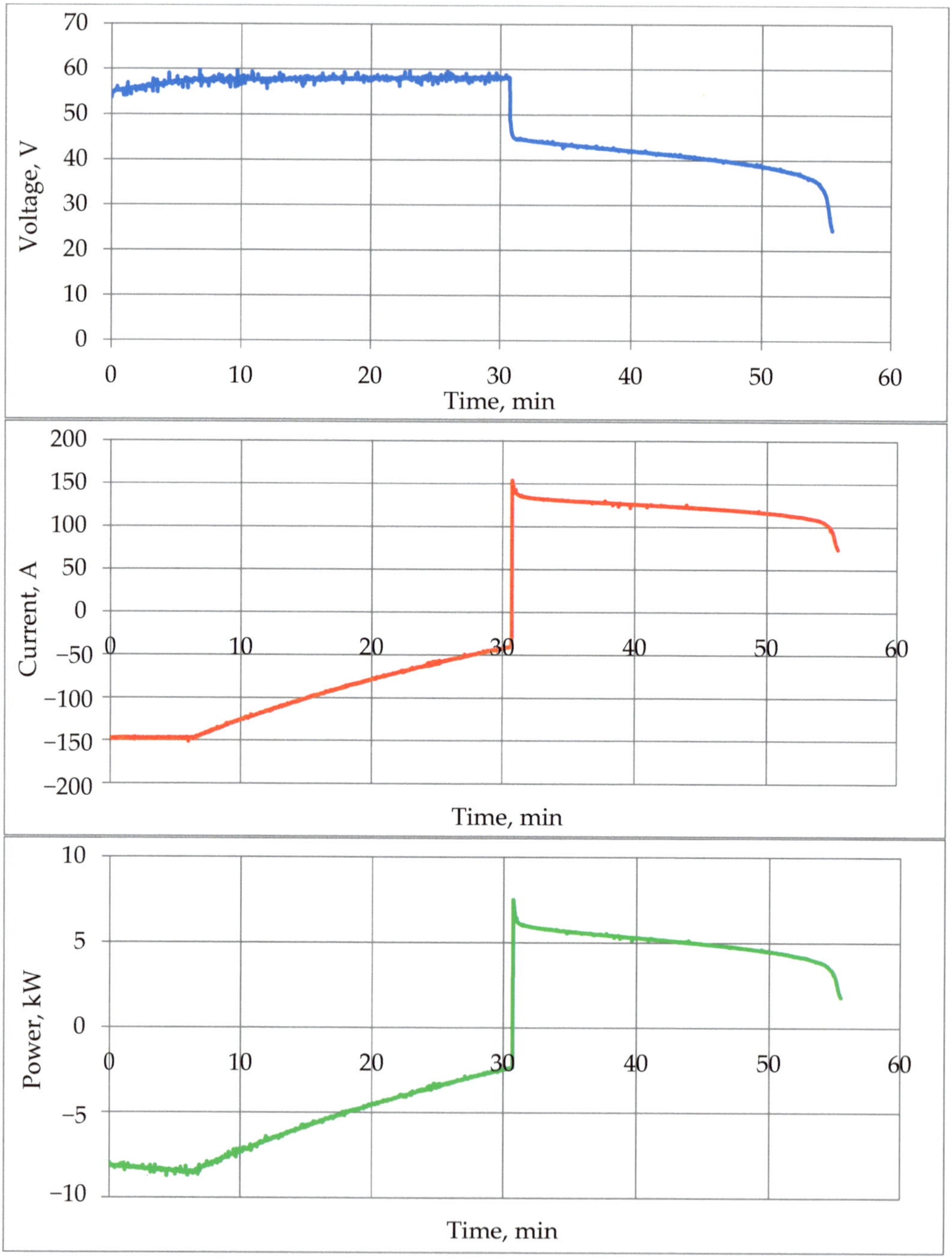

(b)

Figure 10. *Cont.*

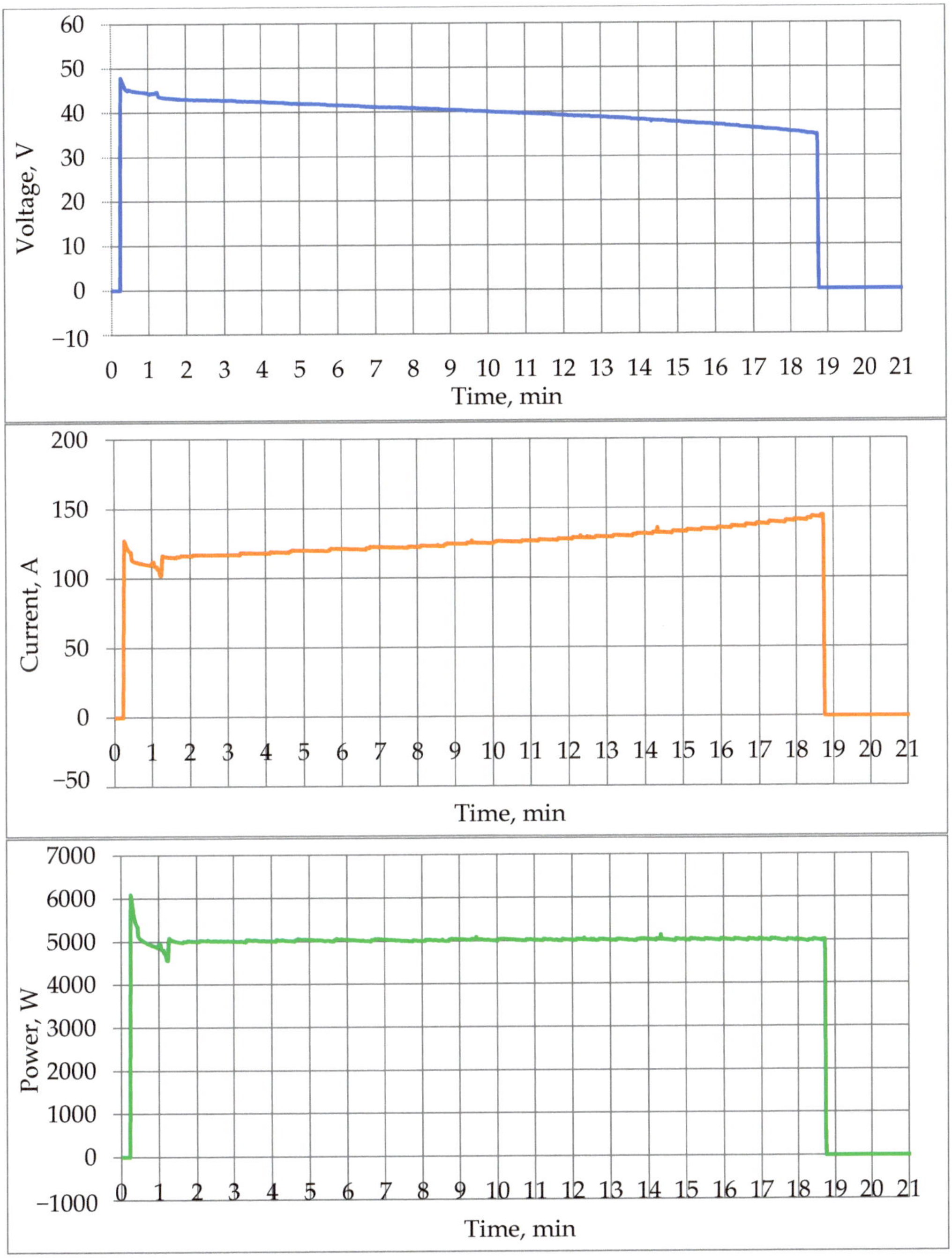

(c)

Figure 10. VRFB test results (voltage, current and power) in the following modes: charge/discharge (**a**); when connecting the load (**b**); when connecting a constant power load (**c**).

The flow battery heats up when the stack is charged and discharged due to the allocation of power losses on the internal resistance, which, as indicated above, is significant. At the same time, electrolyte heating above 45 °C is unacceptable because of the possibility of its decomposition. Therefore, during the tests, the temperatures of the positive and negative electrolytes and the temperature of the VRFB stack itself were constantly monitored, and in the software of the stand, overtemperature protection was implemented, which, if triggered, would turn off the stack. However, during the tests, it turned out that the temperature of the electrolyte increased by only 5.7 degrees compared to the original (27 °C), which is quite acceptable.

Additionally, tests of the VRFB operation were carried out when charging in the same IU mode and then discharging to a load with a constant power consumption (Figure 10c). To implement the latter condition, some of the RB-315 resistive elements were switched on through a load control unit based on a transistor switch. During the discharge, the duty cycle of the control pulses of the key was automatically changed in such a way that the power released at the load was as close as possible to the specified one (5 kW). Since the PACB was charged in a similar way, only a part of the graph corresponding to the discharge mode is shown on Figure 10c.

The presented current, voltage, and power curves illustrate the good dynamic characteristics of the VRFB in charge/discharge mode and load changes.

6. Discussion

Tests of the flow battery stand showed the following results:

(1) The 5 kW VRFB stack provides long-term output of rated power of 5 kW in discharge mode, while short-term output power can reach 8 kW;

(2) VRFB energy efficiency is 69.9%, and the specific capacity of the used electrolyte is 17.11 kWh/m^3, which is at the level of the world's best results;

(3) The load control unit allows one to implement the VRFB discharge mode with constant power;

(4) The increase in the temperature of the electrolyte during the tests did not exceed 5.7 °C, so there is no need to be afraid of unacceptable overheating of the electrolyte during the operation of the VRFB.

The only drawback found during the tests is a significant change in the voltage on the battery during the discharge (from 58 V to 35 V). Therefore, in order to provide a stable voltage to the load, a VRFB-based power system must be equipped with either a regulated DC–DC converter (for supplying a DC load) or a regulated inverter (for supplying an AC load). The implementation of such devices on a modern component base of power electronics does not present any particular difficulties.

7. Conclusions

1. A VRFB hydraulic scheme has been proposed, and a study of the properties of the cells with a change in the electrolyte pumping rate has been carried out. It is shown that in the zone of operating modes of the stack, a change in the electrolyte pumping rate has little effect on the power and efficiency of the stack (±10%).

2. An electrical circuit of the UPS based on VRFB and its control algorithm are proposed.

3. The dynamic characteristics of the UPS in the VRFB charge/discharge modes were studied, and diagrams are given.

4. The presented VRFB can operate even with a 1.5-fold overload without a decrease in efficiency and even with an increase in efficiency without a significant loss in capacity, which proves the good compatibility with the developed electrolyte.

Author Contributions: Conceptualization, A.L. and A.K.; methodology, A.L. and A.K.; software, I.K. and E.O.; validation, I.K. and E.O.; formal analysis, I.K. and E.O.; investigation, A.L., I.K. and E.O.; data curation, I.K. and E.O.; writing—original draft preparation, A.L., A.K., E.O. and I.K.; writing—review and editing, A.L. and A.K.; visualization, I.K. and E.O.; supervision, A.L. and A.K.;

project administration, A.K. and A.L. All authors have read and agreed to the published version of the manuscript.

Funding: The reported study was funded by RFBR, Sirius University of Science and Technology, JSC Russian Railways and Educational Fund "Talent and success", project number 20-38-51016 and Council of the grants of President of the Russian Federation for the state support of Leading Scientific Schools of the Russian Federation (Grant No. NSH-70.2022.1.5).

Conflicts of Interest: The authors declare no conflict of interest.

References

1. Li, C.-P.; Chen, M.-L.; Tseng, C.-C.; Huang, S.-L. Monitoring of thermal effect of redox flow battery by the infrared thermal image technology. In Proceedings of the 6th International Conference on Innovation in Science and Technology, London, UK, 26–28 July 2019.
2. Shibata, T.; Kumamoto, T.; Nagaoka, Y.; Kawase, K.; Yano, K. Redox flow batteries for the stable supply of renewable energy. *SEI Tech. Rev.* **2013**, *76*, 14–22.
3. Huskinson, B.; Marshak, M.P.; Suh, C.; Er, S.; Gerhardt, M.R.; Galvin, C.J.; Chen, X.; Aspuru-Guzik, A.; Gordon, R.G.; Aziz, M.J. A metal-free organic–inorganic aqueous flow battery. *Nature* **2014**, *505*, 195–198. [CrossRef] [PubMed]
4. Shigematsu, T. Redox flow battery for energy storage. *SEI Tech. Rev.* **2011**, *73*, 4.
5. Chen, R.; Kim, S.; Chang, Z. Redox flow batteries: Fundamentals and applications. In *Redox: Principles and Advance Applications*; Khalid, M., Ed.; InTech: London, UK, 2017; pp. 103–118. [CrossRef]
6. Traheya, L.; Brushetta, F.R.; Balsaraa, N.P.; Cedera, G.; Chenga, L.; Chianga, Y.-M.; Hahna, N.T.; Ingrama, B.J.; Minteera, S.D.; Moorea, J.S.; et al. Energy storage emerging: A perspective from the Joint Center for Energy Storage Research. *Proc. Natl. Acad. Sci. USA* **2020**, *117*, 12550–12557. [CrossRef] [PubMed]
7. Alotto, P.; Guarnieri, M.; Moro, F. Redox flow batteries for the storage of renewable energy: A review. *Renew. Sustain. Energy Rev.* **2014**, *29*, 325–335. [CrossRef]
8. Chen, T.-S.; Huang, S.-L.; Chen, M.-L.; Tsai, T.-J.; Lin, Y.-S. Improving electrochemical activity in a semi-V-I redox flow battery by using a C–TiO2–Pd composite electrode. *Hindawi J. Nanomater.* **2019**, *2019*, 7460856. [CrossRef]
9. Cunha, A.; Martins, J.; Rodrigues, N.; Brito, F.P. Vanadium redox flow batteries: A technology review. *Int. J. Energy Res.* **2015**, *39*, 889–918. [CrossRef]
10. Xi, J.; Xiao, S.; Yu, L.; Wu, L.; Liu, L.; Qiu, X. Broad temperature adaptability of vanadium redox flow battery—part 2: Cell research. *Electrochim. Acta* **2016**, *191*, 695–704. [CrossRef]
11. Huang, S.L.; Yu, H.F.; Lin, Y.S. Modification of Nafion® membrane via a sol-gel route for vanadium redox flow energy storage battery applications. *J. Chem.* **2017**, *2017*, 4590952. [CrossRef]
12. Arenas, L.F.; León, C.P.; Walsh, F.C. Mass transport and active area of porous Pt/Ti electrodes for the Zn-Ce redox flow battery determined from limiting current measurements. *Electrochim. Acta* **2016**, *221*, 154–166. [CrossRef]
13. Habekost, A. Vanadium redox flow batteries with different electrodes and membranes. *World J. Chem. Educ.* **2018**, *6*, 8–13. [CrossRef]
14. Wei, Z.; Zhao, J.; Xiong, B. Dynamic electro-thermal modeling of all-vanadium redox flow battery with forced cooling strategies. *Appl. Energy* **2014**, *135*, 1–10. [CrossRef]
15. Cao, L.; Kronander, A.; Tang, A.; Wang, D.-W.; Skyllas-Kazacos, M. Membrane permeability rates of vanadium ions and their effects on temperature variation in vanadium redox batteries. *Energies* **2016**, *9*, 1058. [CrossRef]
16. Cui, Y.; Chen, X.; Wang, Y.; Peng, J.; Zhao, L.; Du, J.; Zhai, M. Amphoteric ion exchange membranes prepared by preirradiation-induced emulsion graft copolymerization for vanadium redox flow battery. *Polymers* **2019**, *11*, 1482. [CrossRef] [PubMed]
17. Arribas, B.N.; Melício, R.; Teixeira, J.G.; Mendes, V.M.F. Vanadium redox flow battery storage system linked to the electric grid. *RE&PQJ* **2016**, *1*, 1025–1030.

energies

MDPI

Article

Modeling of Harmonic Current in Electrical Grids with Photovoltaic Power Integration Using a Nonlinear Autoregressive with External Input Neural Networks

Adán Alberto Jumilla-Corral [1], Carlos Perez-Tello [1], Héctor Enrique Campbell-Ramírez [1], Zulma Yadira Medrano-Hurtado [2], Pedro Mayorga-Ortiz [3] and Roberto L. Avitia [4,*]

1 Instituto de Ingeniería, Universidad Autónoma de Baja California, Mexicali C.P. 21280, Baja California, Mexico; alberto.jumilla@uabc.edu.mx (A.A.J.-C.); carlosperez@uabc.edu.mx (C.P.-T.); hcampbellr@uabc.edu.mx (H.E.C.-R.)
2 Departamento de Ciencias Básicas, Instituto Tecnológico de Mexicali, Mexicali B.C. 21376, Baja California, Mexico; zulmamh@yahoo.com.mx
3 Departamento de Eléctrica-Electrónica, Instituto Tecnológico de Mexicali, Mexicali B.C. 21376, Baja California, Mexico; pmogauss@gmail.com
4 Facultad de Ingeniería, Universidad Autónoma de Baja California, Mexicali C.P. 21280, Baja California, Mexico
* Correspondence: ravitia@uabc.edu.mx

Citation: Jumilla-Corral, A.A.; Perez-Tello, C.; Campbell-Ramírez, H.E.; Medrano-Hurtado, Z.Y.; Mayorga-Ortiz, P.; Avitia, R.L. Modeling of Harmonic Current in Electrical Grids with Photovoltaic Power Integration Using a Nonlinear Autoregressive with External Input Neural Networks. *Energies* **2021**, *14*, 4015. https://doi.org/10.3390/en14134015

Academic Editor: Gabriel Nicolae Popa

Received: 8 June 2021
Accepted: 29 June 2021
Published: 3 July 2021

Publisher's Note: MDPI stays neutral with regard to jurisdictional claims in published maps and institutional affiliations.

Abstract: This research presents the modeling and prediction of the harmonic behavior of current in an electric power supply grid with the integration of photovoltaic power by inverters using artificial neural networks to determine if the use of the proposed neural network is capable of capturing the harmonic behavior of the photovoltaic energy integrated into the user's electrical grids. The methodology used was based on the use of recurrent artificial neural networks of the nonlinear autoregressive with external input type. Work data were obtained from experimental sources through the use of a test bench, measurement, acquisition, and monitoring equipment. The input–output parameters for the neural network were the current values in the inverter and the supply grid, respectively. The results showed that the neural network can capture the dynamics of the analyzed system. The generated model presented flexibility in data handling, allowing to represent and predict the behavior of the harmonic phenomenon. The obtained algorithm can be transferred to physical or virtual systems for the control or reduction of harmonic distortion.

Keywords: model; prediction; inverters; photovoltaic systems; artificial neural networks; nonlinear autoregressive with external input

1. Introduction

Renewable energies are considered clean, abundant, and increasingly competitive energies. They differ from fossil fuels mainly in their diversity, profusion, and harnessing the potential in all regions of the planet, but above all, in that they do not generate greenhouse gases and polluting emissions [1].

Photovoltaic solar (PV) is one of the renewable energy sources, which involves the direct transformation of solar radiation into electrical energy; this transformation is achieved by leveraging the properties of semiconductor materials such as silicon, which can generate electrical power when ionized by solar radiation [2].

PV generation systems are classified into two large groups: isolated and interconnected [3]. In the former, the generation is not connected in any way to the electric supply grid, while in the latter, the energy generated is integrated into the grid by the use of electronic power inverters.

Inverters used in interconnected PV systems have the main function of converting direct waveform power (DC) parameters into sinusoidal alternating waveform (AC) parameters [4]. These devices generate various power quality problems, due to the nonlinear behavior of their components and their operational characteristics [5,6], among which are transients and voltage variations, flickering, harmonic, and interharmonics.

Among some of the harmful effects produced by harmonic distortion in currents and/or voltages are the increase in Joule effect losses in electric feeders, overheating in grounded conductors, in motors, generators, transformers and cables, reducing their service life, vibration in electrical machines, failure of capacitor and transformer banks, resonance effects, as well as operational problems in sensitive electronic devices and interference in telecommunications systems [7].

Due to the imminent growth in the use of PV systems interconnected to power grids, which have led to an increase in energy quality problems [8], specifically those caused by harmonic distortion, it is necessary to characterize and model harmonic behavior due to the integration of the PV power in order to control and/or suppress the content of harmonic distortion at the common coupling point (CCP), as well as in the power supply grid.

1.1. Related Works

Various research studies have been developed on the modeling of harmonic behavior in electrical networks with power integration coming from PV generation systems, both in medium and low voltage. An alternative model of an interactive PV system that can be used to characterize the harmonics of such systems when connected to distorted and undistorted grids, respectively, was presented in [9]. The proposed model was based on the concept of harmonic domain. The real expansion of the Fourier series and the complex Fourier series were used for this model. Flexible harmonic domain models of a three-phase PV system for stable state analysis have been developed [10], based on a selection of dominant harmonics and the reordering of the Toeplitz type matrix involved in frequency convolution. On the other hand, in [11], the authors presented harmonic modeling using a mathematical analysis of a harmonic study of different parallel PV systems connected to a low voltage distribution network to identify the potential resonance between PV inverters and system impedance, while in [12], a novel model was shown in the time domain, using average functions and a novel switching emulator. Despite the results obtained, these models are not suitable when trying to represent the behavior of nonstationary signals due to the limitations presented by mathematical tools such as the transformed and Fourier series.

On the other hand, ANNs have traditionally been used for the detection, classification, and control of energy quality problems, especially those related to harmonic voltage and current distortion, as well as in systems that aim to control and/or eliminate such phenomena. A radial basis function ANN for harmonic load identification was implemented with several of the training and testing data from the combination of 15 different load types to validate the accuracy and efficiency of the proposed algorithm [13]. For the detection and classification of energy quality disturbances [14], the processing of voltage or current signals was used using a phase estimation model, while the classification task was performed using threshold-based rules and an ANN; similar studies [15] that also aim to detect and classify energy quality problems have used the Hilbert–Huang transform and a Perceptron multilayer ANN. With regard to the use of ANN for harmonic control and disposal, the authors of [16] proposed a new ANN structure based on regression by least squares of gradient descent for the control of a PV solar system, integrated into the grid with improved power quality. Another study used a delta–bar–delta ANN-based control [17] to optimally operate by feeding active energy to loads and the remaining power to the grid based on static distribution compensator capabilities, such as harmonic mitigation, load balancing, and power factor improvement. While ANNs have proven their efficiency when used for identification and classification processes, the ANN topologies used have not been suitable for modeling and predicting time-variant systems.

When it comes to modeling and predicting the behavior of load current harmonics injected into power energy micro-networks, self-regressive nonlinear artificial neural networks with exogenous input (NARX neural network) [18] have been used, which performed well in harmonic prediction. Similarly, the authors of [19] described the results of modeling and predicting total harmonic distortion of current and voltage for a nonlinear high power load; modeling was performed using intelligent neural-network-based techniques and diffuse inference, obtaining adequate results with very few errors. Based on the performance of this type of ANN, it is estimated that the application for the analysis of the harmonic behavior of the current in electrical networks with power integration from PV generation systems can deliver results of interest applicable in the field of study of the quality of electrical energy.

1.2. Scope and Contribution

This work shows the modeling and prediction of current behavior in a power grid under critical operating conditions when the CCP integrates powers from this grid with a PV system through a solid-state inverter, jointly powering resistive loads. The methodology used is based on the use of the NARX network, and the results may serve as a future guide for the control and/or suppression of harmonic content in the CCP of such systems.

ANN are computer algorithms that simulate the biological activity of neurons and the processing of human brain information [20]. They are distinguished in fields of science in which the conception of solutions or characteristics of problems analyzed are difficult to determine with conventional programming, such as image and voice processing, pattern recognition, planning, adaptive interfaces for human–machine systems, control and optimization, signal filtering, modeling and prediction [21]. According to their topology, they are classified into monolayers, multilayers, convolutional, radial based, and recurrent.

The recurrent neural networks (RNNs) do not have a layer structure (such as single-layer, multilayer, and convolutional networks) but allow arbitrary connections between neurons, even being able to create loops, thus establishing temporality and allowing the network to have memory [22]. The data that enter at a time "t" to the entry of the network are converted and transferred in it even in the later moments of time, i.e., $t+1$, $t+2$, $t+3\ldots$ The architecture of this type of neural network has become a model implemented in different domains due to its natural ability to process sequential inputs and know their dependencies in the long term [23]. Unlike the conventional ANNs (forward), RNNs are connected to each other in the same hidden layer and a training function is applied repeatedly to hidden states [24]. RNNs are archetypes of deep learning (DL), which are repeatedly fed back.

In this regard, NARX networks have performed adequately in diverse applications, especially in sequential problems, as well as in dynamic system modeling and time series prediction [25–27]. Consequently, this type of architecture gives the training various advantages, such as greater network input accuracy and forward orientation; thus, static reverse propagation can be used [28].

The main contribution of this research lies in determining if an RNN (specifically that of the NARX type) presents an adequate functioning in the modeling and prediction of the behavior of the current in low voltage grids when it is integrated into electrical power from photovoltaic generation systems using electronic inverters; moreover, it contributes to establishing new methodologies based on artificial neural networks (ANNs) for the analysis of the harmonic current phenomenon whose presence is increasingly continuous in the face of the imminent increase in the use of nonconventional renewable energy sources such as solar and wind. In addition, the results obtained may be used in various investigations related to the power quality, as well as in the control and reduction of problems that affect electrical systems, specifically electrical distribution grids.

2. Methodology and Development

2.1. Overview of NARX

When the signals came from time series, some other topologies of ANNs have been successful; here, we can mention the NARXs networks [29]. A regular feedforward neural network contains time series of inputs in the input layer and predicts an output from the output layer [30]. Since RNNs are very complex [31], normally, the training and learning process may take a long period.

Our case is more adapted to the NARX, which is a commonly used discrete nonlinear system and can be mathematically represented as follows:

$$y(k) = f\{u(k-1), u(k-2), \ldots, u(k-l), y(k-1), y(k-2), \ldots, y(k-m)\} \quad (1)$$

where $u(k) \in \mathbb{R}$ and $y(k) \in \mathbb{R}$ denote the inputs and the outputs of the NARX model at the discrete time step, k, respectively. $l \geq 0$ and $m \geq 0$ are the input memory and the output memory used in the NARX model. The unknown function $f(.)$, which is generally nonlinear, can be approximated, for instance, by a regular multilayer feedforward network. The resulting model architecture is called a NARX network or Jordan NARX network.

The network structure of the series–parallel mode Jordan NARX network (Jordan-SP) is presented in Figure 1. v1; v2; ... ; vn are the neurons in the hidden layer, W(i) is the weight matrix from the input layer to the hidden layer, and W(o) is the weight vector from the hidden layer to the output layer. The structure is a regular feedforward neural network (FNN) structure, in which the output's regressor is formed only by the actual data of the system's output.

$$\hat{y}(k) = f\{u(k-1), u(k-2), \ldots, u(k-l), d(k-1), d(k-2), \ldots, d(k-m)\} \quad (2)$$

where $d(k)$ is the desired or actual output data, and $\hat{y}(k)$ is the network's estimated output at the time step, k. The state of the p-th neuron and its output ($x_p(k)$ are defined as

$$\begin{array}{c} x_p(k) = \sum_{q=1}^{l+m+1} w_{pq}^{(i)} u_q(k) \\ v(k) = f(x_p(k)) \end{array} \quad (3)$$

where the subscripts stand for the index of the element in a vector or a matrix, $w_{pq}^{(i)}$ is the weight connecting the q-th input and the p-th neuron in the hidden layer. Then, the output of the network is

$$y(k) = \sum_{p=1}^{n} w_p^{(o)} v_p(k) \quad (4)$$

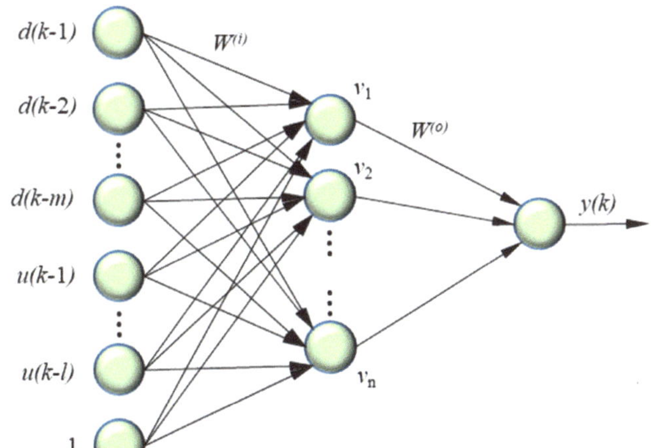

Figure 1. Network structure of series–parallel mode Jordan NARX network.

The hyperbolic tangent sigmoid function is applied to the hidden layer and the linear function is applied to the output layer.

$$f(x) = \frac{2}{1 + e^{-2x}} - 1 \tag{5}$$

In order to let the neural network model track the desired outputs, the cost function mean squared error (MSE) method is utilized here, which should be minimized between the desired outputs and the network's estimated output. Thus, the cost function (J) for N steps is

$$J = \frac{1}{N} \sum_{k=1}^{N} (\hat{y}(k) - d(k))^2 \tag{6}$$

which is a function of W(i) and W(o).

In the training process, for example, by using the Levenberg–Marquardt backpropagation algorithm [32], error functions in each time step can be calculated in parallel in one epoch as a regular FNN. Normally, an accurate training result is quickly achieved based on this network. However, since this type of network is generally only a one-step-forward predictor, it cannot be utilized in pure numerical simulations or as a reference model, which involves a long-term time prediction. Manual feedback of the output signal to the regressor without additional training may accumulate errors in every single step and finally cause instability.

In the equation for the NARX model, the next value of the dependent output signal $y(k)$ is regressed on previous values of the output signal and previous values of an independent (MSE) input signal. A diagram of the resulting network is shown below (Figure 2), where a two-layer feedforward network is used for the approximation. This implementation also allows for a vector autoregressive with exogenous terms (ARX) model, where the input and output can be multidimensional.

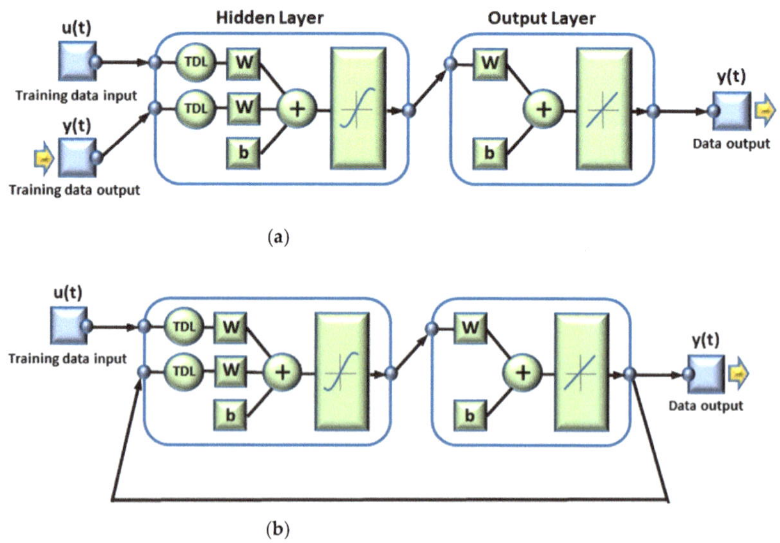

Figure 2. Jordan NARX network architecture: (**a**) open-loop network (parallel–series); (**b**) closed-loop network (parallel).

You can consider the output of the NARX network to be an estimate of the output of some nonlinear dynamic system that you are trying to model. The output is fed back to the input of the feedforward neural network as part of the standard NARX architecture.

As the true output is available during the training of the network, you could create a series–parallel architecture [33], in which the true output is used instead of feeding back the estimated output.

In this configuration, the errors are very small. However, because of the series–parallel configuration, these are errors for only a one-step-ahead prediction. A more stringent test would be to rearrange the network into the original parallel form (closed loop) and then to perform an iterated prediction over many time steps [33].

Now, the NARX (and other) networks must be converted from the series–parallel configuration (open loop), which is useful for training, to the parallel configuration (closed loop), which is useful for multi-step-ahead prediction.

All of the training is performed in open loop (also called series–parallel architecture), including the validation and testing steps. The typical workflow is to fully create the network in open loop, and only when it has been trained (which includes validation and testing steps) is it transformed to closed loop for multistep-ahead prediction.

You can now use the closed-loop (parallel) configuration to perform an iterated prediction in N time steps. In this network, you need to load the two initial inputs and the two initial outputs as initial conditions.

It is important to remark that each time a neural network is trained, it can result in a different solution due to different initial weight and bias values and different divisions of data into training, validation, and test sets. To ensure that a neural network of good accuracy has been found, retrain several times [33–35].

2.2. Experimental Setup

For the modeling and behavior prediction of current in the 220 V, 60 Hz power supply grid, at the CCP, where the powers of the grid and a PV system are integrated through electronic inverters, a test bench consisting of 6 monocrystalline solar panels of 250 W each was used in an arrangement of 1500 W. A central inverter interconnected to the grid, of the voltage-controlled type, with a full-bridge configuration of 3000 W capacity, 220 V_{AC}, 2 phases, and 3 wires was used. The CCP was located on a single-phase, 220 V, 100 A electric board, through which 530 W at 127 V resistive loads were fed (the configuration of the test bench—inverter type, load type, and generating power—was determined by applying an experimental design based on variance analysis and complete factorial designs). The harmonic current extraction stage was carried out by using Hall effect current sensors, model ACS712-20, and a USB-6008 NI data acquisition card, processing the information in SignalExpress software and finally generating the data to power the neural networks in a MATLAB environment. Table 1 shows the ACS712 current sensor characteristics.

Table 1. Current sensor characteristics.

ACS712 Current Sensors			
Imax	**Sensitivity**	**Vout**	**Resolution**
±5 A	185 mV/A	1575 V to 3425 V	26 mA
±20 A	100 mV/A	0.5 V to 4.5 V	49 mA
±30 A	66 mV/A	0.52 V to 4.48 V	74 mA

The experimental methodology for data acquisition and modeling of dynamic system behavior using ANN is as follows:

1. Design and construction of the test bench;
2. Connection and configuration of acquisition equipment;
3. Data acquisition for references;
4. Data acquisition for modeling;
5. Determination of the ANN to be used;
6. Structure of the ANN to be used;
7. Application of ANN (training, validation, and testing of ANN);

8. Obtaining results and conclusions.

Figure 3 shows the arrangement of the test bench elements.

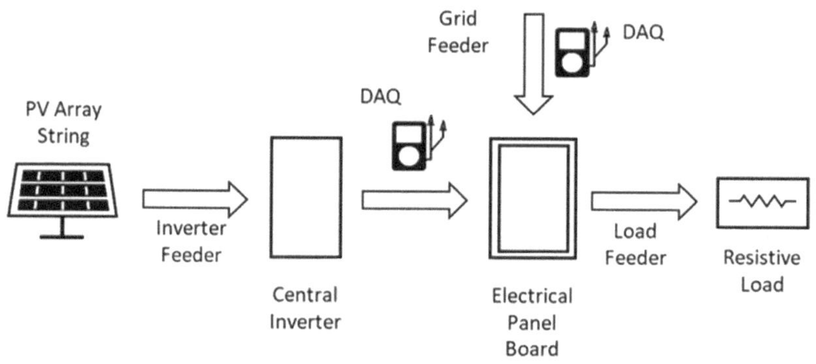

(a)

(b)

Figure 3. Arrangement of elements in test bench: (**a**) schematic diagram; (**b**) physical arrangement of equipment.

The phase conductors in each feeder (inverter and supply grid) were connected in series to the ACS712 current sensors, between their terminals: source (IP+) and load (IP−), with a polarization voltage of 5 V_{DC} between the Vcc and ground (GND) terminals of each sensor. Figure 4 shows the connection of data acquisition devices.

2.3. Data Acquisition

First, the data acquisition of the current parameters was performed in the grid feeder without power inputs from the PV system for 5 s, with a sample rate of 10,000 samples/second, to obtain reference information on the background harmonic distortion in the feeder, under the referred operating conditions. Second, data acquisition was carried out simultaneously of the current parameters in both feeders (supply grid and inverter), at CCP, with a sample rate of 5000 samples/second for each feeder for 5 s. Table 2 shows the measurement points and signal acquisition conditions.

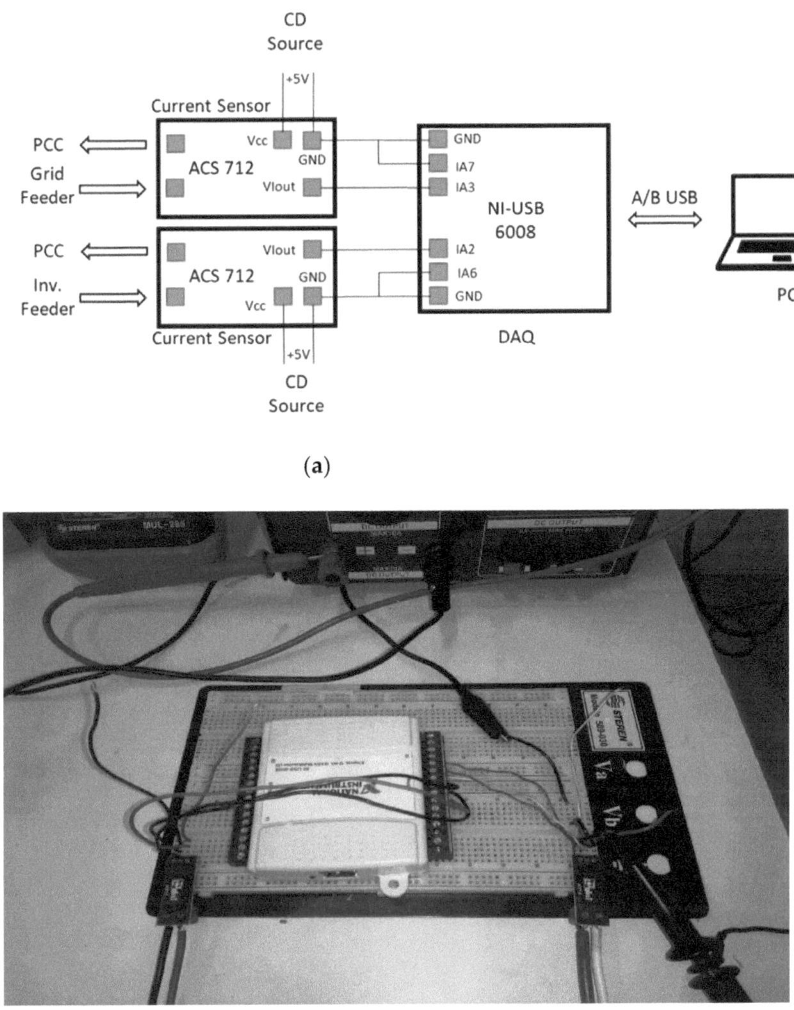

Figure 4. Connecting devices for data acquisition: (**a**) schematic diagram; (**b**) physical arrangement of data acquisition equipment.

Table 2. Description and conditions for data acquisition.

Data Acquisition		
No.	**Description**	**Conditions of Acquisition**
1	Supply grid feeder	No current input from the inverter and charging at CCP
2	Supply grid feeder	With input of current by the inverter and charging at CCP
3	Inverter feeder	With charging at CCP

The figure above (Figure 5) shows a sinusoidal signal with a positive peak magnitude of 5.98 A, with little background harmonic distortion (THDi= 1.07%) and nondynamic (invariant in time) behavior.

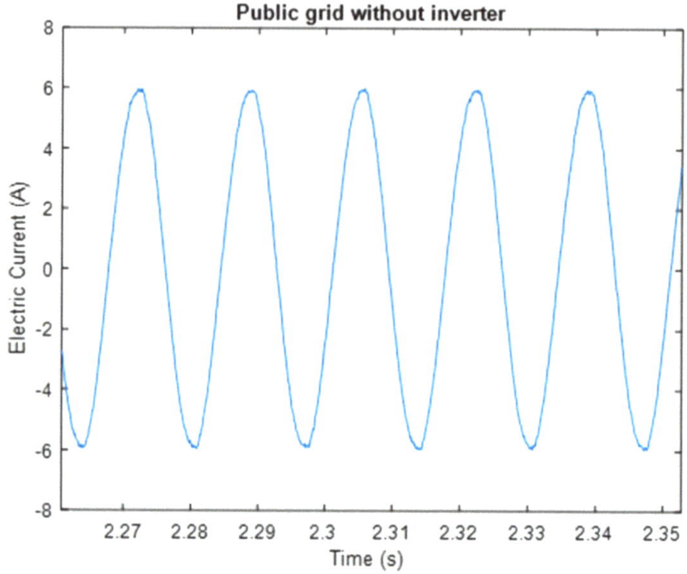

Figure 5. Current in grid feeder without inputs from the PV system.

Figure 6 corresponds to the current in both feeders (supply grid and inverter), jointly feeding the resistive load. This graph shows the change in the waveform of the current signal in the supply grid in the face of the power integration of the PV system, using the electronic inverter. Both signals show distortion due to the presence of harmonics and nonlinear dynamic behavior.

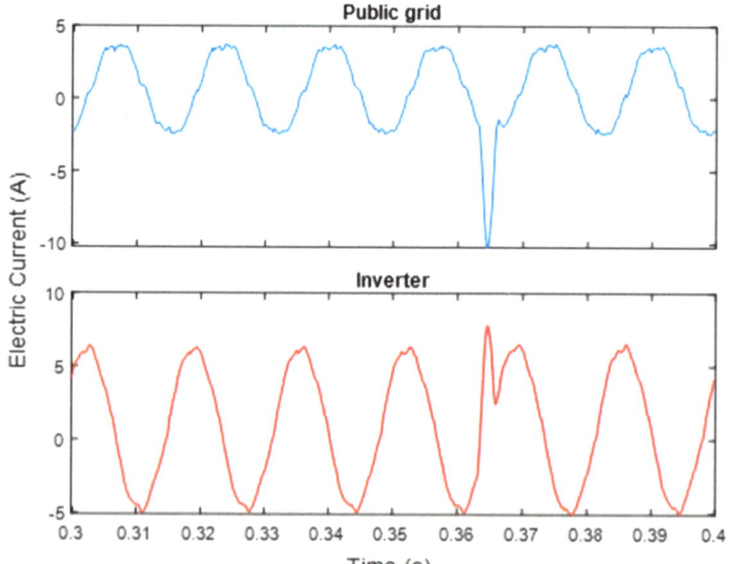

Figure 6. Currents in supply grid and inverter, feeding the resistive load.

2.4. NARX's Structure

Since one of the characteristics of ANN is not having a single structure, this work used NARX networks configured with a series–parallel architecture (open loop) for the modeling network and in parallel (closed loop) for the prediction network, which was used to check the proper functioning of the resulting model [28]. Both networks are powered by input signals at the start of the network, containing two hidden layers with 10 neurons each, activated by sigmoid functions, and an output layer with a single neuron activated by a linear function. Figure 7 shows the structures of the NARXs networks used. These networks have four variable offsets (shown by the 1:4 ratios in the tapped delay line (TDL)), which indicates that the input signals are made up of $x(t)$, $x(t-1)$, $x(t-2)$, $x(t-3)$, and $x(t-4)$, and have input variables $y(t-1)$, $y(t-2)$, $y(t-3)$, and $y(t-4)$ for the series–parallel network, and $z(t-1)$, $z(t-2)$, $z(t-3)$, and $z(t-4)$ for the network in parallel, with "y" being the actual output and "z" the estimated output. The structure of the NARX network (the number of hidden layers, the number of neurons per layer, and the number of offsets) was determined by iteration of values to obtain the best performance of the network.

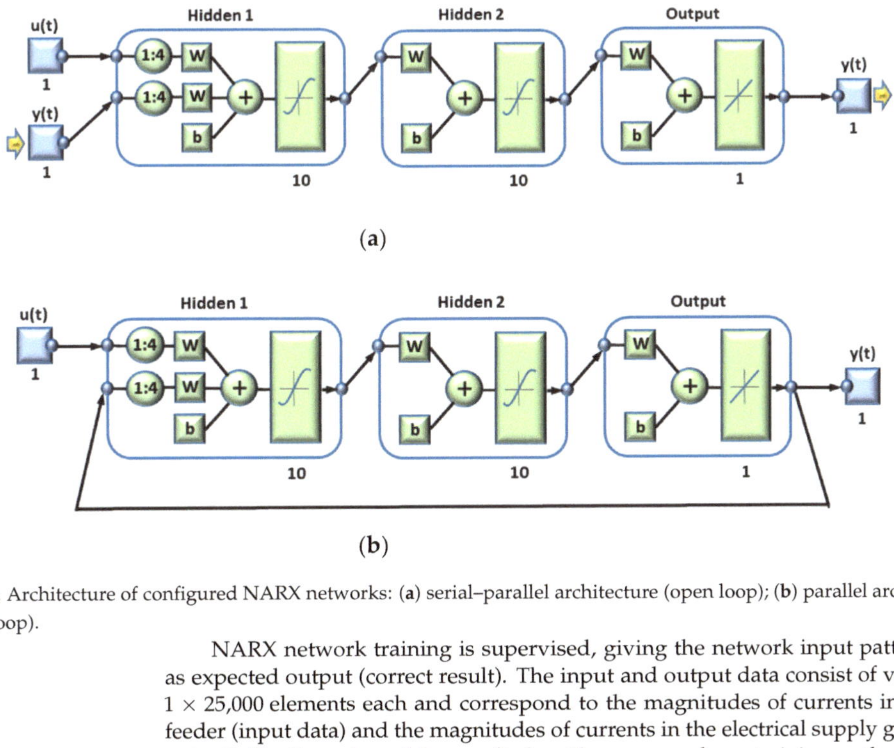

Figure 7. Architecture of configured NARX networks: (**a**) serial–parallel architecture (open loop); (**b**) parallel architecture (closed loop).

NARX network training is supervised, giving the network input patterns, as well as expected output (correct result). The input and output data consist of vectors line of $1 \times 25{,}000$ elements each and correspond to the magnitudes of currents in the inverter feeder (input data) and the magnitudes of currents in the electrical supply grid (expected output). The Levenberg–Marquardt algorithm was used as a training method, along with the MSE performance function, with a total of a thousand iterations (epochs, i.e., the number of times that the algorithms will be executed). The method used for calculating gradients was dynamic backpropagation.

The criteria for stopping training were mainly defined by the number of epochs (1000 iterations), gradient (less than 10^{-5}), and 6 as the number of validation checks.

The general procedure in both networks consisted of the introduction of inputs (where the input neurons are activated); next, the information was propagated through the networks and outputs were generated, and then the outputs of the networks were compared to the desired outputs and the errors were calculated; finally, corrections were made to the weights that were based on these errors until they were minimized. During training, the dataset was randomly divided (to avoid an overfit effect), into three subsets for each network. The first subset was the training set, with this dataset network learning was carried out through weight adjustment and corresponded to 70% of the total data. The second subset (with 15% of the data) was the validation set that served to monitor the error during the training process. The last subset with the remaining 15% of the data was the test set, it was not used during training but was subsequently used to evaluate network performance [28].

3. Results and Discussion

After the parameterization of the series–parallel neural network, training was conducted using the Levenberg–Marquardt method using 269 iterations of the 1000 available epochs to obtain the lowest performance value before stopping the algorithm because six or more times, there were no changes in that performance during testing with validation data. Figure 8 exhibits a very fast drop in MSE, before 10 iterations.

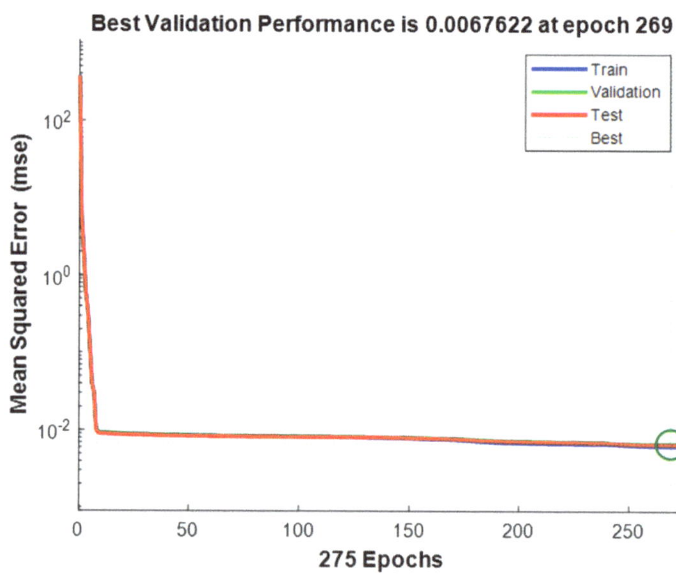

Figure 8. Development of training, validation, and testing, with reference to the number of iterations used, as well as the MSE obtained.

The figure above shows how errors in training, validation, and test data follow the same trends during the algorithm run, finally achieving an MSE less than 10^{-2}; in addition, it shows the minimum value of the network performance obtained in iteration 269, as 0.0067622; the similar behavior of the trends of the three data groups, the rapid decrease of the MSE, coupled with the fast ending of the algorithm, determine an adequate configuration and effectiveness of the NARX network.

Figure 9 shows the histogram containing the errors in the three datasets; it is determined that on the centerline when the error is zero, the data ratio of the training set is higher, and this behavior is constant during the analysis of the data furthest from the null error, while always retaining the higher proportion of the training data in reference to validation and test data. This trend is also observed in the behavior of the data shown in Figure 8.

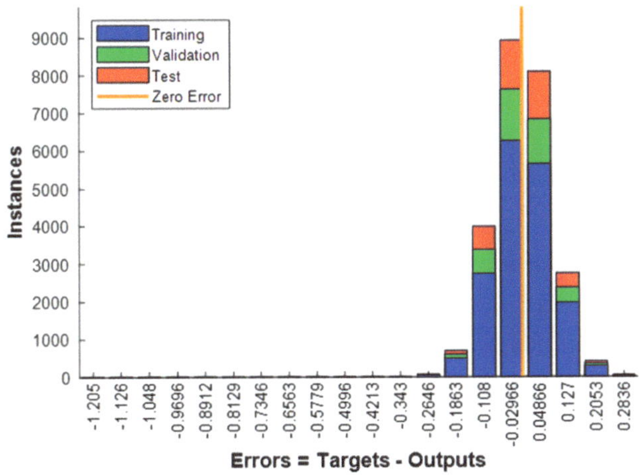

Figure 9. Histogram of the error of outputs against the target.

Another way to analyze the results of significant errors is using the linear regression method for each dataset (training, validation, and testing). Figure 10 shows the values between the achieved output and the required targets, which are far from the ideal linear adjustment (where the x-axis represents the target—in this case, the distorted signal measured in the network feeder—while on the y-axis represents the output value generated by the neural network); the best conditions are those of training data that have a 99.938% effectiveness in regression. This result can be attributed to a greater amount of data (70% of the total), with which the training was carried out. There is also a lot of proximity to the other data groups regarding the effectiveness of regression.

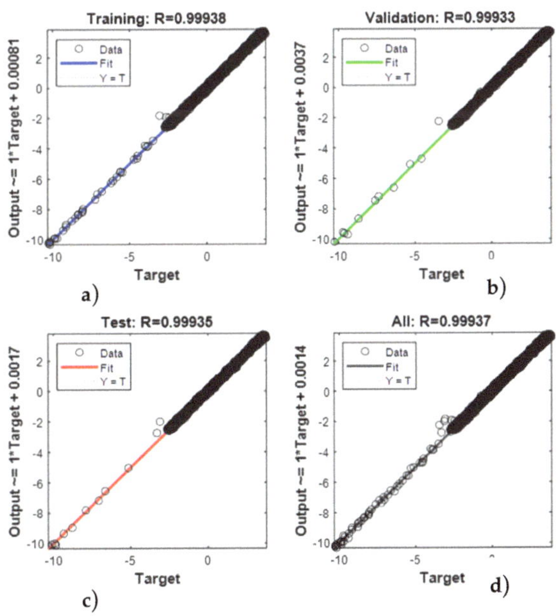

Figure 10. Datasets linear regressions: (**a**) training data regression; (**b**) validation data regression; (**c**) test data regression; (**d**) total data regression.

Subsequently, the temporary response obtained was evaluated by graphically relating the result of the NARX network, against the required values. Figure 11 shows the comparison between the results generated by each of the datasets against the target, in addition to the error present in each of moments of time. The response exhibits the characteristics of the difference between the current signal in the grid feeder in presence of power supplied by the PV system, and the response obtained by the NARX when the current signal enters it from the electronic inverter to the CCP. From the errors' graph, it is observed that the differences between the target values and those resulting from the three training datasets are very close to zero, showing the biggest errors when signal behavior exhibits rapid and abrupt changes in its waveform.

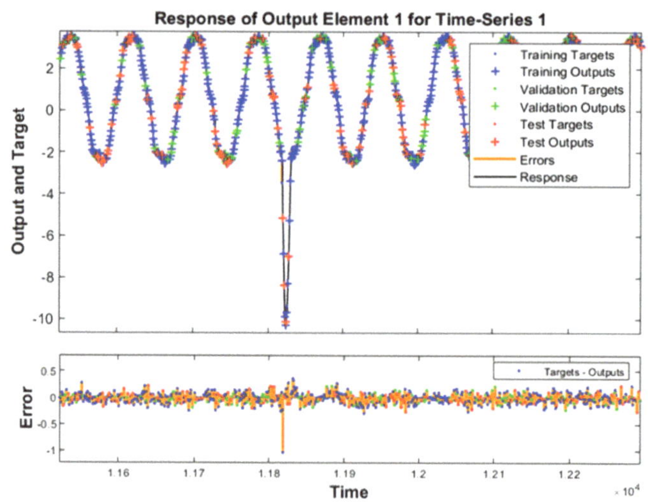

Figure 11. The temporary response obtained and error, with respect to the objective.

Finally, the correlation plots of errors with respect to time and inputs are displayed. Figure 12 presents error–time autocorrelation; it can be seen how the training is correct because the central correlation (MSE) with zero value is greater, while the rest are within the expected confidence limits.

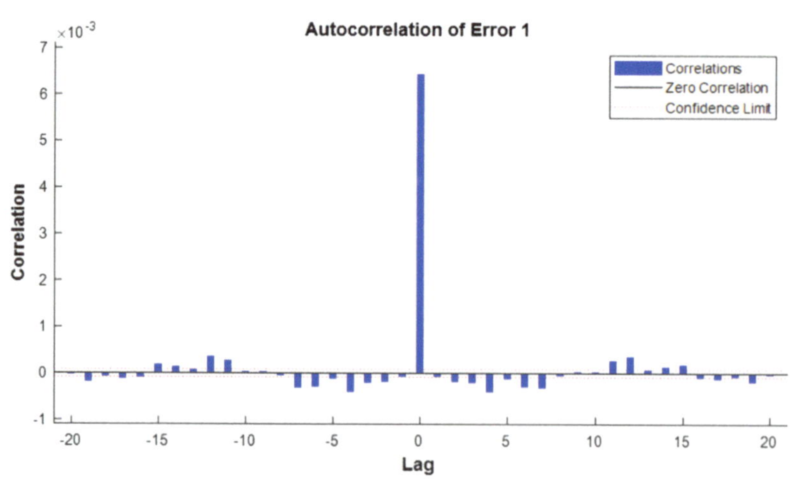

Figure 12. Error–time autocorrelation.

Figure 13 shows the existence of a large number of correlations between inputs and error, which are within the limits, mostly concentrated in the zero value; thus, the training has optimal performance.

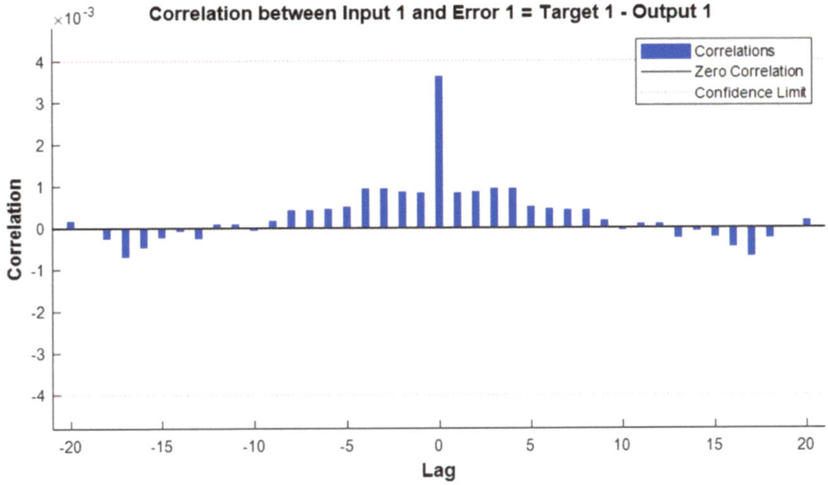

Figure 13. Error–input correlation.

The configuration of the open-loop neural network is replaced with a closed-loop neural network (Figure 7b) in such a way that from the target inputs, the network uses its own predictions to check whether the model obtained through the open-loop network, has adequately defined the behavior of the current signal in the feeder of the electrical supply grid. This architecture produces results that differ from the temporal response of Figure 11 and depend on how appropriate the training has been. Now, the target values will be unknown, and the temporary response obtained will be the one shown in Figure 14.

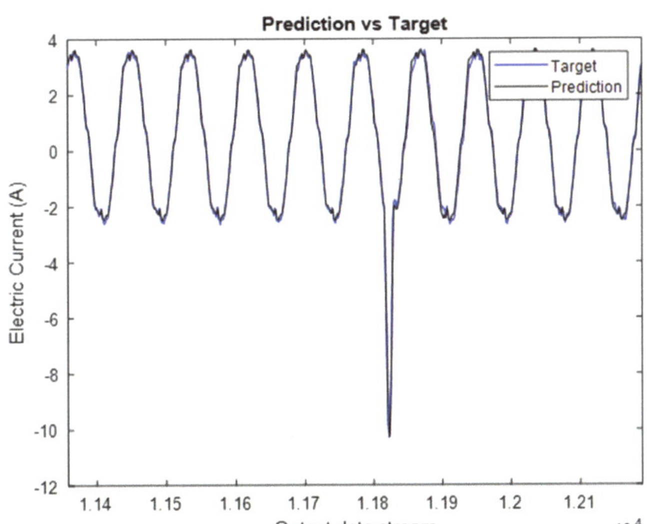

Figure 14. Comparison of the temporal response of closed-loop network against target output.

Figure 15 shows the comparison of the output of the open-loop network (model) against the target output. Similarity to Figure 14 can be seen, with little difference being observed between the temporal response of the closed-loop network and the open-loop network.

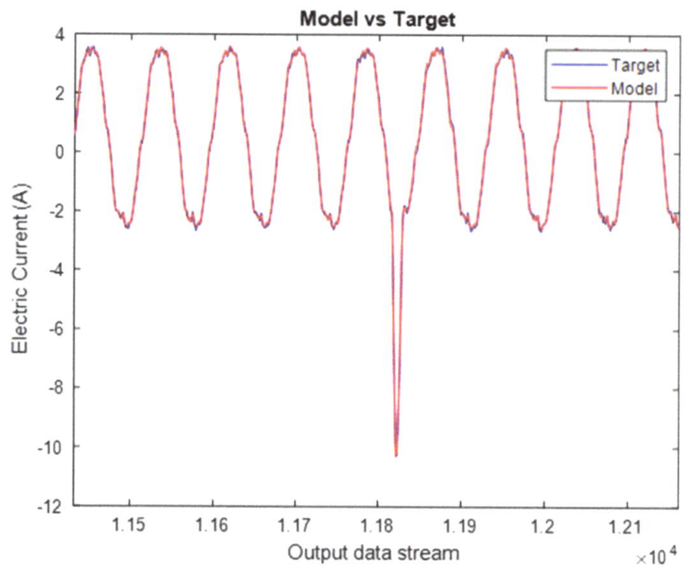

Figure 15. Comparison of the temporary response of open-loop network against target output.

Table 3 shows the MSE for each of the algorithms.

Table 3. MSE in modeling and prediction.

No.	Algorithm	MSE
1	Open loop NARX network (parallel-series)	0.0067
2	Closed loop NARX network (parallel)	0.0094

This article handled real harmonic current values from a PV generation test bench, whose power is integrated into the public power grid using a centralized electronic inverter. The configuration of the test bench was determined by applying an experimental design based on variance analysis and complete factorial designs to obtain the conditions of greatest harmonic current distortion. The factors considered were the type of inverter, the type of load, and the generating power of the photovoltaic system. The acquired data were used to train a neural network of the NARX type, aiming to model and predict the dynamic behavior of the system.

The percentage of background harmonic distortion in the supply network (without PV power integration) was determined with the aim of establishing a comparison point for harmonic current distortion at the time of PV power input (Figure 5). Figure 6 makes it clear that the integration of PV power distorts the current in the supply network.

For NARX network analysis, the current signal generated by the inverter was considered to be input to the neural network, while the current signal in the electrical network was considered to be the target to be determined. The NARX network was configured in such a way that it was able to model (using an open-loop network) and predict (with a closed-loop network) the dynamic behavior of the distorted current signal in the supply network in an efficient and accurate manner, achieving results with low computational load by using few iterations in the execution of the algorithm and obtaining minimum values of the MSE in modeling and prediction. In addition, the analysis of linear regressions (Figure 10) shows an almost ideal fit for the best case (which corresponds to the training data).

With the application of the NARX network, the disadvantages that occur when analyzing nonstationary systems with deterministic methods based on the frequency domain were overcome, as well as extending the scope of such computational algorithms commonly used in the identification and classification of electrical loads and the harmonic distortion they produce.

On the other hand, the results of the application of the NARX network for modeling and predicting distorted current behavior in the supply network of dynamic characteristics are similar to the results obtained in the research studies that have used this same type of resource in modeling and predicting the behavior of load current harmonics injected into energy micro-networks, as well as in the study of nonlinear high-power loads [18].

Among the main advantages of this modeling methodology over others are the following:

1. Ability and ease to represent the nonlinear dynamics of the system;
2. Convergence with a small amount of data with reduced computational time;
3. Suitability to represent internal dynamics through few input variables to the network;
4. Robustness for training-like conditions;
5. Simplicity in the use of the neural network.

4. Conclusions and Future Works

At the end of the analysis of results are as follows:

- It was determined that the sinusoidal waveform of current in the electrical supply grid is affected when integrated into the CCP, powers coming from PV system through electronic inverters.
- Modeling the dynamic and nonlinear behavior of that signal using NARX networks resulted in the establishment of a highly efficient pattern in terms of execution times, as well as computational resources, presenting an MSE of 0.0067 with respect to the actual behavior of the signal, which indicates a high performance of the neural network.
- The validity of the model was checked by forecasting the results, obtaining an MSE of 0.0094 when using the closed-loop NARX, which shows a strong correlation between inputs and error values within the confidence limits.

In view of obtained results, the following observations can be made:

- It can be stated that a neural network with the appropriate characteristics may be considered suitable for capturing the dynamics of such systems.
- The obtained model presents great flexibility in terms of variety and amount of data that can be managed, allowing to represent and predict the behavior of the system under study in long periods of time and under various operating conditions of the system. The resulting algorithm can be used for the generation of physical or virtual systems that can be used for the control or reduction of harmonic phenomena affecting electrical grids.

Based on the results, for future work, it is proposed to apply the methodology developed in current signals from electrical grids of greater capacity in medium and high voltage, which contain power integration of larger PV systems and with different topologies in the inverters. In addition, it is intended to analyze the sensitivity and precision of NARXs networks in the study of the behavior of these distorted currents in order to find the optimal and appropriate parameters for the configuration of the neural network.

Author Contributions: Conceptualization, A.A.J.-C. and C.P.-T.; methodology, A.A.J.-C., C.P.-T. and Z.Y.M.-H.; software, A.A.J.-C. and P.M.-O.; validation, P.M.-O.; formal analysis, P.M.-O.; investigation, Z.Y.M.-H. and A.A.J.-C.; resources, H.E.C.-R. and C.P.-T.; writing original draft preparation, A.A.J.-C. and C.P.-T.; writing review and editing, A.A.J.-C. and R.L.A.; supervision, Z.Y.M.-H. and R.L.A.; project administration, C.P.-T. and H.E.C.-R. All authors have read and agreed to the published version of the manuscript.

Funding: This research received no external funding.

Institutional Review Board Statement: Not applicable.

Informed Consent Statement: Not applicable.

Data Availability Statement: The data that support the findings of this study are available from the corresponding author: upon reasonable request.

Acknowledgments: The authors acknowledge the help and resources provided by Consejo Nacional de Ciencia y Tecnologia (CONACyT), Instituto Tecnologico de Mexicali (ITM) and Universidad Autonoma de Baja California (UABC).

Conflicts of Interest: The authors declare no conflict of interest.

References

1. Javaid, N.; Hafeez, G.; Iqbal, S.; Alrajeh, N.; Alabed, M.S.; Guizani, M. Energy Efficient Integration of Renewable Energy Sources in the Smart Grid for Demand Side Management. *IEEE Access* **2018**, *6*, 77077–77096. [CrossRef]
2. Reinders, A.H.; Debije, M.G.; Rosemann, A.A. Measured Efficiency of a Luminescent Solar Concentrator PV Module Called Leaf Roof. *IEEE J. Photovoltaics* **2017**, *7*, 1663–1666. [CrossRef]
3. Singh, S.; Kewat, S.; Singh, B.; Panigrahi, B.K.; Kushwaha, M.K.; Seema, S.S. Seamless Control of Solar Pv Grid Interfaced System with Islanding Operation. *IEEE Power Energy Technol. Syst. J.* **2019**, *6*, 162–171. [CrossRef]
4. Gupta, A.K.; Pawar, V.; Joshi, M.S.; Agarwal, V.; Chandran, D.; Jimeno, J.C.; Gutiérrez, J.R.; Fano, V.; Del Cañizo, C.; Habib, A.; et al. A Solar PV Retrofit Solution for Residential Battery Inverters. In Proceedings of the 2017 IEEE 44th Photovoltaic Specialist Conference (PVSC), Washington, DC, USA, 25–30 June 2017; pp. 2986–2990.
5. Jose, B.K. Grid integration of PV systems-issues and requirements. In Proceedings of the 2017 IEEE International Conference on Circuits and Systems (ICCS), Thiruvananthapuram, India, 20–21 December 2017; pp. 215–219.
6. Plangklang, B.; Thanomsat, N.; Phuksamak, T. A verification analysis of power quality and energy yield of a large scale PV rooftop. *Energy Rep.* **2016**, *2*, 1–7. [CrossRef]
7. Liu, Y.-W.; Rau, S.-H.; Wu, C.-J.; Lee, W.-J. Improvement of Power Quality by Using Advanced Reactive Power Compensation. *IEEE Trans. Ind. Appl.* **2018**, *54*, 18–24. [CrossRef]
8. Sangwongwanich, A.; Blaabjerg, F. Mitigation of Interharmonics in PV Systems with Maximum Power Point Tracking Modification. *IEEE Trans. Power Electron.* **2019**, *34*, 8279–8282. [CrossRef]
9. Nduka, O.S.; Pal, B.C. Harmonic characterisation model of grid interactive photovoltaic systems. In Proceedings of the 2016 IEEE International Conference on Power System Technology (POWERCON), Wollongong, NSW, Australia, 28 September–1 October 2016; pp. 1–6.
10. Vargas, U.; Ramirez, A.; Lazaroiu, G.C. Flexible harmonic domain model of a photovoltaic system for steady-state analysis. In Proceedings of the 2017 International Conference on ENERGY and ENVIRONMENT (CIEM), Bucharest, Romania, 19–20 October 2017; pp. 311–315.
11. Deng, Z.; Rotaru, M.D.; Sykulski, J.K. Harmonic Analysis of LV distribution networks with high PV penetration. In Proceedings of the 2017 International Conference on Modern Power Systems (MPS), Cluj-Napoca, Romania, 6–9 June 2017; pp. 1–6.
12. Todeschini, G.; Balasubramaniam, S.; Igic, P. Time-Domain Modeling of a Distribution System to Predict Harmonic Interaction Between PV Converters. *IEEE Trans. Sustain. Energy* **2019**, *10*, 1450–1458. [CrossRef]
13. Mubarok, A.F.; Octavira, T.; Sudiharto, I.; Wahjono, E.; Anggriawan, D.O. Identification of harmonic loads using fast fourier transform and radial basis Function Neural Network. In Proceedings of the 2017 International Electronics Symposium on Engineering Technology and Applications (IES-ETA), Surabaya, Indonesia, 26–27 September 2017; pp. 198–202.
14. Mejia-Barron, A.; Amezquita-Sanchez, J.P.; Dominguez-Gonzalez, A.; Valtierra-Rodriguez, M.; Razo-Hernandez, J.R.; Granados-Lieberman, D. A scheme based on PMU data for power quality disturbances monitoring. In Proceedings of the IECON 2017—43rd Annual Conference of the IEEE Industrial Electronics Society, Beijing, China, 29 October–1 November 2017; pp. 3270–3275.
15. Rodriguez, M.A.; Sotomonte, J.F.; Cifuentes, J.; Bueno-Lopez, M. Classification of Power Quality Disturbances using Hilbert Huang Transform and a Multilayer Perceptron Neural Network Model. In Proceedings of the 2019 International Conference on Smart Energy Systems and Technologies (SEST), Porto, Portugal, 9–11 September 2019; pp. 1–6.
16. Kumar, N.; Singh, B.; Panigrahi, B.K. Framework of Gradient Descent Least Squares Regression-Based NN Structure for Power Quality Improvement in PV-Integrated Low-Voltage Weak Grid System. *IEEE Trans. Ind. Electron.* **2019**, *66*, 9724–9733. [CrossRef]
17. Shukl, P.; Singh, B. Delta-Bar-Delta Neural-Network-Based Control Approach for Power Quality Improvement of Solar-PV-Interfaced Distribution System. *IEEE Trans. Ind. Inform.* **2020**, *16*, 790–801. [CrossRef]
18. Hatata, A.; Eladawy, M. Prediction of the true harmonic current contribution of nonlinear loads using NARX neural network. *Alex. Eng. J.* **2017**, *57*, 1509–1518. [CrossRef]
19. Panoiu, M.; Panoiu, C.; Ghiormez, L. Neuro-fuzzy modeling and prediction of current total harmonic distortion for high power nonlinear loads. In Proceedings of the 2018 Innovations in Intelligent Systems and Applications (INISTA), Thessaloniki, Greece, 3–5 July 2018; pp. 1–7.

20. Alhroob, E.; Mohammed, M.F.; Lim, C.P.; Tao, H. A Critical Review on Selected Fuzzy Min-Max Neural Networks and Their Significance and Challenges in Pattern Classification. *IEEE Access* **2019**, *7*, 56129–56146. [CrossRef]
21. Shrestha, A.; Mahmood, A. Review of Deep Learning Algorithms and Architectures. *IEEE Access* **2019**, *7*, 53040–53065. [CrossRef]
22. Rezk, N.M.; Purnaprajna, M.; Nordstrom, T.; Ul-Abdin, Z. Recurrent Neural Networks: An Embedded Computing Perspective. *IEEE Access* **2020**, *8*, 57967–57996. [CrossRef]
23. Xia, W.; Zhu, W.; Liao, B.; Chen, M.; Cai, L.; Huang, L. Novel architecture for long short-term memory used in question classification. *Neurocomputing* **2018**, *299*, 20–31. [CrossRef]
24. Salas, J.; Vidal, F.D.B.; Martinez-Trinidad, F. Deep Learning: Current State. *IEEE Lat. Am. Trans.* **2019**, *17*, 1925–1945. [CrossRef]
25. Li, Y.; Cao, H. Prediction for Tourism Flow based on LSTM Neural Network. *Proc. Comput. Sci.* **2018**, *129*, 277–283. [CrossRef]
26. Cortez, B.; Carrera, B.; Kim, Y.-J.; Jung, J.-Y. An architecture for emergency event prediction using LSTM recurrent neural networks. *Expert Syst. Appl.* **2018**, *97*, 315–324. [CrossRef]
27. Liu, F.; Chen, Z.; Wang, J. Video image target monitoring based on RNN-LSTM. *Multimedia Tools Appl.* **2018**, *78*, 4527–4544. [CrossRef]
28. Hudson, M.; Hagan, M.; Demuth, H. *Matlab Deep Learning Toolbox User's Guide*; MATHWORKS: Natick, MA, USA, 2019.
29. Dzielinski, A. Difference inequalities and approximate NARX models. In Proceedings of the 1999 European Control Conference (ECC), San Diego, CA, USA, 2–4 June 1999; pp. 4784–4788.
30. Yazdani-Asrami, M.; Taghipour-Gorjikolaie, M.; Song, W.; Zhang, M.; Yuan, W. Prediction of Nonsinusoidal AC Loss of Superconducting Tapes Using Artificial Intelligence-Based Models. *IEEE Access* **2020**, *8*, 207287–207297. [CrossRef]
31. Zhang, X.; Luo, T. A RNN Decoder for Channel Decoding under Correlated Noise. In Proceedings of the 2019 IEEE/CIC International Conference on Communications in China (ICCC), Changchun, China, 11–13 August 2019; pp. 30–35.
32. Anggriawan, D.O.; Satriawan, A.L.; Sudiharto, I.; Wahjono, E.; Prasetyono, E.; Tjahjono, A. Levenberg Marquardt Backpropagation Neural Network for Harmonic Detection. In Proceedings of the 2018 International Electronics Symposium on Engineering Technology and Applications (IES-ETA), Bali, Indonesia, 29–30 October 2018; pp. 129–132. [CrossRef]
33. Nazaripouya, H.; Wang, B.; Wang, Y.; Chu, P.; Pota, H.; Gadh, R. Univariate time series prediction of solar power using a hybrid wavelet-ARMA-NARX prediction method. In Proceedings of the 2016 IEEE/PES Transmission and Distribution Conference and Exposition (T&D), Dallas, TX, USA, 3–5 May 2016; pp. 1–5.
34. Mohanty, S.; Patra, P.K.; Sahoo, S.S. Prediction of global solar radiation using nonlinear auto regressive network with exogenous inputs (narx). In Proceedings of the 2015 39th National Systems Conference (NSC), Greater Noida, India, 14–16 December 2015; pp. 1–6.
35. Gautam, A.; Singh, V. Comparison of different NN training functions of NARX architecture for financial time series. In Proceedings of the 2017 IEEE International Conference on Industrial and Information Systems (ICIIS), Peradeniya, Sri Lanka, 15–16 December 2017; pp. 1–6.

Short Biography of Author

A.A. Jumilla-Corral, electrical engineer, professor at the Instituto Tecnológico de Mexicali (Tecnológico Nacional de México), in Mexicali City, Baja California, Mexico. He earned a master's degree in electrical engineering in the area of power quality from the Universidad Autónoma Baja California (2017). PhD student in electrical engineering in the area of power quality at the Instituto de Ingeniería of the Universidad Autónoma de Baja California. His areas of interest: Energy utilization, electrical machines, power quality, electrical maintenance.

C. Pérez-Tello, Chemical Engineer graduated from the Instituto Tecnólogico de Sonora, Master of Science from the Instituto Politécnico Nacional. Doctor of Science from the Instituto Politécnico de Celaya. He has experience as a process engineer in copper ore benefit metallurgical plants. Associate researcher at the Instituto Mexicano del Petróleo attached to the Department of Physical Processes, and since 1990 is a researcher in the Energy Systems Laboratory of the Instituto de Ingeniería of the Universidad Autónoma de Baja California, Mexico, specialized in the area of energy efficiency and energy saving, applied thermodynamics, experiment design, and thermal simulation. He is the author of several specialized articles and co-author of a book related to artificial intelligence applied to energy systems and several chapters in e-books specialized in energy saving and efficient use.

H.E. Campbell-Ramirez earned a master's degree in Thermodynamic Engineering and a Doctor of Engineering degree from the Universidad Autónoma de Baja California. Leader of the academic body of energy systems, coordinator of the thermal and electrical systems laboratory of the chemical engineering area of the Instituto de Ingeniería, and president of the chemical-energy subcommittee in the chemistry area of the graduate program. His areas of expertise are energy-saving and efficient use, energy system planning, economic engineering, fluid mechanics, and heat transfer. He has participated in 49 research and development projects, 37 of them linked to the municipal, state, and federal public sectors.

Z.Y. Medrano-Hurtado, electrical engineer, professor at the Instituto Tecnológico de Mexicali (Tecnológico Nacional de México), in Mexicali Baja California Mexico. master's degree in electrical engineering in the area of metrology and instrumentation (2007) from the Universidad Autónoma de Baja California engineering faculty, Doctor of electrical engineering (2014) from the Instituto de Ingeniería of the Universidad Autónoma de Baja California. Areas of interest: metrology and instrumentation, Data Acquisition, Signal Processing, Analysis and Diagnostics of electrical machines.

Pedro Mayorga-Ortiz earned his bachelor's science degree in physics at the Universidad Autónoma de Baja California, Mexico. He earned a master's degree in digital systems, later he went on to earn a doctorate from INPG (Grenoble, France). Finally, he made a postdoctoral stay in California State University Long Beach (CSULB) working in lung sounds signals. From 1993 up to now he has been working as a full-time professor researcher at the Instituto Tecnológico de Mexicali. His main interest is in topics related to signal processing and pattern recognition, specifically acoustic signals such as speech, lung sounds, or heart sounds. Additionally, he worked with acoustic signals extracted from the lungs sounds and the heart sounds applying some machine learning methodologies to reinforce diagnostic processes in medicine.

Roberto Lopez-Avitia received the BS degree in biomedical electronic engineering from the National Institute of Technology of Mexico in 2004, the MS degree in bioelectronics from the Center for Research and Advanced Studies of the National Polytechnic Institute in 2006, and the PhD degree in biomedical engineering from Autonomous University of Baja California (UABC) in 2013. His past employment includes industrial at Hirata Engineering and hospital experience at the National Institute of Cardiology. He joined the Department of Bioengineering and Environmental Health at UABC in 2007 by applying dynamic programming and machine learning algorithms in high-resolution electrocardiography (HRECG). He was the creator of the bioengineering undergraduate program at UABC. Dr. L. Avitia is currently a full-time professor of the bioengineering program at the UABC and a member of the National System of Researchers. His research interests include machine learning algorithms applied to biomedical systems, the development of medical instrumentation, and pattern recognition.

energies

MDPI

Article

Cost of Industrial Process Shutdowns Due to Voltage Sag and Short Interruption

Édison Massao Motoki [1,2,*], José Maria de Carvalho Filho [2], Paulo Márcio da Silveira [2], Natanael Barbosa Pereira [3] and Paulo Vitor Grillo de Souza [2,4]

1 Power Systems Department, Mackenzie Presbyterian University, São Paulo 01302-907, SP, Brazil
2 Institute of Electrical Systems and Energy, Federal University of Itajubá, Itajubá 37500-903, MG, Brazil; jmariacarvalho@gmail.com (J.M.d.C.F.); pmsilveira@gmail.com (P.M.d.S.); paulo.grillo@ufla.br (P.V.G.d.S.)
3 Energias de Portugal—EDP, São José dos Campos 12210-010, SP, Brazil; natanael.pereira@edpbr.com.br
4 Automatic Department, Federal University of Lavras, Lavras 37200-000, MG, Brazil
* Correspondence: edison.motoki@gmail.com; Tel.: +55-11-98943-3975

Abstract: The objective of this work is to propose and apply a methodology to obtain the cost of industrial process shutdowns due to voltage sag and short interruption. A field survey, aided by a specific questionnaire, was carried out in several industries connected to medium voltage networks, in the states of Espírito Santo and São Paulo in Brazil. The results obtained were the costs per event and the costs per demand in a total of 33 companies in 12 different types of activities. It is noteworthy that this survey brings a relevant technical contribution to the electricity sector, helping to fill, even partially, an existing gap in both national and international literature.

Keywords: voltage sag; equipment sensitivity; costs; production losses

Citation: Motoki, É.M.; Filho, J.M.d.C.; da Silveira, P.M.; Pereira, N.B.; de Souza, P.V.G. Cost of Industrial Process Shutdowns Due to Voltage Sag and Short Interruption. *Energies* **2021**, *14*, 2874. https://doi.org/10.3390/en14102874

Academic Editor: Andrea Mariscotti

Received: 7 April 2021
Accepted: 11 May 2021
Published: 16 May 2021

1. Introduction

1.1. Relevance

Voltage sags and short interruptions are the power quality disturbances that most affect industrial consumers and are caused by the occurrence of short circuits, especially in aerial transmission and distribution networks that are exposed to various climatic events and environmental effects, such as atmospheric discharges, gales, burning, and vandalism, which can cause temporary or permanent outages in the network.

Voltage sags and short interruptions can influence the operation of various equipment and result in serious consequences, especially for those consumer units that have sensitive loads.

Due to the advancement of the electronic area, the equipment brings in its technological support devices with great sensitivity so that the voltage sags and short interruptions can impact the operation of the industries. The effects result in losses due to production stoppages, losses of raw material, damage to equipment, delays, and fines for non-delivery of products, and even lost profits [1].

In this context, the main objective of this work is to propose and apply a survey methodology to obtain the cost estimate of industrial process shutdowns due to the occurrence of voltage sags and short interruptions.

1.2. State of Art

A search for published articles on the costs of process shutdowns due to voltage sags and short interruption was carried out, and a summary of the main works found is presented below.

In [1], research showed the possibilities to mitigate the voltage sags with some applications of technologies for the industry and presented the cost per voltage sag of US $5000.00 for industry in general.

In [2], a survey was carried out by sending a questionnaire in three districts in Italy, to which 93 industries from 18 different types of activity responded. The results indicated that the processes were susceptible to certain power quality disturbances, these being: 84% due to voltage sags and short interruption, 8% due to transient overvoltage, 6% due to voltage fluctuation, and finally 2% due to harmonics. Qualitatively, 89% of the companies answered about the annual economic impact due to power quality problems, these being: 8% despised it, 28% as a minimum, 20% as low, 25% as medium, 16% as high, and 3% as very high. Only 10 industries in the sample reported the average annual cost, with 7 of them indicating an average economic loss of up to US $20,000.00; in another, the cost varied between US $20,000.00 and US $50,000.00, and for 2 of them, the value varied between US $50,000.00 and US $250,000.00 per year.

In the research [3], a method was presented to estimate the number of voltage sags and the resulting costs through a study of five electricity utilities in Finland. For this purpose, the number of occurrences of voltage sag was estimated using the fault position method. Data regarding permanent faults in the network were used, as well as a correction factor being applied to consider temporary faults of the different types (single-phase, two-phase, three-phase). The economic consequences were assessed considering the number of process shutdowns, the costs involved and the number of customers affected. Annual costs due to voltage sags were €530,000.00 for 500 industries in Rural area 1; €270,300.00 for 150 industries in Rural area 2; €8,474,700.00 for 1950 industries in Rural area 3; €1,038,800.00 for 1400 industries in the Urban area 1 and €519,400.00 for 700 industries in the Urban area 2.

In [4], a method was used to calculate the cost of process shutdowns for small industrial plants caused by voltage sags and interruptions in Italy. The formulation of the cost of process shutdowns includes loss of production, loss of raw material, imperfections in production, damaged equipment, extra maintenance, and finally, the process stop times. The methodology was applied in a plastic company supplied at medium voltage and which experienced an annual average of 25 to 30 process shutdowns due to voltage sags. The analysis was performed on four types of process machines that are sensitive: injection, compression, polymerization, and molds. Considering their average production resumption times and with their cost parameters linked to each type of machine, a cost per voltage sags was reached with a variation of €300 to €550, between the stop times (0 sec to 1 h).

In the work of [5], a study was carried out in a generic distribution network consisting of 295 buses, 296 transmission lines, and a large number of switches and circuit breakers in order to allow changing the network topology to obtain better system reliability. The methodology used to assess the financial losses suffered by the consumer as a result of voltage sags was the fault position method. Various types of fault were applied at different points in the network according to the probability of occurrence of each one of them. Thirty-seven industries with high sensitivity to voltage sags were chosen and the Monte Carlo method was used to perform the simulations. Ten thousand tests were simulated to verify the variation in the number of process shutdowns, considering different categories of industries, classified into 3 groups: group I for high loads (>2 MW); group II for medium loads (1 MW to 2 MW) and group III for small loads (up to 1 MW). The costs obtained for the industrial processes shutdowns were £1000.00 for small loads, £16,300.00 for medium load industries, and £581,000.00 for large industrial loads.

In [6,7], a methodology was proposed to estimate financial losses due to supply interruptions and voltage sags by means of a probabilistic assessment, applicable both for the evaluation of individual consumer losses and of the total system losses. For economic losses, it was considered that every shutdown of the plant requires 24 h to resume normal production and that the costs of various types of activities are those of interruption due to voltage sags. The estimated cost for the class of industrial consumers was US $19,594.00 per event.

In article [8], a practical implementation of a stochastic assessment of financial losses due to voltage sags was used. The study characterized the process shutdown by a cumulative probabilistic equation based on the study of [6,7] and took into account the sensitivity and the interconnection of equipment present in a given process, as well as the types of consumers and their location in the distribution network. The simulated network consisted of a distribution system of 29 bars, 28 lines, and 11 transformers, using the Monte Carlo method. The loads were classified into three categories: group I (>20 MW); group II (between 5 and 20 MW); group III (up to 5 MW). For all categories, costs per process shutdown due to voltage sags were from Rs 46,300.00 to Rs 781,000.00 (US $648.00 to US $10,934.00) per event.

Reference [9] presents the methodology used and the main results of a survey carried out in Norway on short-term interruption costs, based on questionnaires about direct costs and the consumer's willingness to pay to avoid voltage variations or to accept them. The results are normalized and standardized by the energy not supplied, in kWh, in the case of interruptions lasting more than three minutes, and by the power, in kW, for short interruptions and voltage sags less than or equal to three minutes. The results showed that the cost per interrupted demand due to voltage sags in industrial consumers was between US $3.34/kW and US $5.18/kW.

In [10], the study presented the cost estimate due to the occurrence of voltage sags based on the momentary reduction of the power flow and of the energy not supplied. As a case study, real measurements made in a transmission system in Brazil were used. The costs obtained were due to 37 events considered significant and ranged from US $510.00/event to US $772,800.00/event, considering that the voltage sag was proportional to the energy not supplied and that the costs were more concentrated in values below US $50,000.00 per event.

In work [11], a hybrid method was used to assess risks linked to interruptions in sensitive processes due to faults in the electrical distribution system. For each consumer, indices related to voltage sags are determined, such as the frequency of occurrences classified in ranges of magnitude and duration, as well as the impacts on industrial processes fed by the distribution system used as a test case. The average annual cost due to voltage sags for industrial consumers was U$ 64,417.00.

In [12], researchers presented a methodology for cost-benefit analysis aimed at mitigating short-term voltage variation in a cement factory, considering the probability of process shutdown. The cost per voltage sag reported by the cement industry was Rs 750,000.00 or US $13,385.00 per event.

In [13], a study was carried out to assess the level of power quality in Malaysia, as well to estimate the cost associated with the occurrence of voltage sags. The events were captured by meters installed at appropriate locations on the power grid. The costs per event vary according to the type of activity of the consumer unit, with the lowest cost being found at US $24,124.00/event (commercial sector) and the highest at US $723,729.00/event (semiconductors sector); glass/stone/clay/cement, ceramic and tiles the cost was US $96,500.00/event; metal/aluminum/copper products, US $168,700.00/event. Also, in this reference is shown the cost per voltage sag of US $5000.00 at the plastics industry in the USA.

In [14], the impact of voltage sags on consumer units in China was discussed and the method of calculating losses resulting from voltage sags was through questionnaires or personal interviews. The results show that there is a great variation in the cost per voltage sags according to the industrial sector, mainly due to the added value of the final product; for the chemical fiber industry they ranged from US $29,000.00 to US $172,000.00, while for the semiconductor industry the costs were between US $574,000.00 and US $3,585,000.00.

In [15], costs were assessed due to the occurrence of voltage sags in a distribution system typical of a chemical industry, considering five types of sensitive equipment installed in the industrial process. The cost considered was US $1893.00 per event (voltage sag), including labor costs, unfinished product costs, losses of raw materials, among others. The results obtained with the simulations vary with the bus where the sensitive loads are installed. The number of estimated shutdowns ranges from 24 to 344 per year, resulting in losses ranging from US $380.80 to US $12,128.20 per year.

In [16], an investment analysis study was carried out to minimize losses due to the occurrence of voltage sags and short interruptions. Using the Monte Carlo simulation method for stochastic simulation of voltage sags for a period of 30 years, the Modified RBTS Bus 2 test system was used. To validate the simulation data, the costs considered were obtained from reference [13], in the plastic and automotive industries as US $48,920.00 and US $56,142.00, respectively, representing an average of US $52,531.00 per event. For the analysis, two types of systems were considered, overhead lines and underground cables. The best result was obtained when you chose to change a total of 4 overhead lines to underground cables.

In [17], the importance of assessing financial losses in industrial consumers due to voltage sag and short interruptions was mentioned. The authors proposed a new model for assessing the impact of voltage sags using voltage tolerance curves (VTC) associated with a truth table to characterize the logical relationship of sensitive equipment with the production process. The model is applied in a production process of TLT-LCD, whose estimated annual loss due to voltage sags was US $776,439.00 per year. The case study also showed that the proposed model had better adaptability in relation to the conventional model.

1.3. Contributions

The bibliographic research carried out included about two decades of work and aimed to verify the existence of works related to process shutdown costs due to voltage sags and short interruption. Directly related to the theme, several procedures were used to obtain cost estimates, namely: conducting surveys through forms, conducting stochastic simulations of voltage sags, and conducting measurements. Among the main countries where the costs were assessed, the following stand out: Italy, China, India, Finland, Brazil, Malaysia, and Norway. In general, cost estimates were presented on an annual basis, per event and per interrupted kW. Despite the significant number of types of activities surveyed in the publications, it was observed that the costs obtained in different countries were very different for the same productive sector. It was also found that most publications were old and few of them were in journals. In the specific case of Brazil, the works found were generic and did not quantify costs by type of activity. In this context, this work aims to fill the gaps identified by conducting direct surveys, in a total of 33 companies in 12 different business areas, supplied at medium voltage (11.9/13.8 kV) by the company EDP in the states of Espírito Santo and São Paulo in Brazil.

2. Theory

2.1. Voltage Sag and Short Interruption

The IEEE 1159-2009 defines voltage sag as being a decrease between 0.1 and 0.9 p.u. in rms voltage and with a duration from 0.5 cycles to 1 min [18]. Figure 1 shows the representation of voltage sag and is characterized by sag magnitude and duration. The sag magnitude is defined as the lowest rms voltage of the three phase voltages during the sag event, and its duration is the time that the voltage is lower than the 0.9 p.u. threshold in all three phases. Usually, voltage sags and short interruptions occurred due to power system faults.

Table 1 shows the category, the type of event, and the duration of each one. The information contained in the table was adopted from [18].

2.2. Overview of Equipment and Process Sensitivity

Each piece of equipment has a behavior in terms of voltage sags, which can be represented by a tolerance curve, which is usually rectangular and presents the thresholds of magnitude and duration. For example, Figure 2 shows the rectangular voltage tolerance curve indicating that when the voltage sags are longer than the duration threshold

(T_{max}) and deeper than the voltage magnitude threshold (V_{min}), the equipment will trip (or malfunction).

Table 1. Category of event, duration, and voltage magnitude.

Category		Transitory	Duration	Voltage Magnitude
Short-time duration	Instantaneous	Voltage sag	0.5→to 30 cycles	0.1 to 0.9 pu
	Momentaneous	Interruption Voltage sag	0.5→to 30 cycles 30 cycles to 3 s	<0.1 pu 0.1 to 0.9 pu
	Temporary	Interruption Voltage sag	3 s to 1 min 3 s to 1 min	<0.1 pu 0.1 to 0.9 pu

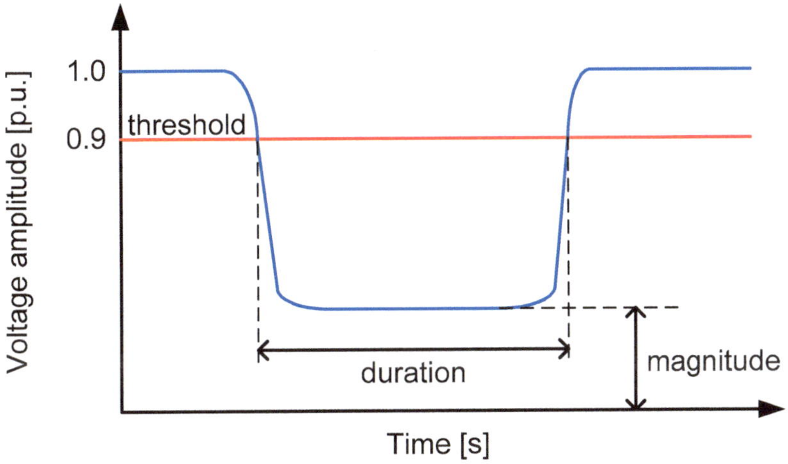

Figure 1. Voltage sag.

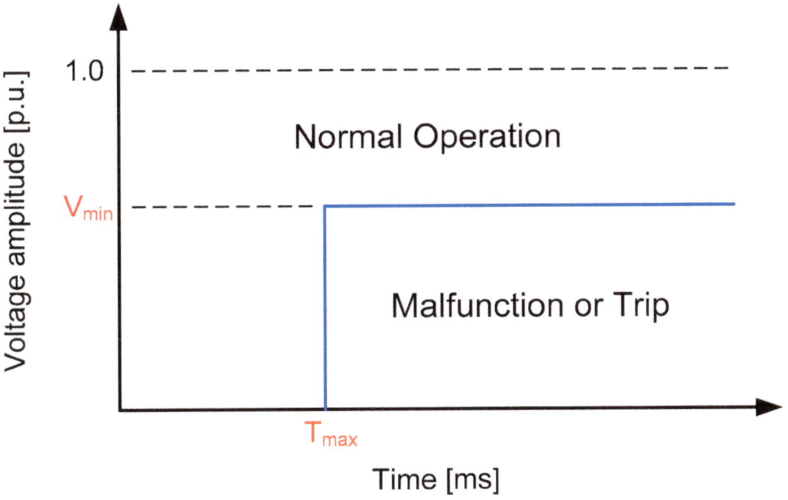

Figure 2. Equipment voltage tolerance curve.

Sensitivity curves are often used to assess the impact of voltage sags on industrial loads and processes. However, these curves do not have a single pattern of behavior, and there may be significant variations due to differences in manufacturer, model, hardware topology, software configurations, loading, among others [5,19]. Two others important characteristics of the voltage sags, in addition to the magnitude and duration of the voltage sags already mentioned are:

- point on wave—which corresponds to the phase angle of the instantaneous voltage at the sag initiation.
- phase–angle shift—the difference between the voltage phase angle at the pre sag moment and during the event.

The equipment most used in the industries and which were also found during the field survey carried out were: programmable logic controller—adjustable speed drive—ASD [20], PLC [21], contactor and [22–24].

2.2.1. Adjustable Speed Drive—ASD

ASDs are equipment widely used in industries to drive induction motors for better process control. In addition to reducing thermal and mechanical stress during starting and breaking the motors, they optimize the use of energy in applications that require variable torque or reduced speeds [20].

To evaluate the performance of a three-phase equipment, it must be taken into account that different types of voltage sags results and their different effects on their operation [20,25]. Figure 3 shows sensitivity curves obtained from ASDs, referring to the type I event (most severe reduction in voltage in one of the phases). It can be seen in Figure 3 that the ASDs had different levels of sensitivity.

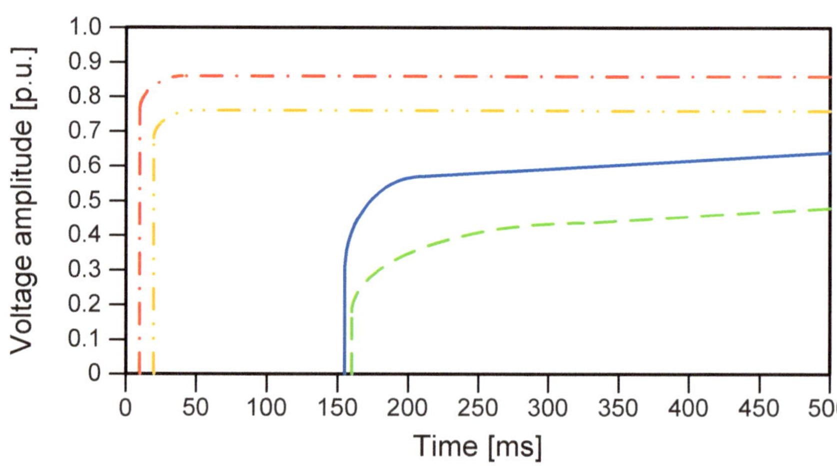

Figure 3. ASD sensitivity curves type I.

2.2.2. Programmable Logic Controller—PLC

The programmable logic controller—PLC, is a digital system that performs control and monitoring functions through a set of instructions previously defined by the user. The basic structure of its circuit is formed by: power supply, processing unit (CPU), and signal input and output modules, which can be digital or analog [21].

Figure 4 shows the sensitivity curves obtained through tests on several PLCs, illustrating the "worst case" and "best case". In the worst case, the sensitivity of the PLC was 0.75 p.u.

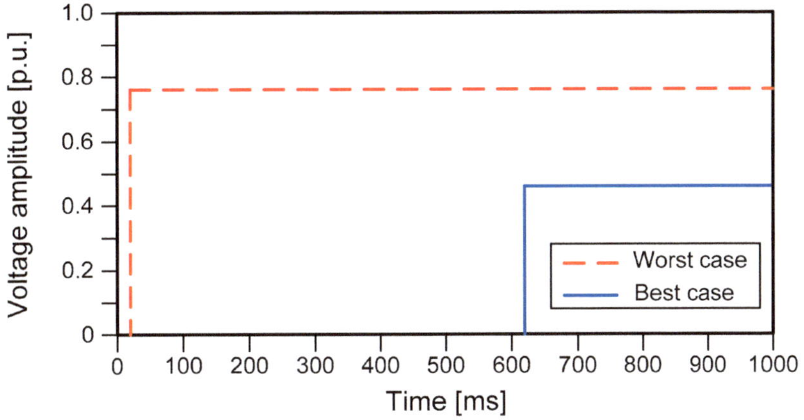

Figure 4. Sensitivity curves referring to the PLC.

2.2.3. Contactor

The contactor is an electromechanical device that controls loads from a command circuit, it is one of the most used devices in the industry [22–24]. The starting point of the event is also an important parameter in evaluating the performance of this equipment [19,22–24], as shown in the sensitivity curve of Figure 5.

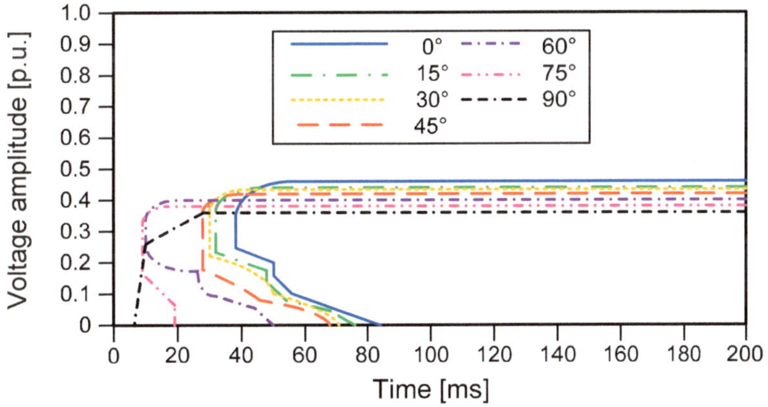

Figure 5. Contactor sensitivity curves for different starting points.

3. Methods

The methodology used in the research adopted a procedure similar to that used in [2,9,14] and was divided into four stages:

- Preparation of the survey form;
- Selection of companies;
- Field survey;
- Analysis of the results.

3.1. Preparation of the Survey Form

The form was prepared with the objective of obtaining detailed data on:

- Type of industrial activity, contracted demand, supply voltage, installed loads;
- Industry opening hours, number of employees;

- Critical parts of the process, identification of sensitive equipment, number of process shutdowns, types of losses, and critical period of the day;
- Overtime to recover lost production, lost raw material, investments to mitigate shutdowns.

The proposed form contained 19 questions, the content of which was prepared on the basis of Annex 2 of [26,27]. The questions contained in the survey form and their justifications are shown in Appendix A.

Therefore, it is noteworthy that this form was intended to obtain, in a more realistic way, the data necessary for the survey of process shutdown costs for each industrial activity due to the occurrence of voltage sag or short interruption.

3.2. Selection of Companies

The energy distributor that financed the research project (EDP), together with the researcher, generated a list of companies that could participate in the survey, from a total of 70 industries.

The selection criteria were: companies that "most complain" about process shutdowns due to the occurrence of events in the concessionaire's network; to contemplate in the research the largest possible number of types of activities; geographic location of the consumer units and to select from the sample companies located in two different states, the state of São Paulo and Espírito Santo.

Among the 70 selected industries, the established goal is to obtain at least 30 responses, considering that some would not be interested in participating in the survey due to confidentiality issues or because they do not have data available to be provided.

3.3. Field Survey

As a general strategy for conducting the visits, the following procedure was adopted:

- Schedule visits in advance with the assistance of the electric utility;
- Clarify the client about the survey objectives and send the survey form in advance to be evaluated and filled out.

As an operational strategy, visits should be carried out as follows:

- The researcher was always accompanied by an EDP representative. On the day of the visit, the objectives of the survey were explained again and its importance clarified, both in relation to the academic sphere, as well as the business aspect of customer service;
- When possible, the plants of the companies' production process were visited, in order to understand the manufacturing process, identifying the critical parts of the process, in order to better understand the causes of the shutdowns due to the occurrence of voltage sags and short interruptions.

An important point of the survey was to have the survey form filled out by the technician responsible for the visited company. If the questions were not answered previously, the researcher must ask the questions verbally in the meeting on the day of the visit, in order to obtain the necessary data to complete the form.

3.4. Analysis of the Results

The results obtained with the application of the survey forms were treated statistically and organized into tables and graphs in order to facilitate the display, interpretation, and analysis.

4. Case Study

4.1. Results

Among the 70 pre-selected companies, 11 withdrew from the survey, so 59 were visited between May and November 2019 and 33 responded in full to the survey form.

Table 2 shows the list of companies containing the type of activity, the size of the company, and a column of observations that presents some additional information.

Table 2. Company identification.

Number	Activity Type	Company Size	Observation
1	Plastic	Big (1000 employees)	Plastic film for food area
2	Automotive	Medium (560 employees)	Locks and door handles
3	Tires	Big (1000 employees)	Tires for cars
4	Tires	Big (1000 employees)	Tires for cars
5	Commercial	Medium (300 employees)	Space lease
6	Commercial	Small (150 employees)	Space lease
7	Wallpaper	Medium (300 employees)	Wallpaper
8	Metallurgical	Small (170 employees)	Aluminum profiles
9	Toys	Small (190 employees)	Toys
10	PVC Plastic	Medium (400 employees)	Plastic film for food area
11	PVC Plastic	Medium (800 employees)	Plastic film for food area
12	Electroplating	Small (35 employees)	Pieces and parts
13	Extrusion	Small (49 employees)	Metallurgical
14	Textile	Small (120 employees)	Wool fiber
15	Textile	Small (70 employees)	Dyeing clothes
16	Glass	Small (80 employees)	Glass for boxing, doors of residences
17	Extrusion	Medium (500 employees)	Use in deep sea waters
18	Plastic	Medium (300 employees)	Packaging for food area
19	Foundry	Small (50 employees)	Freight train parts—Vale do Rio Doce
20	Rock mining	Small (45 employees)	Granules for civil industry
21	Glass	Small (60 employees)	Glass for boxing, doors of residences
22	Food	Small (250 employees)	Pulp juices
23	Furniture	Medium (300 employees)	MDF boards
24	Furniture	Medium (300 employees)	MDF boards
25	Animal food	Small (50 employees)	Animal food
26	Stones and granites	Small (250 employees)	Manufacturing granite and marble sinks
27	Chips cards	Small (200 employees)	Card for banks and general trade
28	Plastic	Small (100 employees)	Plastic packaging for food industry
29	Metallurgical	Medium (300 employees)	Vertical, horizontal movements
30	Commercial	Big (1000 employees)	Space lease
31	Fertilizer	Small (100 employees)	Packaging
32	Automotive	Medium (500 employees)	Bus assembler
33	Petroleum	Big (1000 employees)	Onshore oil prospecting

Table 3 shows, for each company, the sensitive equipment used in its processes, the average number of monthly shutdowns, and the average time to resume production when the process stops.

Table 3. Sensitive equipment, event history and breakdown time.

Number	Activity Type	Sensitive Equipment	Event History	Breakdown Time
1	Plastic	ASD, PLC	2 events/month	120 min
2	Automotive	ASD, PLC, Robot	4 events/month	60 min
3	Tires	ASD, PLC	2 events/month	120 min
4	Tires	ASD, PLC, Electronic cards	3 events/month	60 min
5	Commercial	Computer, No break	7 events/month	15 min
6	Commercial	Building Manag. System, No break, Computer	4 events/month	20 min
7	Wallpaper	ASD, PLC	5 events/month	30 min
8	Metallurgical	ASD, PLC	2 events/month	60 min
9	Toys	ASD, PLC	8 events/month	45 min
10	PVC Plastic	ASD, PLC, electronic cards	10 events/month	30 min
11	PVC Plastic	ASD, PLC, electronic cards	10 events/month	30 min
12	Electroplating	Contactor	1 event/month	10 min
13	Extrusion	ASD, PLC	3 events/month	15 min
14	Textile	ASD, Contactor	2 events/month	30 min
15	Textile	ASD, PLC, Electronic cards	6 events/month	40 min

Table 3. *Cont.*

Number	Activity Type	Sensitive Equipment	Event History	Breakdown Time
16	Glass	ASD, PLC	not available	10 min
17	Extrusion	ASD, PLC, RX SCANNER	30 events/month	60 min
18	Plastic	ASD, PRINTER	25 events/month	60 min
19	Foundry	ASD, PLC	1 event/month	60 min
20	Rock mining	ASD, PLC	not available	20 min
21	Glass	ASD, PLC	not available	20 min
22	Food	ASD, PLC	not available	360 min
23	Furniture	ASD, PLC, UV lamp	8 events/month	60 min
24	Furniture	ASD, PLC, UV lamp	8 events/month	30 min
25	Animal food	ASD, PLC	8 events/month	30 min
26	Stones and granites	ASD, PLC	4 events/month	30 min
27	Chips cards	Printers machines	8 events/month	120 min
28	Plastic	ASD, PLC	4 events/month	15 min
29	Metallurgical	ASD, PLC	4 events/month	15 min
30	Commercial	Computers, Servers, Elevators, Escalators	5 events/month	10 min
31	Fertilizer	ASD, PLC, Contactor, Elevator	not available	10 min
32	Automotive	Laser, Tube Bender, Welding Robot	30 events/month	20 min
33	Petroleum	ASD, PLC	18 events/month	60 min

Table 4 shows, for each company surveyed, the demand, the cost per event, the cost per demand, and other information about additional costs such as payment of fines, rework, and repairs of equipment damaged by the occurrence of voltage sag or short interruption.

Table 4. Cost per event and cost per kW.

Number	Activity Type	Demand (kW)	Cost/Event (US $)	Cost/kW (US $/kW)	Other Information
1	Plastic	3000	4190.30	1.40	US $3286.50 equipment damage
2	Automotive	2500	6847.00	2.74	Damage to some equipment
3	Tires	2900	6003.00	2.07	US $12,005.00/month (scrap reprocessed)
4	Tires	1700	6000.00	3.53	US $18,000.00/month (scrap reprocessed) + 2096.00 maintenance
5	Commercial	1200	2717.00	2.26	US $76,070.00/month of penalty (12 h no energy)
6	Commercial	1300	6851.00	5.27	US $261.00 in damaged equipment and 1 h no monitoring Investment—US $227,718.00 in 8 generators, 2018/Jan.
7	Wallpaper	1700	9667.00	5.69	US $2370.00 equipment damage
8	Metallurgical	1400	2071.00	1.48	-
9	Toys	1200	3657.00	3.05	US $740.00 equipment damage
10	PVC Plastic	500	1640.00	3.28	US $2368,00 equipment damage
11	PVC Plastic	1350	6056.00	4.50	US $7130,00 equipment damage
12	Electroplating	350	556.00	1.60	US $470,00 equipment damage
13	Extrusion	500	1811.00	3.62	-
14	Textile	950	1501.00	1.67	-
15	Textile	450	1509.45	3.35	US $2841,00 equipment damage
16	Glass	410	1437.00	3.51	Loss of 1 oven due to an 8-h interruption
17	Cables Extrusion	850	52,800.00	62.18	Loss of umbilical tubes up to 2 km
18	Plastic	2400	5690.00	2.37	-
19	Foundry	850	4773.00	5.61	Scraps are reprocessed
20	Rock mining	2000	4085,00	2.04	Loss of particle size
21	Glass	405	2161.00	5.33	Scraps are destined for beverage factories
22	Food	700	5145.00	7.35	5 no breaks—800 kVA
23	Furniture	760	7238.00	9.52	Damage of UV Lamps
24	Furniture	800	7153.00	8.94	Damage of Transformer and track rollers
25	Animal food	1000	650.00	0.65	Refused feed is recycled in the process
26	Stones and granites	3150	11,130.00	3.53	Imperfect granite slabs are reprocessed
27	Chips cards	500	6613.00	13.26	-
28	Plastic	400	2031,00	5.08	Material is recycled and reprocessed
29	Metallurgical	500	2715.00	5.43	-
30	Commercial	6000	9311.00	1.55	Investment in harmonic filter
31	Fertilizer Chemical	250	4565.00	18.26	Refuse is reprocessed
32	Automotive	2500	13,049.00	5.22	Investment in no breaks, generator
33	Petroleum	2500	41,414.00	16.57	-

4.2. General Analysis

Considering all the companies surveyed, boxplots were generated for the number of employees, number of events per month, breakdown time, demand for electricity, cost per event, and cost per demand.

Figure 6 shows the boxplot of the number of employees; in this case, there were no data considered to be outliers. The average value of the number of employees found was 349 and the boxplot demonstrates that, in general, the research was focused on small and medium-sized companies.

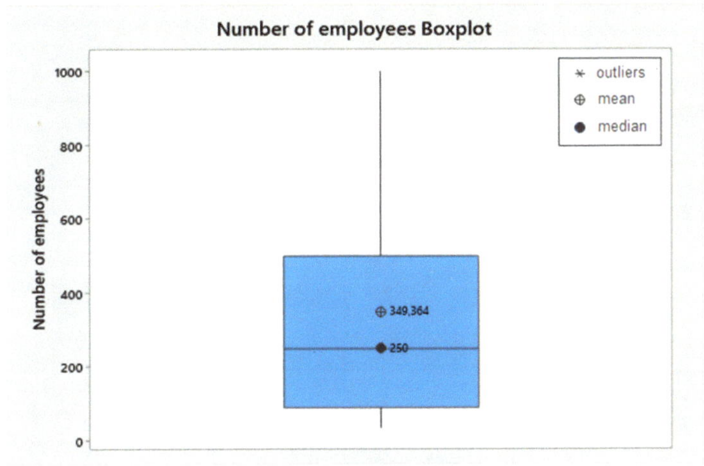

Figure 6. Boxplot (Number of employees).

Figure 7 presents the boxplot of the number of voltage sags or short interruptions per month, as noted, there are 4 companies with a number of events considered to be outliers. In average terms, there are approximately 8 events per month, which can be considered a high value, since depending on the type of product manufactured, a single process shutdown can generate major losses.

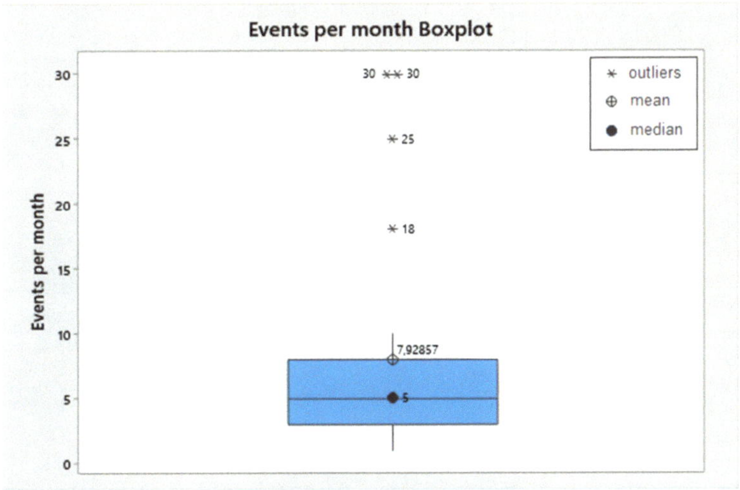

Figure 7. Boxplot (Events per month).

Figure 8 shows the boxplot for the breakdown time, which corresponds to the time necessary for the company to resume production when a process shutdown occurs. There is only one outlier, most companies had a breakdown time between 10 min and 60 min, and the average was 50.76 min.

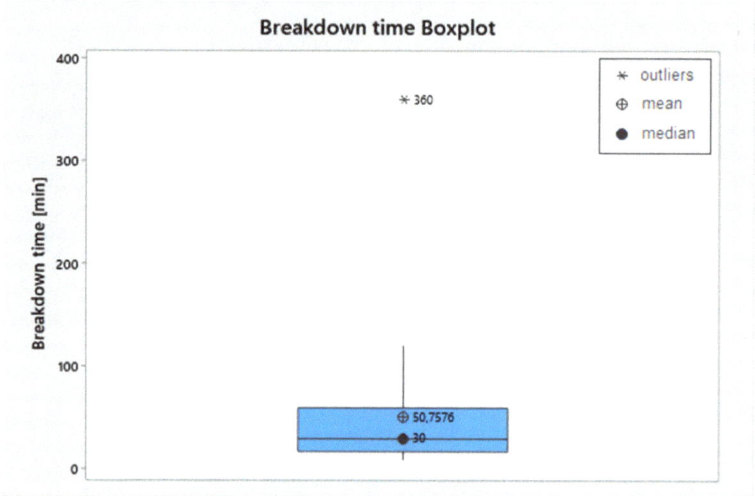

Figure 8. Boxplot (Breakdown time).

The graph in Figure 9 corresponds to the boxplot of electricity demand, which presented a discrepant value (company 30). The average demand for electricity is 1423.48 kW, with a large concentration of companies in the range from 500 kW to 2500 kW, reinforcing that the majority of companies surveyed were small and medium-sized.

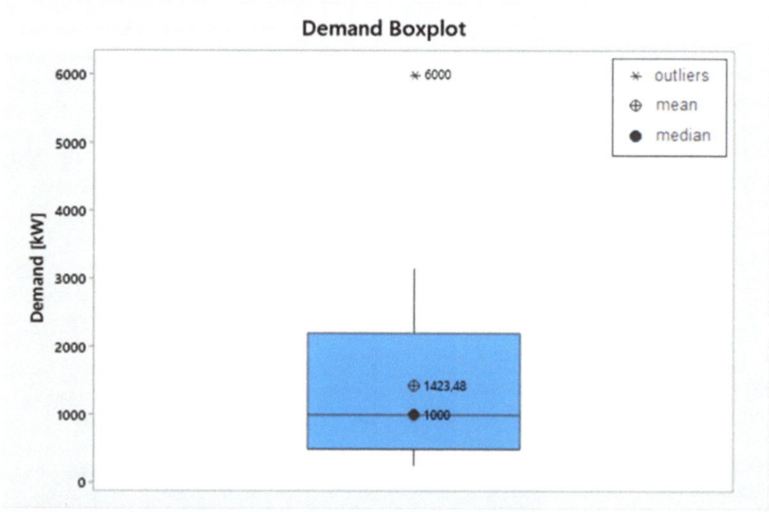

Figure 9. Boxplot (Demand).

Figure 10 shows the boxplot of cost per event, it was observed that there were two outliers, companies 17 and 33, that presented a cost per event much higher than the other companies surveyed due to the high added value of the product. It was also noted that the average cost related to the occurrence of voltage sag and short interruption was US $7364.75, with the minimum and maximum values being US $556.00 and $13,049.00, respectively.

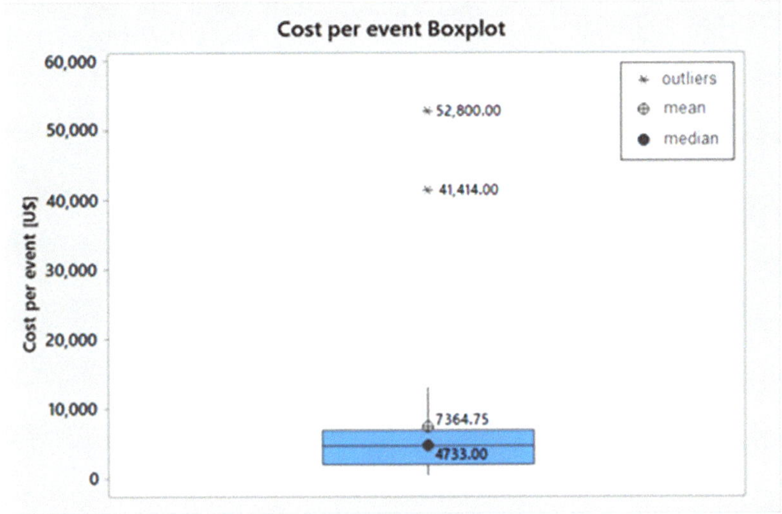

Figure 10. Boxplot (Cost per event).

Figure 11 presents the boxplot of costs per demand, where companies 17, 27, 31, and 33 were characterized as outliers. It was observed that the average found was US $6.72/kW and that there was a large concentration of companies in the range of US $2.00/kW to US $6.00/kW.

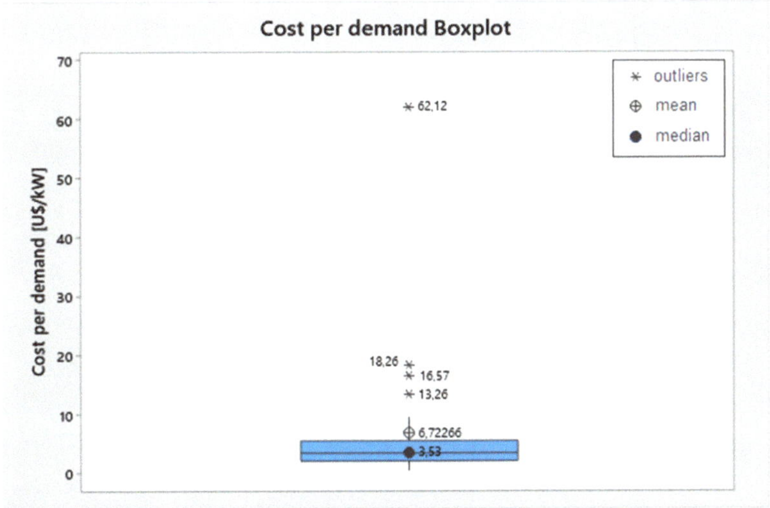

Figure 11. Boxplot (Cost per demand).

It is worth mentioning that considering the average values found for the number of events per month (8) and the average cost per event (US $7364.75), there was an estimate of the average annual cost due to voltage sags and short interruption of US $707,016.00.

The magnitude of this value shows the importance of assessing the economic impacts of these events and when viable to use energy conditioners to mitigate these costs.

Another analysis that can be done is whether the company's electricity demand correlates with the cost per event or the cost per demand. The graph in Figure 12 shows the dispersion and the correlation coefficient between cost per event and demand, while Figure 13 shows the same information between the variables cost per demand and demand.

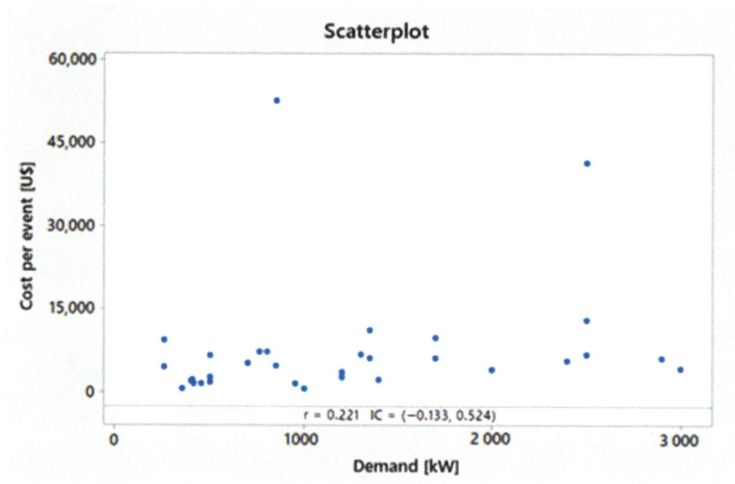

Figure 12. Correlation (Cost per event × demand).

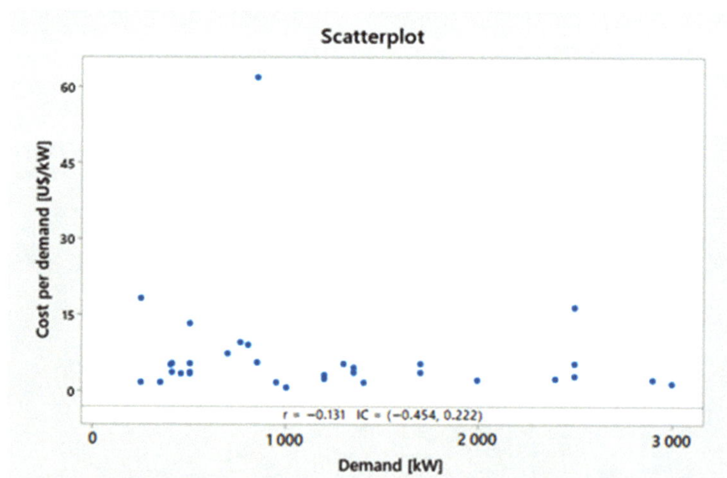

Figure 13. Correlation (Cost per demand × demand).

Due to the low correlation coefficient found in both cases (r = 0.22 and r = −0.13 respectively), it can be concluded that demand did not affect the cost per event or cost per demand, probably the main variable that influences costs was the added value of the product generated.

4.3. Analyzes by Type of Activity

Among the 33 companies analyzed, 12 different types of activities were identified. Table 5 presents the demand, cost per event, and cost per demand, data obtained for the different types of activity. The foundry, galvanizing, aluminum extrusion, and umbilical cable industries were included in the metallurgical activity, since they produce products designed for this purpose. The toy industry was considered in the activity of the plastic industry, since they are manufactured by polymer injection machines.

Table 5. Costs by type of activity.

Number	Demand (kW)	Cost/Event (US $)	Cost/kW (US $/kW)
	Plastic—Total Companies: 8—24.24%		
1	3000	4190.00	1.40
7	1700	9667.00	5.69
9	1200	3657.00	3.05
10	500	1640.00	3.28
11	1350	6056.00	4.50
18	2400	5690.00	2.37
27	500	6613.00	13.26
28	400	2031.00	5.08
Range	400 to 3000	1640.00 to 9667.00	1.40 to 13.26
Average	1381.25	4943.00	4.83
	Automotive—Total Companies: 4—12.12%		
2	2500	6847.00	2.74
3	2900	6003.00	2.07
4	1700	6000.00	3.53
32	2500	13,049.00	5.22
Range	1700 to 2900	6000.00 to 13,049.00	2.07 to 5.22
Average	2400	7974.75	3.39
	Commercial—Total Companies: 3—9.09%		
5	1200	2717.00	2.26
6	1300	6851.00	5.27
30	6000	9311.00	1.55
Range	1200 to 6000	2717.00 to 9311.00	1.55 to 5.27
Average	2833.33	6293.00	3.03
	Glass—Total Companies: 2—6.06%		
16	410	1437.00	3.51
21	405	2161.00	5.33
Range	405 to 410	1437.00 to 2161.00	3.51 to 5.33
Average	408	1799.00	4.42
	Metallurgical—Total Companies: 6—18.18%		
8	1400	2071.00	1.48
12	350	556.00	1.60
13	500	1811.00	3.62
17	850	52,800.00	62.18
19	850	4773.00	5.61
29	500	2715.00	5.08
Range	350 to 1400	556.00 to 52,800.00	1.48 to 62.18
Average	741.67	10,787.67	13.26
	Mining—Total Companies: 1—3.03%		
20	2000	4085.00	2.04

Table 5. *Cont.*

Number	Demand (kW)	Cost/Event (US $)	Cost/kW (US $/kW)
	Food—Total Companies: 2—6.06%		
22	700	5145.00	7.35
25	1000	650.00	0.65
Range	700 to 1000	650.00 to 5145.00	0.65 to 7.35
Average	850	2897.50	4.00
	Furniture—Total Companies: 2—6.06%		
23	760	7238.00	9.52
24	800	7153.00	8.94
Range	760 to 800	7153.00 to 7238.00	8.97 to 9.52
Average	780	7195.50	9.23
	Stones and Granites—Total Companies: 1—3.03%		
26	3150	11,130.00	3.53
	Chemical—Total Companies: 1—3.03%		
31	250	4565.00	18.26
	Textile—Total Companies: 2—6.06%		
14	950	1501.00	1.67
15	450	1509.45	3.35
Range	450 to 950	1501 to 1509.45	1.67 to 3.35
Average	700	1505.23	2.51
	Petroliferous—Total Companies: 1—3.03%		
33	2500	41,414.00	16.57

Table 5 shows that the largest number of visits were made to the plastics industry, corresponding to 24.24% of the total number of companies visited whose costs per event and cost per demand ranged from US $1640.00 to US $9667.00 and US $1.40 to US $13.26, respectively. This wide range of values is due to the fact that the types of products produced have different aggregate values. The average cost per event found was US $4943.00 and the cost per demand was US $4.79.

The automotive industries represented 12.12% of the total visited companies and their costs were very close to each other, ranging from US $6000.00 to US $6847.00 for auto parts, although each company manufactures a specific product, and US $13,048.00 for the automaker. Average costs per event and demand were US $7947.50 and US $3.39, respectively.

The commercial activity sector had three companies visited, representing 9.09% of the companies, and the costs per event obtained ranged from US $2717.00 to US $9311.00, presenting an average of US $6293.00, while the average cost per demand was US $3.03. In this activity, 2 shopping centers and 1 storage shed were grouped.

The two glass industries had an average cost per event of US $1799.00 and the cost per demand of US $4.42, remembering that the glass companies visited were destined to the construction sector.

Metallurgical industries accounted for 18.18% of the visits and presented cost values per event from US $556.00 to US $52,800.00, with an average of US $10,787.67. In this activity, small metal smelting and galvanizing companies were grouped, which produce raw materials for metallurgical industries. The extruded umbilical optical cable industry was also grouped, for its use in deep waters, for the extraction of oil, whose cost of production shutdown is high.

In the food industries, it was observed that the costs per event were very different. The factory dedicated to animal feeding had a lower value, while the beverage manufacturer had a higher cost. In the latter case, the operation of resuming the unit was complex, since there was a need to wash the production lines, increasing the time to resume the process.

The two companies in the furniture sector had the cost per event values very close with an average of US $7195.50 and cost per demand of US $9.23, demonstrating that possibly this value is typical in this type of activity.

The textile industries also had very close costs per event, ranging from US $1501.00 to US $1509.45. The average cost per demand obtained was US $2.51.

The rest of the activities surveyed, that is, mining, chemistry, stones and granites, and oil prospecting, had only one company in the sample and it was not possible to make a comparative analysis. However, it is noted that there was great variability in the values of cost per event in these industries, mainly due to the added value of the product that each company produces.

Figure 14 shows a graphical comparison between the average values of the cost per event of the types of activities surveyed. It should be noted that, with the exception of the petroliferous sector, the other sectors had an average cost close to each other and around US $6000.00. Therefore, the size of the company is probably a variable that also affects the cost per event.

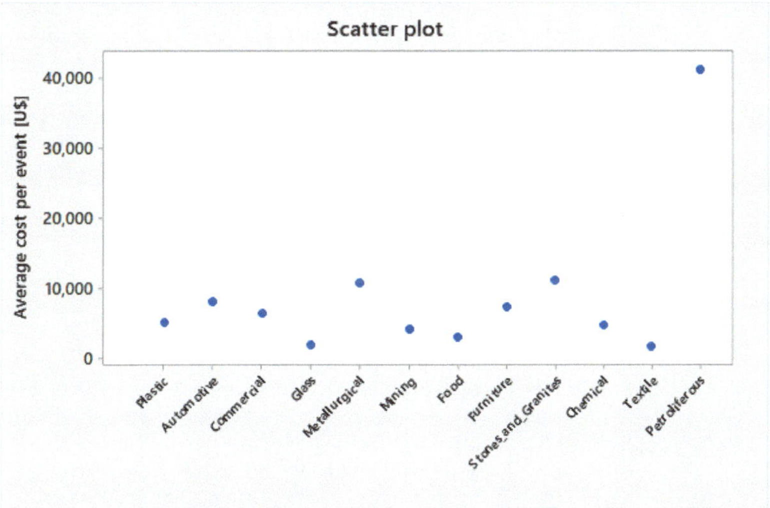

Figure 14. Scatter plot (Type of activity).

4.4. Difficulties

The main difficulties faced in carrying out this survey were:

- Planning visits, scheduling difficulties, and delay in responding to the interest in participating in the survey;
- Difficulty in logistics and locomotion to visit companies. The distance between the city of São Paulo, where the researcher resides, and the companies chosen in this state was around 150 km. Regarding the state of Espírito Santo, travel is only viable by air, since the distance between São Paulo and Espírito Santo is 1100 km.

Besides, during the field survey, it was noted the concern of the interviewees in not being able to provide their costs in detail, possibly due to fear of speculation from competing companies in the market. In general, companies consider costs to be a strategic part of business, even though the researcher had guaranteed data confidentiality and that the name of the company would not be disclosed.

Of the 59 companies visited, 13 did not respond to the questionnaire claiming that they had not received authorization from the board, in 9 companies it was not possible to obtain the data requested in the questionnaire during the visit, and 4 companies gave up on the survey. In summary, 26 companies did not participate in the survey, which represented 44% of the total visited.

Of the 33 companies visited, 10 medium-sized companies requested that their names be kept confidential. In small companies, almost all agreed to have their names revealed, including allowing costs to be disclosed. However, for reasons of consistency, the authors chose not to disclose the name of any researched company.

5. Conclusions

The literature review presented in Section 1.2 proves that in fact, several types of industries installed in different countries face problems of loss of production due to the occurrence of voltage sags and short interruptions. It was found that several procedures are used to obtain an estimate of these costs, which are: conducting research through forms, conducting stochastic simulations, and conducting measurements. Cost estimates are presented on an annual basis, per event, and per interrupted kW. Despite the significant number of branches of activity surveyed in the publications, it was observed that the costs obtained in different countries for the same productive sector are very disparate.

This work presented and applied a survey methodology with the objective of evaluating the financial impact of voltage sags and short interruptions on industrial and commercial consumers. The methodology is based on conducting on-site visits and applying a specific questionnaire containing a total of 19 questions.

The survey results made it possible to quantify the losses involved in 33 small and medium-sized companies, with an average number of 349 employees, grouped in 12 business areas, all supplied in medium voltage networks (11.9 kV and 13.8 kV). The average cost obtained was US $7364.75 per event and US $6.72/kW interrupted.

We conclude that this work achieved the objectives initially established, representing an effective contribution to the electricity sector, especially for the Brazilian sector.

As a proposal to continue this line of research, it is recommended to contemplate other segments of activities and expand the sample size of the segments already researched (food industry, furniture, mining, stones and granites, oil).

Regarding the success rate of response to the form, considering that the initial sample was 70 companies, with 33 responses to the survey form, the rate was 47.1%, however, considering the sample of 59 companies due to withdrawal of 11 companies, the rate becomes 55.9%.

As improvements to the procedure used, it is proposed to take actions to publicize the survey in advance, such as the holding of a workshop with wide participation of companies, with the objective of attracting and motivating the participation of consumers. With this attitude, it is believed that the percentage of participation could be improved.

Author Contributions: Conceptualization, É.M.M., J.M.d.C.F. and P.M.d.S.; data curation, É.M.M. and P.V.G.d.S.; formal analysis, É.M.M., J.M.d.C.F. and P.M.d.S.; funding acquisition, N.B.P.; investigation, É.M.M. and J.M.d.C.F.; methodology, É.M.M., J.M.d.C.F. and P.M.d.S.; project administration, J.M.d.C.F. and P.M.d.S.; supervision, J.M.d.C.F. and P.M.d.S.; validation, É.M.M. and J.M.d.C.F.; writing—original draft, É.M.M. and P.V.G.d.S.; writing—review and editing, É.M.M., J.M.d.C.F., P.M.d.S. and P.V.G.d.S. All authors have read and agreed to the published version of the manuscript.

Funding: The authors gratefully acknowledge the financial support of Capes, CnPq, Fapemig, and EDP.

Institutional Review Board Statement: Not applicable.

Informed Consent Statement: Not applicable.

Data Availability Statement: The data presented in this study are contained in the article.

Acknowledgments: The authors thank the support of EDP in providing the data used in this study and are also grateful for the technological support provided by the Federal University of Itajubá.

Conflicts of Interest: The authors declare no conflict of interest.

Appendix A

Following is the survey form:

1. **Company name.**
2. **Main type of activity.**
3. **Average monthly demand, consumption and supply voltage.**

Questions 1 to 3 aim to identify the company, its type of activity, as well as its demand, consumption, and supply tension.

4. **Opening hours of the company.**
5. **Number of employees of the company.**

Questions 4 and 5 seek to obtain the production regime, the number of work shifts and to identify the size of the company.

6. **Types of loads and processes installed.**
7. **How many processes shutdowns have occurred due to voltage sags and short interruptions in the last 12 months?**

Questions 6 and 7 seek to know the purpose of energy use and the types of loads installed in the industry. It also seeks to identify the number of process stops due to voltage sag or short interruption.

8. **Considering voltage sag and short interruption identify the events that have the greatest impact (instantaneous, momentary, and temporary).**

Question 8 aims to know, in a qualitative way, the interviewees' perception regarding process shutdowns, their frequencies and their impacts on the process and to identify the types of most impactful events.

9. **What is the critical period of the day that the occurrence of a voltage sag or short interruption causes more losses?**
10. **Which equipment in your company is most sensitive to voltage sag or short interruption?**

Questions 9 and 10 seek to verify, in a qualitative way, if there is a period of the day that presents a higher frequency of voltage sag or short interruption that directly affects the company in terms of process shutdowns. In addition, they aim to identify sensitive loads and the interviewees' perception of whether these loads have or not correlations with the process shutdowns.

11. **How long would it take to resume production or operation if there was an unexpected process shutdown?**

Question 11 seeks to determine the time taken to resume the production process.

12. **When there is an unexpected process shutdown due to voltage sag or short interruption, what type of damage or loss does the company have, and to what extent?**

This question aims to verify whether it is possible to detail each item that makes up the total cost of the loss according to the specifics of each industrial activity.

13. **Considering a day and a period of high production, what is the estimated cost of restarting production/operation and the loss of production/operation/billing that your company would have in that period if there was an unexpected process shutdown due to voltage sag and short interruption?**

Question 13 is complementary to the previous one, in order to verify whether there is also a lost profit.

14. **Does the company pay overtime to employees in order to recover production or billing losses, even to end interrupted production, due to an unexpected process shutdown due to voltage sag and short interruption? If so, could the expenditure on the payment of such overtime be estimated?**

This question aims to verify, in qualitative and quantitative terms, whether there is payment of overtime to employees due to the occurrence of voltage sag and short interruption.

15. **Does your company have raw material, product in process, or finished product that would be discarded due to an unexpected process shutdown? If so, could you estimate the cost related to these losses?**

This question aims to verify, depending on the company's activity, if there is a loss of raw material, finished product or in processing, in addition to obtaining the associated costs.

16. **Does the company have expenses for repairs and replacement of damaged equipment due to an unexpected process shutdown? If so, could you estimate the cost related to this?**

Question 16 is intended to verify if there are expenses with repairs and replacement of equipment damaged due to voltage sag and short interruption.

17. **Would the company have, in addition to the factors mentioned above, any other type of loss if an unexpected process shutdown occurred due to voltage sag and short interruption? If so, which one? Could you estimate the cost related to this other factor?**

The question aims to verify if there is another type of cost due to an unexpected process shutdown.

18. **Does the company have procedures or equipment to monitor voltage sags and short interruptions? If yes, detail the type of monitoring.**

This question is intended to find out if the company has voltage sag or short interruption monitoring procedures or equipment.

19. **What does your company do to mitigate voltage sag and short interruption? And what is the investment for that?**

This question aims to find out if the company invests in equipment to mitigate voltage sag or short interruption.

References

1. MCGranaghan, M.; Roettger, B. Economic Evaluation of Power Quality. *IEEE Power Eng. Rev.* **2002**, *22*, 8–12. [CrossRef]
2. LaMedica, R.; Esposito, G.; Tironi, E.; Zaninelli, D.; Prudenzi, A. A survey on power quality cost in industrial customers. In Proceedings of the 2001 IEEE Power Engineering Society Winter Meeting. Conference Proceedings (Cat. No.01CH37194), Columbus, OH, USA, 28 January–1 February 2001.
3. Heine, P.; Pohjanheimo, P.; Lehtonen, M.; Lakervi, E. A method for estimating the frequency and cost of voltage sags. *IEEE Trans. Power Syst.* **2002**, *17*, 290–296. [CrossRef]
4. Quaia, S.; Tosato, F. A method for the computation of the interruption costs caused by supply voltage dips and outages in small industrial plants. In Proceedings of the IEEE Region 8 EUROCON 2003, Computer as a Tool, Ljubljana, Slovenia, 22–24 September 2003; Volume 2, pp. 249–253. [CrossRef]
5. Gupta, C.P.; Milanovic, J.V. Costs of Voltage Sags: Comprehensive Assessment Procedure. In Proceedings of the IEEE Russia Power Tech, St. Petersburg, Russia, 27–30 June 2005; pp. 1–7. [CrossRef]
6. Milanovic, J.V.; Gupta, C.P. Probabilistic Assessment of Financial Losses due to Interruptions and Voltage Sags-Part I: The Methodology. *IEEE Trans. Power Deliv.* **2006**, *21*, 918–924. [CrossRef]
7. Milanovic, J.V.; Gupta, C.P. Probabilistic Assessment of Financial Losses due to Interruptions and Voltage Sags—Part II: Practical Implementation. *IEEE Trans. Power Deliv.* **2006**, *21*, 925–932. [CrossRef]
8. Goswami, A.K.; Gupta, C.P.; Singh, G.K. Assessment of Financial Losses due to Voltage Sags in an Indian Distribution System. In Proceedings of the 2008 IEEE Region 10 and the Third international Conference on Industrial and Information Systems, Kharagpur, India, 8–10 December 2008; pp. 1–6. [CrossRef]
9. Kjølle, G.H.; Samdal, K.; Singh, B.; Kvitastein, O.A. Customer Costs Related to Interruptions and Voltage Problems: Methodology and Results. *IEEE Trans. Power Syst.* **2008**, *23*, 1030–1038. [CrossRef]
10. Carvalho Filho, J.M.; Leborgne, R.C.; Oliveira, T.C.; Oliveira, J.F.; Watanabe, G.T. Voltage Sag Cost Assessment Based on Power Flow Reduction and Non Supplied Energy. In Proceedings of the 2009 IEEE Power & Energy Society General Meeting, Calgary, AB, Canada, 26–30 July 2009.
11. Cebrian, J.C.; Kagan, N. Hybrid Method to Assess Sensitive Process Interruption Costs Due to Faults in Electric Power Distribution Networks. *IEEE Trans. Power Deliv.* **2010**, *25*, 1686–1696. [CrossRef]

12. Goswami, A.K.; Gupta, C.P.; Singh, G.K. Cost-benefit analysis of voltage sag mitigation methods in cement plants. In Proceedings of the 2014 16th International Conference on Harmonics and Quality of Power (ICHQP), Bucharest, Romania, 25–28 May 2014; pp. 866–870.
13. Salim, F.; Nor, K.M.; Said, D.M.; Rahman, A.A.A. Voltage sags cost estimation for malaysian industries. In Proceedings of the 2014 IEEE International Conference on Power and Energy (PECon), Kuching, Sarawak, 1–3 December 2014; pp. 41–46.
14. Chen, W.; Ding, C.; Wang, L.; Zhu, X. Economic analysis of voltage sag loss and treatment based on on-site data. In Proceedings of the 2016 China International Conference on Electricity Distribution (CICED), Xi'an, China, 10–13 August 2016; pp. 1–4.
15. Behera, C.; Banik, A.; Nandi, J.; Dey, S.; Reddy, G.H.; Goswami, A.K. Assessment of Financial Loss Due to Voltage Sag in an Industrial Distribution System. In Proceedings of the 2019 IEEE 1st International Conference on Energy, Systems and Information Processing (ICESIP), Chennai, India, 4–6 July 2019; pp. 1–6.
16. Somrak, T.; Tayjasanant, T. Minimized Financial Losses Due to Interruptions and Voltage Sags with Consideration of Investment Cost. In Proceedings of the 2019 IEEE PES GTD Grand International Conference and Exposition Asia (GTD Asia), Bangkok, Thailand, 19–23 March 2019; pp. 29–34.
17. He, H.-Y.; Zhang, W.-H.; Wang, Y.; Xiao, X.-Y. A Sensitive Industrial Process Model for Financial Losses Assessment Due to Voltage Sag and Short Interruptions. *IEEE Trans. Power Deliv.* **2020**. [CrossRef]
18. *Institute of Electrical and Electronics Engineers IEEE Std 1159-2009: IEEE Recommended Practice for Monitoring Electric Power Quality*; IEEE: New York, NY, USA, 2009. [CrossRef]
19. Escribano, A.H.; Gómez-Lázaro, E.; Molina-García, A.; Fuentes, J. Influence of voltage dips on industrial equipment: Analysis and assessment. *Int. J. Electr. Power Energy Syst.* **2012**, *41*, 87–95. [CrossRef]
20. Djokic, S.; Stockman, K.; Milanovic, J.; Desmet, J.; Belmans, R. Sensitivity of AC adjustable speed drives to voltage sags and short interruptions. *IEEE Trans. Power Deliv.* **2005**, *20*, 494–505. [CrossRef]
21. Wu, Y.; Li, C.; Xu, Y.; Wei, P. Characterizing the tolerance performance of PLCs to voltage sag based on experimental research. In Proceedings of the 2016 IEEE PES Asia-Pacific Power and Energy Engineering Conference (APPEEC), Xi'an, China, 25–28 October 2016; pp. 496–501.
22. Djokic, S.; Milanovic, J.; Kirschen, D. Sensitivity of AC Coil Contactors to Voltage Sags, Short Interruptions, and Undervoltage Transients. *IEEE Trans. Power Deliv.* **2004**, *19*, 1299–1307. [CrossRef]
23. Kanokbannakorn, W.; Saengsuwan, T.; Sirisukprasert, S. The modeling of AC magnetic contactor for immunity studies and voltage sag assessment. In Proceedings of the The 8th Electrical Engineering/ Electronics, Computer, Telecommunications and Information Technology (ECTI) Association of Thailand—Conference 2011, Khon Kaen, Thailand, 17–19 May 2011; pp. 621–624.
24. Shareef, H.; Marzuki, N.; Mohamed, A.; Mohamed, K. Experimental Investigation of ac contactor ride through capability during voltage sag. In Proceedings of the 2010 9th International Conference on Environment and Electrical Engineering, Institute of Electrical and Electronics Engineers (IEEE), Cappadocia, Turkey, 16–19 May 2010; pp. 325–328.
25. *Cigre/Cired/Uie Joint Working Group C4.110 Voltage Dip Immunity of Equipment and Installations*; Cigre Technical Brochure 412; CIGRE: Paris, France, 2010.
26. Magalhaes, C.H.N. Recursos Operativos no Planejamento de Expansão de Sistemas de Potência. Doctoral Thesis, Universidade de São Paulo–USP, São Paulo, Brazil, 23 March 2009. (In Portuguese).
27. Pelegrini, M.A.; Almeida, C.F.M.; Kondo, D.V.; Magalhaes, C.H.; Silva, F.T.; Baldan, S.; Filho, F.C.S.; Garcia, V.V. Survey and applications of interruption costs in large customers. In Proceedings of the 2012 IEEE 15th International Conference on Harmonics and Quality of Power, Hong Kong, China, 17–20 June 2012; pp. 860–864.

energies

Article

Harmonic Mitigation Using Passive Harmonic Filters: Case Study in a Steel Mill Power System

Byungju Park [1], Jaehyeong Lee [2], Hangkyu Yoo [1] and Gilsoo Jang [2,*]

[1] PQ Tech Incorporation, Suwon 16690, Korea; bjpark@pqtech.co.kr (B.P.); hkyoo@pqtech.co.kr (H.Y.)
[2] School of Electrical Engineering, Korea University, Seoul 02841, Korea; bluesky6774@korea.ac.kr
* Correspondence: gjang@korea.ac.kr; Tel.: +82-2-3290-3246

Abstract: In this study, we mitigated the harmonic voltage in a power system that contained the roughing mill (RM) and finishing mill (FM) motor drives. AC/DC converter type RM drive is a non-linear, large-capacity varying load that adversely affects power quality, e.g., a flicker, voltage distortion, etc. The voltage drop can be compensated within a certain limit by using the proper capacity of a power capacitor bank. In addition, the voltage distortion can be controlled as per the guidelines of IEEE Std. 519 using the passive harmonic filter corresponding to the characteristic harmonics of the motor drive load. The passive harmonic filter can provide an economical solution by mitigating the harmonic distortion with a proper reactive power supply. However, at the planning level, attention should be paid to avoid system overvoltage that is caused by the leading power under light load conditions and also the problem of parallel resonance between the harmonic filter and the step-down transformer. In addition, when designing the filter reactor, the K-factor and peak voltage must be considered; the filter capacitor also requires a dielectric material that considers the harmonic peak voltage. The purpose of this study was to acquire a better understanding of the filter applications as well as verify the field measurement, analysis, and design of harmonic filters together with its performance.

Keywords: converter; easy power system software; harmonic distortion; hi-pass filter; IEEE Std. 519; passive harmonic filter; single tune filter

Citation: Park, B.; Lee, J.; Yoo, H.; Jang, G. Harmonic Mitigation Using Passive Harmonic Filters: Case Study in a Steel Mill Power System. *Energies* **2021**, *14*, 2278. https://doi.org/10.3390/en14082278

Academic Editor: Gabriel Nicolae Popa

Received: 18 March 2021
Accepted: 15 April 2021
Published: 18 April 2021

Publisher's Note: MDPI stays neutral with regard to jurisdictional claims in published maps and institutional affiliations.

1. Introduction

The plate, in this paper, refers to an iron plate that is made by rolling an intermediate slab obtained through the iron-manufacturing process or the steel-manufacturing process. It is usually a steel plate with a thickness of at least 6 mm that is difficult to process and is generally used in the manufacturing of ships, bridges, boiler pressure vessels, etc. Owing to the characteristics of speed and torque control, a Ward Leonard, mercury arc, and thyristor converter have been used in the past. In recent years, this development process has been extended to the cyclo-converter, rectifier-pulse width modulation (PWM) inverter, and active rectifier-PWM inverter [1].

The rolling mill for plate processing is operated in the acceleration speed, pass, and idling modes. To satisfy these characteristics, precise control performance of constant torque and braking characteristics are required in response to a wide speed control range and whirlwind overload. Generally, the voltage drop and the voltage distortion of the mill motor drive converter system reduce the output torque of the motor drive and adversely affect the control performance. Therefore, proper reactive power compensation and harmonic mitigation of the mill motor drive system can reduce the system losses and increase the productivity and quality of the plate [1,2]. Figure 1 shows the power fluctuation of the mill motor drive system. The measured apparent (S), active (P), and reactive powers (Q) of the plate mill system are red, green, and blue, respectively.

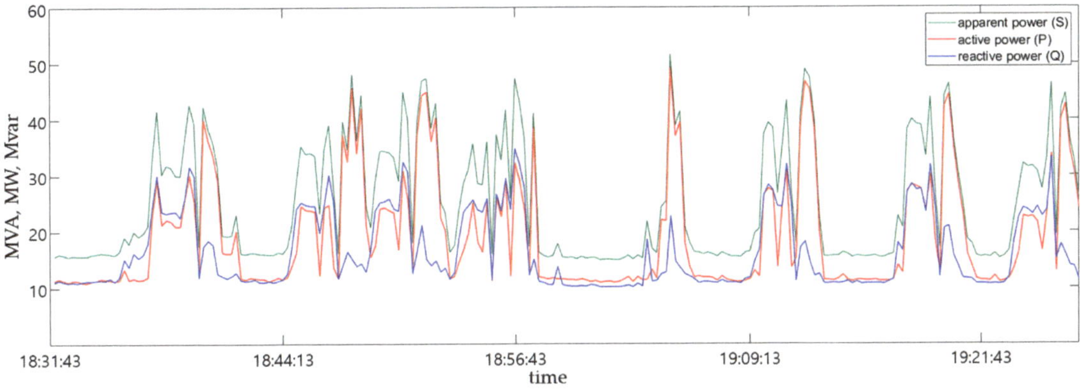

Figure 1. Measured P, S, and Q on plate mill system.

Harmonics generated in the power system have various adverse effects. First, the effects of harmonics on electrical equipment are generally as follows: disturbance, increased losses, extra neutral current, improper working of metering devices, resonance problems [3,4]. In addition, the effects of harmonics on the power system are as follows: the possibility of amplification of harmonic levels, efficiency reduction, ageing of electrical plant components, and malfunction [4,5]. To solve these problems, various studies have been conducted to mitigate harmonics of distribution systems using filters [6–11]. In particular, the effect is evaluated by sharing application case study on the real power system as in this paper [12–15].

Harmonic distortion can be suppressed using two methods: a passive power and an active power filter. A passive filter is the conventional solution to reduce harmonic distortion [16–18]. While it is simple, a passive filter does not always respond correctly to dynamics behavior [19]. This type of filter has been continuously developed over the past years, and notch filter (which bypasses a specific harmonic current) and high pass filter (which allow a large percentage of all harmonics above the corner frequency to pass through) are typical [17,18]. With the development of power electronics (PE) technology, active power filter has been used to mitigate harmonics. The basic principle is to use PE to create a specific current component that cancels the harmonic current component. Active power filter is better than passive filter in terms of technical aspects such as resonance, the range of harmonic mitigation range, and occupied bulky space, but has a disadvantage of being expensive [20].

In this paper, system modeling, harmonic calculation, filter design, construction, and field measurement were performed for the steel mill power system, and the results were evaluated based on IEEE Std. 519 [21]. Conventionally, to minimize voltage fluctuation, SVC (static var compensator) combined with TCR (thyristor-controlled reactor) and passive tuned filter, or STATCOM combined with PWM inverter module and high-pass filter is used [2,22]. However, depending on the system conditions, a passive filter can be a sufficient solution. In this case, the space required for the TCR can be saved and the power loss on TCR can be reduced [1].

1.1. *Brief Description of the Plate Mill System*

1.1.1. Simplified Single Line Drawing

The outline view of the plate mill system is shown in Figure 2. As seen from Figure 2, the short-circuit capacity (SCC) of the system at the grid side is 2500 MVA, X/R 10. On the left portion of Figure 2, the main system loads, roughing mill (RM) TOP, and BOT are in the form of a 5.5 MW DC motor drive with a 12-pulse thyristor rectifier. The finishing mill (FM) TOP and BOT are in the form of an 8.8 MW synchronous motor with an injection enhanced gate transistor (IEGT)-PWM inverter. (TOP and BOT mean motors and inverters located at the upper and lower side of RM and FM, respectively.) There are three 22/3.3 kV, 3 phase, 20 MVA, step-down transformers in the middle that are used for the 3.3 kV mill line, 3.3 kV 2 pre-leveler roll (PLR) power, and 3.3 kV shear/finishing line.

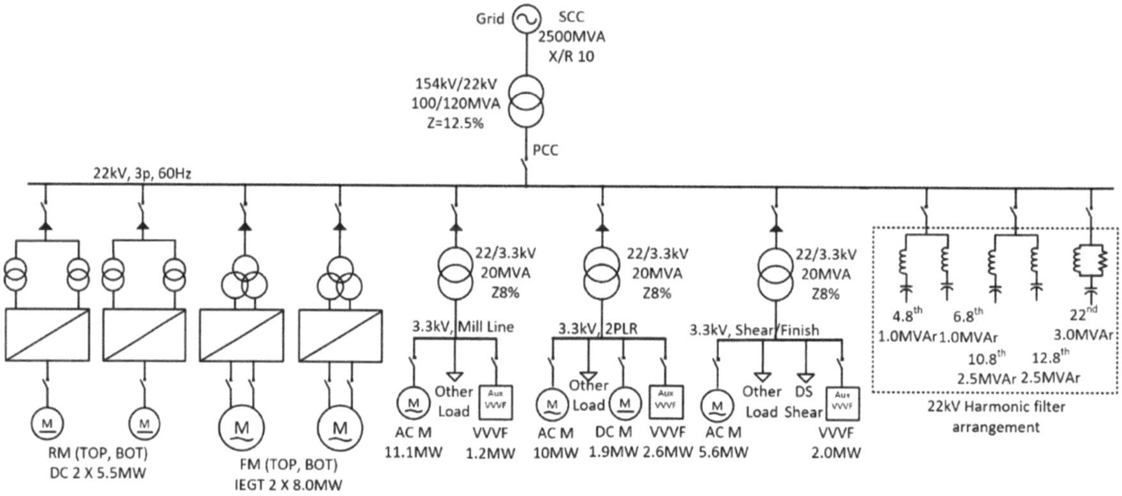

Figure 2. Simplified plate mill system single line drawing.

1.1.2. Summation of Input Data

As listed in Table 1, the SCC of the power supply utility is 2500 MVA, X/R 10, and specifications of the step-down transformers are 154/22.9 kV, 100 MVA, and Z = 12.5%. For RM TOP and BOT, 167% of the rating was applied considering the overload characteristics and 107% for FM TOP and BOT. In the 3.3 kV mill line, 2PLR, and finishing line, the AC motor, FM field, variable voltage variable frequency of the thyristor rectification and other loads were included.

Table 1. Summation of major input data.

Clarification	Description
Utility	25,000 MVA, X/R 10
Step down trans.	154 kV/22 kV, 100 MVA, Z = 12.5%
RM TOP	9.167 MW, −9.352 MVAR (167% of rating), DC motor
RM BOT	9.167 MW, −9.352 MVAR (167% of rating), DC motor
FM TOP	8.521 MW, 0.852 MVAR (107% of rating), synchronous motor
FM BOT	8.521 MW, −0.852 MVAR (107% of rating), synchronous motor
Step down trans.	22 kV/3.3 kV, 20 MVA, Z = 8%, 3 set
Mill line 3.3 kV	16.4 MW, −11.1 MVAR (AC motor, FM field, VVVF, and others)
2PLR 3.3 kV	19.1 MW, −12.8 MVAR (AC motor, DC motor, VVVF, and others)
Shear/finishing 3.3 kV	17.8 MW, −11.6 MVAR (AC motor, shear, VVVF, and others)

Tables 2 and 3 list the harmonic current spectrum of the RM TOP/BOT and FM TOP/BOT mill motor drives. RM TOP/BOT represents the typical characteristic harmonic current of the 12-pulse thyristor rectifier, and as listed in Table 1, the reactive power injection is also large. On the other hand, FM TOP/BOT is a PWM inverter drive with a large capacity IEGT, and the harmonic current and reactive power injection are low.

Table 2. Harmonic injection on RM and FM drives.

	Harmonic Current I_h(A), 22 kV Bus Base			
h Order	**RM TOP**	**RM BOT**	**FM TOP**	**FM BOT**
5	0	0	1.81	1.81
7	0	0	0.31	0.31
11	16.66	16.66	4.6	4.6
13	14.1	14.1	3.53	3.53
17	0	0	0.4	0.4
19	0	0	1.25	1.25
23	6.97	6.97	1.25	1.25
25	6.42	6.42	0.65	0.65
29	0	0	0.21	0.21
31	0	0	0.14	0.14
35	3.93	3.93	0.49	0.49
37	3.72	3.72	0.65	0.65
41	0	0	0.07	0.07
43	0	0	0.13	0.13
47	1.95	1.95	0.44	0.44
49	1.87	1.87	0.64	0.64

Table 3. Harmonic injection on 3.3 kV buses.

	Harmonic Current I_h(A), 22 kV Bus Base		
h Order	**Mill Line**	**2PLR**	**Shear Finishing**
5	40.7	152.7	149.4
7	28.1	105.5	103.2
11	20	74.8	73.2
13	12.8	48	47
17	10.5	38.4	38.5
19	6.7	25.2	24.6
23	4.2	15.7	15.4
25	3.2	11.8	11.3

2. Harmonic Filter Design

There are various effects of harmonics, such as overheating, electromagnetic noise of wires, noise of transformers, and malfunction of power devices. In addition, it increases the system losses and causes reliability problems [21,23,24]. A harmonic filter (HF) is a device used to reduce the harmonic current or harmonic voltage of the system to protect the power equipment from these problems [25,26]. HFs usually comprise capacitors, inductors, and resistors to provide a lower impedance than the system impedance at one or more specific frequencies as per requirement [23]. Thus, the filter can be used to lower the impedance of the corresponding order, absorb the harmonic current of the order, and reduce the harmonic voltage. In general, the filter tuning is set slightly lower than the corresponding order, which allows only a positive error when determining the manufacturing error of the filter capacitor and reactor [26,27]. Due to manufacturing error, if the tuning is at a higher frequency than that of the corresponding order, the harmonic current of the order acts in the leading phase, thereby increasing the harmonic voltage.

2.1. Capacitor Banks

The capacitor bank of the high-voltage system is mainly in the form of a wye connection, and it is a double wye type when the capacity is large. The neutral is classified as grounded and ungrounded. In the case of a grounded neutral, the recovery voltage can be significantly reduced [28]. On the other hand, there is a disadvantage that the current flows to the ground, which can cause communication failure and ground relay malfunction. In this study, an unground double wye connection was adopted that was easy to configure for open delta unbalance voltage sensing protection in a 22 kV system. For the capacitor, the voltage rating should be determined by considering the harmonic voltage synthesis due to the inflow of harmonic current (for reference, the V_{SUM} of the 5th harmonic filter capacitor was considered as 1.26 pu) and hazed polypropylene film and phenyl xylyl ethane (PXE) oil, which have excellent performance, were used [27].

$$V_{SUM} = V_{1pu-sys} + \sum_{h \neq 1}^{h} V_{hpu-sys}\ (\mathrm{pu}) \tag{1}$$

where V_{SUM} is the summation of the capacitor voltage in pu, $V_{1pu-sys}$ is the system fundamental voltage of the capacitor expressed in pu, $V_{hpu-sys}$ is the system harmonic voltage of the capacitor expressed in pu, and h is the harmonic order.

2.2. Series Reactors

In the design of a series reactor (SR), the derating factor (DF) should be applied considering the loss caused by the inflow of harmonic current. The DF is related to the *K-factor* [23,29] and the harmonic loss factors (F_{HL}) [30].

2.2.1. DF Using *K-Factor*

In the case of SR of 5th harmonic filter, the loss was 11.6 kW (the fundamental current loss was 2.7 kW), *K-factor* was 38, and DF was 0.26 pu (eddy current loss was 0.08). Compared to the case where only the fundamental current flows, this case should consider the loss and current tolerance of 3.5 to 4.0 times.

$$DF = \frac{(1 + P_{EC-R})}{\left(1 + P_{EC-R} \times K_{factor}\right)}\ (\mathrm{pu}) \tag{2}$$

$$K_{factor} = \frac{\sum_{h=1}^{h} (I_h)^2 (h)^2}{I_{Rating}^2} \tag{3}$$

where DF is expressed as a per unit, P_{EC-R} is the eddy current loss factor (8–12%), K_{factor} is a weighting of the harmonic currents, I_h is the harmonic current of h order into the SR, and I_{Rating} is the rating current of SR.

2.2.2. DF Using F_{HL}

In addition, the derating can be calculated by applying the F_{HL} instead of *K-factor*. In this case, F_{HL} was 15.91. The DF to which F_{HL} was applied in P_{LL-R}(0.563 pu) and P_{EC-R}(0.08 pu) was 0.497 pu. Compared to the former method, it shows about twice the difference [31].

$$F_{HL} = \frac{P_{EC-R}}{P_{EC-O}} = \frac{\sum_{h=1}^{h} \left(\frac{I_h}{I_1}\right)^2 (h)^2}{\sum_{h=1}^{h} \left(\frac{I_h}{I_1}\right)^2} \tag{4}$$

$$DF = \frac{(P_{LL-R})}{(1 + P_{EC-R} \times F_{HL})}\ (\mathrm{pu}) \tag{5}$$

$$P_{LL-R} = I_R^2 R_{DC} + F_{HL} P_{EC-R} + F_{HL-STR} P_{OSL-R} \tag{6}$$

where P_{LL-R} is the loss of the load, P_{EC-O} is the measure eddy current loss of winding under current and frequency conditions, R_{DC} is the DC resistance, F_{HL-STR} is the harmonic loss factor of other stray losses of transformer winding, and P_{OSL-R} is the rated other stray loss.

The harmonic filter reactor loss can be calculated by the coil, core, and gap losses. (7) shows the calculation formula for the total coil loss.

$$P_C = \sum [I(h)^2 R_{ac}(h) + P_{eddy}(h) + P_{stray}(h) \quad (7)$$

where P_C is the total coil loss (W), $I(h)$ is the hth harmonic current (Arms), $R_{ac}(h)$ is the conductor resistance at hth (Ω), $P_{eddy}(h)$ is the hth eddy current loss (W), and $P_{stray}(h)$ is the hth stray loss (W).

The core loss and gap loss are strongly influenced by the harmonic order and magnitude. The core and gap losses should be designed to be less than 40% and 20% of the total coil loss, respectively. In this case, the core and gap losses are 27% and 1.05% of the total losses, respectively. Moreover, in the design process, several parameters must be considered. Among them, since the voltage at the point where the filter is connected is low, saturation was not considered. In addition, inductance is determined by the cross-sectional area of the core and the width of the air gap [27].

There are two types of cores: the iron and air core. The iron core is efficient in terms of the installation space but has disadvantages such as magnetic saturation due to core hysteresis characteristics, noise between core layers, and partial heat generation in steel structures. The air core has a good stability because of its excellent linearity and low noise, but it has disadvantages that it requires a large installation space to maintain a distance considering the effect of the magnetic field and is vulnerable to external pollutants. In this study, an SR with an oil-filled magnetic shield air core type was selected, which took advantage of both these types due to the limited installation space.

2.3. Harmonic Filter System

The RM TOP/BOT and FM TOP/BOT rolling mill drives were connected to the secondary bus of a 154/22 kV, 100 MVA transformer. From this bus, power was supplied to the auxiliary devices such as a mill line, 2PLR, and shear/finishing line load via three transformers (22/3.3 kV, 20 MVA). The filter banks were connected based on this main bus, and the harmonic filter was designed to perform harmonic filtering and reactive power compensation [32]. The total reactive power compensation capacity of the filter was 10 MVAR and consisted of a total of 5 filter banks, including 4 single-tuned filters and 1 high-pass filter.

The reactive power compensation capacity Q_C, capacitance C, SR inductance L, and quality factor Q_f can be calculated, respectively, as illustrated in (8)–(12) [27,32].

$$C = \frac{1}{\omega}\frac{Q_C}{V_L^2}\left(1 - \frac{1}{h^2}\right) \quad (8)$$

$$L = \frac{1}{\omega}\frac{V_L^2}{Q_C}\frac{h}{h^2 - 1} \quad (9)$$

$$R_d = \gamma \frac{V_L^2}{Q_C}\frac{h}{h^2 - 1} \quad (10)$$

$$\gamma = \frac{R_d}{\omega L h} = \frac{R_d}{Z_0} \quad (11)$$

$$Q_f = \frac{\omega L h}{R_s} = \frac{Z_0}{R_d} \quad (12)$$

where Q_C is the compensation capacity of the fundamental frequency, V_L is the system line voltage, C is the capacitance of the filter bank, L is the inductance of the filter bank, R_d is the damping resistor of the high-pass filter, Z_0 is the characteristic impedance, γ is the quality factor of the high-pass filter, R_s is the resistance of the SR, and Q_f is the quality factor of the band-pass filter. Table 4 illustrates the electrical design parameters of the 5th, 7th, 11th, 13th, and 22nd HF.

Table 4. Harmonic filter design parameters.

Description	Unit	4.8th HF Band Pass	6.8th HF Band Pass	10.8th HF Band Pass	12.8th HF Band Pass	22nd Hi-Pass
System VLL	kV	22	22	22	22	22
Bank capacity	kVAr	1000	1000	2500	2500	3000
Dimensional capacity	kVAr	1046	1022	2522	2516	3006
Bank capacitance	µF	5.24	5.36	13.58	13.62	16.41
Capacitor string voltage	kV	23.01	22.49	22.19	22.14	22.05
Bank current	A1	26.24	26.24	65.61	65.61	78.73
SR% reactance	%	4.38	2.18	0.87	0.62	0.21
SR inductance	mH	58.86	28.67	4.49	3.19	0.89
SR reactance	Ω	22.19	10.81	1.69	1.2	0.34
SR ohmic resistance	Ω	1.31	0.91	0.22	0.19	0.03
SR X/R ratio	-	16.9	11.9	7.5	6.3	10.1
Quality factor Q	-	80.9	80.8	80.9	80.6	5
Damping R	Ω	-	-	-	-	37
Reactor voltage	V1	582.3	283.6	110.9	78.8	26.6

2.4. Distribution Network Topology and Current Divider $H_{id}(s)$

The step-down transformer, main mill motor load, and harmonic filter banks can be expressed as shown in Figure 3, where $H_{id}(s)$ is the current divider transfer function between the power system, and $I_h(s)$ is the harmonic current injected from the load. $I_h(s)$ is calculated as in (13) [33].

$$H_{id}(s) = \frac{I_{hs}(s)}{I_h(s)} = \frac{Z_0}{R_d} \tag{13}$$

Figure 4 shows the attenuation of the filtering system as a function of the impedance of the power system. The filtering effect was insufficient with $h_5 = -1.63$ dB. If the 5th harmonic had a large proportion, better results could be obtained by moving the tuning order to h = 4.9 or increasing the capacity. The 11th, 13th, 23rd, and 25th order harmonics, which were of primary concern, demonstrated smooth filtering performance. The transfer function ($H_{id}(s)$) was obtained from an equivalent circuit substituted with a single phase based on the systematic equilibrium.

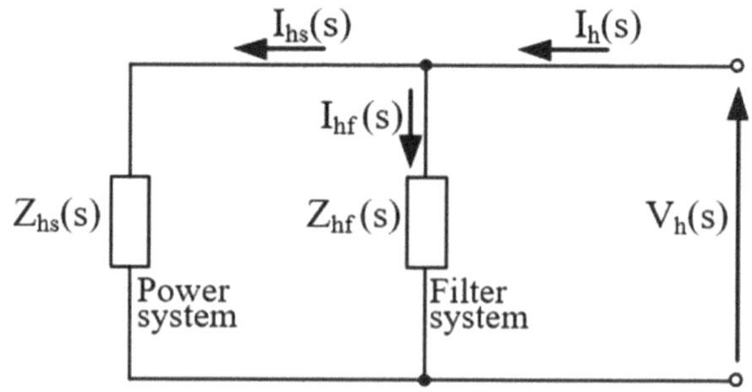

Figure 3. Conceptual $H_{id}(s)$ divider.

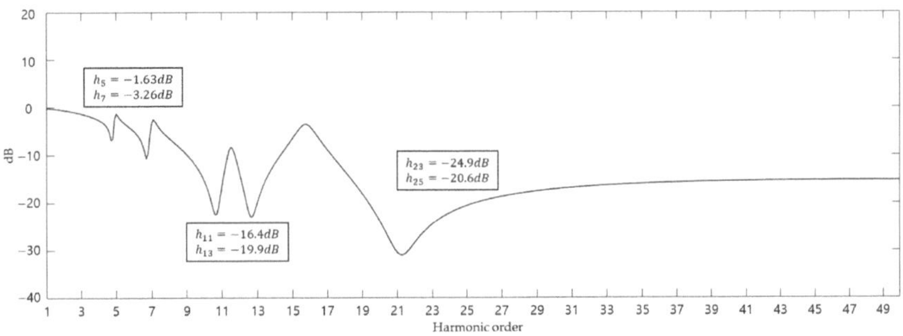

Figure 4. Frequency response of $H_{id}(s)$.

3. Easy Power System Software (ESA) Calculation Results

3.1. Without Filters

Prior to harmonic current and voltage distortion calculation, the harmonic phase angle was assumed to be zero. Depending on the phase angle, the calculation result is drastically changed [6], and in this paper, the worst conditions were considered.

Table 5 lists the calculation results of the ESA assuming that there are no filters and shows the load flow, voltage distortion, and power factor. From the table it is seen that the voltage drops for 3.3 kV mill line, 2PLR, shear, and finishing line are 5%, the voltage distortion V_{thd} also exceeds both the KEPCO limits (154 kV 1.5%, 22 kV 3%) and IEEE std. 519 (PCC 22 kV 5%). Tables 6 and 7 list the current report and voltage report of the IEEE Std. 519 [23]. From Table 6 it is seen that the harmonic current for Harmonic Number 11–16 is 5.21%, exceeding the limit of 4.5%, and for Harmonic Number 23–34 is 1.72%, exceeding the limit of 1.5%. From Table 7 it is seen that the voltage distortion V_{thd} is 9.27%, exceeding the limit of 5%, and the expected maximum individual voltage distortion exceeds the limit of 3% and is listed as 4.72%. As shown in the simulation results to determine the planning level, installation of the harmonic filter is indispensable to mitigate the harmonic distortion and voltage compensation.

3.2. With Filters

Table 8 lists the results of the load flow, voltage distortion V_{thd}, and power factor, assuming that the designed harmonic filter banks are connected to the 22 kV bus. The load (S) on the 154 kV bus was reduced from 47.65 to 42.6 MVA, and the voltage phase and voltage drop were improved on both the 22 and 3.3 kV buses. The voltage distortion V_{thd} was improved from 2.24 to 0.68% based on the 154 kV bus, satisfying the KEPCO limit of 1.5% and also the power factor was improved from 0.82 to 0.91. Tables 9 and 10 list the current distortion and voltage distortion at the 22kV bus, the current distortion was improved from 8.75% to 4.75%, and the V_{thd} improved from 9.267 to 2.819%, satisfying both the KEPCO limit and the IEEE Std. 519 guideline.

Table 5. ESA calculation results for load flow, V_{thd}, and power factors without filter.

Bus	Nominal Voltage (kV)	Voltage (pu)	Angle (deg)	V_{thd} (%)	pf	S (MVA)
154 kV	154	1.0	0.0	2.24 *	0.82	47.65
22 kV main	22	0.97	−2.8	9.27 *	0.84	46.01
RM TOP, BOT	22	0.96	−2.9	9.36 *	0.70	15.72
FM TOP, BOT	22	0.96	−2.9	9.36 *	0.99	10.28
ML 3.3 kV	3.3	0.95 *	−4.2	10.23 *	0.82	6.65
2PLR 3.3 kV	3.3	0.95 *	−4.4	12.75 *	0.82	7.74
S/F 3.3 kV	3.3	0.95 *	−4.3	12.68 *	0.82	7.14

* violation of standards.

Table 6. ESA calculation results for IEEE Std. 519 current report without filter-1.

Harmonic Current Distortion in Percent of Plant Loading						
Harmonic Number	**3–10**	**11–16**	**17–22**	**23–34**	**35–50**	**ITDD (%)**
Odd Harmonics	4.2	5.21 *	1.08	1.72 *	0.74 *	8.75
IEEE Limits	10	4.5	4	1.5	0.7	12.00
Even Harmonics	0	0	0	0	0	0.00
IEEE Limits	2.5	1.13	1	0.38	0.17	3.00
Plant Load kVA = 50,000						
PCC I_{sc}/I_{load} = 55.0						

* violation of standards.

Table 7. ESA calculation results for IEEE Std. 519 current report without filter-2.

Harmonic Voltage Distortion in Percent of PCC Base Voltage		
	Max Individual	V_{thd} **(%)**
PCC Bus	4.720 *	9.267 *
IEEE Limits	3	5.00
PCC Base kV = 22.000		

* violation of standards.

Table 8. ESA calculation results for load flow, V_{thd}, and power factors with filter.

Bus	Nominal Voltage (kV)	Voltage (pu)	Angle (deg)	V_{thd} (%)	pf	S (MVA)
154 kV	154	1.0	0.0	0.68	0.91	42.60
22 kV main	22	0.98	−2.8	2.82	0.93	41.67
RM TOP, BOT	22	0.98	−2.8	2.87	0.7	15.72
FM TOP, BOT	22	0.98	−2.8	2.87	0.99	10.28
ML 3.3 kV	3.3	0.96	−4.1	3.49	0.82	6.65
2PLR 3.3 kV	3.3	0.96	−4.3	5.91	0.82	7.74
S/F 3.3 kV	3.3	0.96	−4.2	5.83	0.82	7.14

Table 9. ESA calculation results for IEEE Std. 519 current report with filter-1.

Harmonic Current Distortion in Percent of Plant Loading						
Harmonic Number	**3–10**	**11–16**	**17–22**	**23–34**	**35–50**	**ITDD (%)**
Odd Harmonics	3.58	0.88	1.06	0.16	0.13	4.75
IEEE Limits	10	4.5	4	1.5	0.7	12.00
Even Harmonics	0	0	0	0	0	0.00
IEEE Limits	2.5	1.13	1	0.38	0.17	3.00
Plant Load kVA = 50,000						
PCC I_{sc}/I_{load} = 55.0						

Table 10. ESA calculation results for IEEE Std. 519 current report with filter-2.

Harmonic Voltage Distortion in Percent of PCC Base Voltage		
	Max Individual	V_{thd} **(%)**
PCC Bus	1.484	2.819
IEEE Limits	3	5.00
PCC Base kV = 22.000		

4. Harmonic Filter Configuration and Measured Power Profile

4.1. Harmonic Filter Configuration

Figure 5 shows a part of the harmonic filter banks installed in the field. The SR and the capacitor were located at the top, and an unbalanced voltage detector for filter bank protection was located in between them. The 5th and 7th HFs consisted of 6 cans/banks with a single wye 1-parallel connection, and the 11th, 13th, and 22nd HFs consisted of 12 cans/banks with a single wye 2-parallel connection. Figure 6 shows a single line diagram of the 22nd harmonic filter bank. The unbalanced voltage detector had a discharge coil function discharging the residual voltage in the capacitor to less than $50V_{DC}$ within 5 s after being disconnected from the system. It also detected the unbalanced voltage between the upper and lower can groups simultaneously when an element failure occurred inside the can [28].

Figure 5. A part of harmonic filter banks.

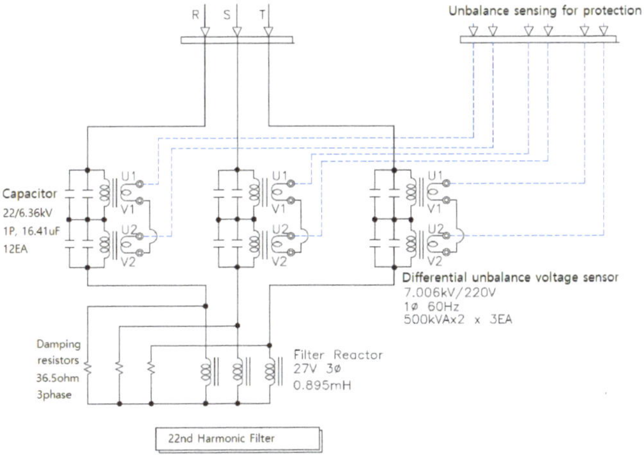

Figure 6. Single line drawing of 22nd HF.

4.2. Validation of Results

4.2.1. Variations in Power and Voltage

Figure 7 shows the apparent (S), active (P), and reactive powers (Q) of the plate mill system with and without the filter banks. From 15:53:54, 2nd June, when filter banks were closed, the reactive power was significantly reduced, and it was slightly leading at light loads. Without harmonic filter banks, *S* and *P* changed differently. However, it can be seen from Figure 7 that *S* and *P* change to approximately the same values as *Q* decreases when the filter is closed.

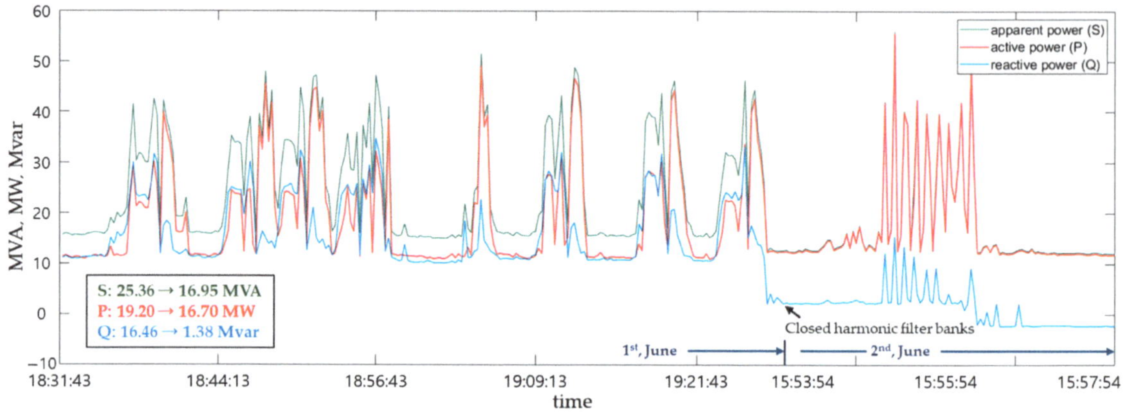

Figure 7. Measured *S*, *P*, and *Q* on plate mill system.

Figure 8 shows the maximum line voltage of the 22 kV bus. By comparing the results before and after applying the filter, the range of voltage amplitude improved from 5.3–2.3% to 3.3–1.5%, variance from 23,434 to 6894 V, and standard deviation from 153.1 V to 83.0 V (because of the tap adjustment of the transformer, only the voltage fluctuation depth was considered).

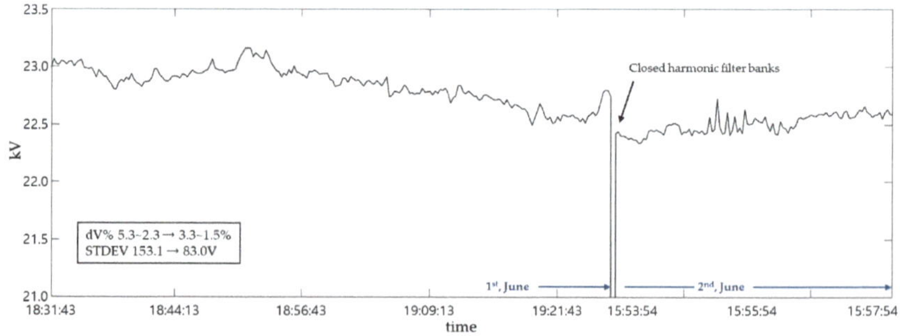

Figure 8. Measured V_{max} for 22 kV plate mill system.

4.2.2. Voltage Distortion

Figure 9 shows the measured values of V_{thd}. The maximum value, average value, and standard deviation improved from 2.60 to 1.05%, 1.78 to 0.70%, and 0.38 to 0.08%, respectively. Contrary to the previous simulation, it showed a significant difference, and the cause of the overall low level was due to the fact that the harmonic generation spectrum data applied during the simulation was based on the worst case scenario. However, it was confirmed that the tendency of the voltage distortion V_{thd} to improve was similar to the previous simulation.

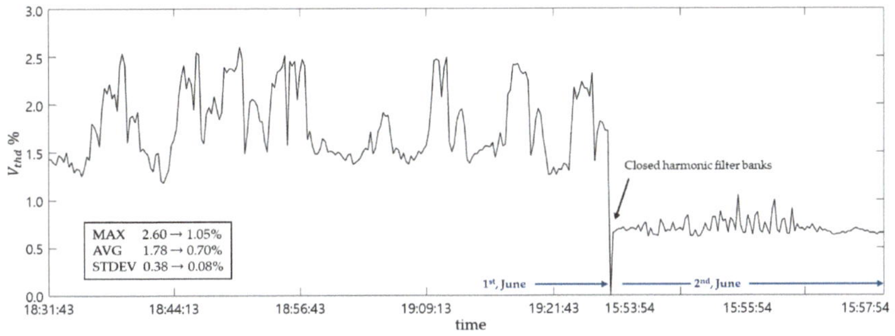

Figure 9. Measured V_{thd} for 22 kV plate mill system.

4.2.3. Voltage Harmonic Spectrum V_5, V_7, V_{11}, and V_{13}

Figure 10 shows the measured harmonic voltage spectra V_5, V_7, V_{11}, and V_{13}. Figure 10 shows that for the major harmonic voltage V_{11}, the size is significantly reduced from 418 to 73 V. From Figure 10, we can see that V_7 and V_{13} are decreased from 180 to 94 V and 169 to 26 V, respectively, but V_5 is increased from 170 to 194 V. When the tuning frequency of the 5th HF was 4.8th (288 Hz), the attenuation of the simulation value was −1.63 dB, but in the actual measurement, V_5 increased slightly. This is because the curve from the point of the lowest impedance to the peak impedance is steep, as shown by the green curve in Figure 11. To overcome this problem, the two following solutions were proposed. First, it could be improved by changing Q. If Q is changed from 80 to 20, as shown in the red in Figure 11, a relatively smooth curve is shown, and the attenuation is improved to −2.73 dB. The other solution is to move the tuning frequency to the right. If the tuning frequency is considered as 4.9th (294 Hz), as shown in blue in Figure 11, the attenuation can be improved by −2.71 dB.

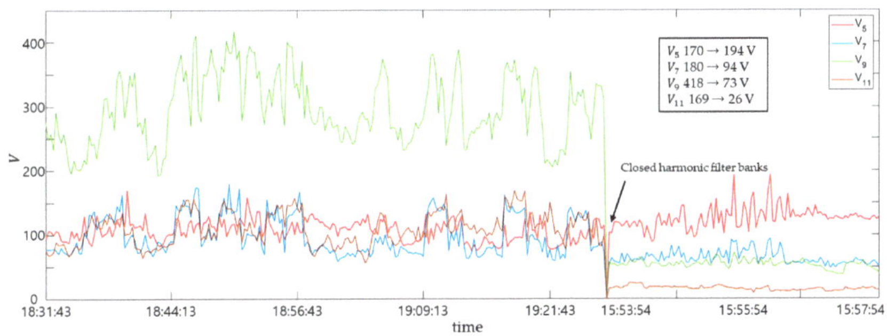

Figure 10. Measured $V_{h(5,\,7,\,11,\,13)}$ for 22 kV system.

4.2.4. Voltage Harmonic Spectrum V_{17}, V_{19}, V_{23}, and V_{25}

Figure 12 shows the measured harmonic voltage spectra V_{17}, V_{19}, V_{23}, and V_{25}. As shown in Figure 12, the major harmonic voltage V_{23} is improved from 304 to 9 V, and V_{17}, V_{19}, and V_{25} are decreased from 113 to 26 V, 79 to 15 V, and 255 to 8 V, respectively.

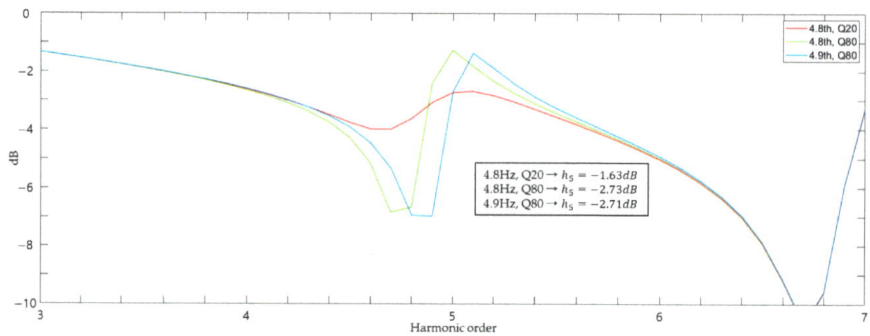

Figure 11. Attenuation with 5th HF parameters changed.

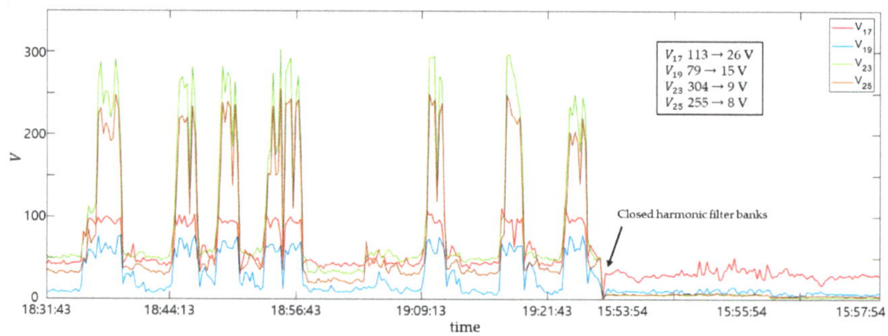

Figure 12. Measured $V_{h(17,\,19,\,23,\,25)}$ on 22 kV system.

4.2.5. Voltage Harmonic Spectrum V_{29}, V_{31}, V_{35}, and V_{37}

Figure 13 shows the measured harmonic voltage spectra V_{29}, V_{31}, V_{35}, and V_{37}. As shown in Figure 13, the largest harmonic voltage V_{37} is improved from 164 to 15 V, and V_{29}, V_{31}, and V_{35} are decreased from 82 to 5 V, 98 to 5 V, and 119 to 9 V, respectively.

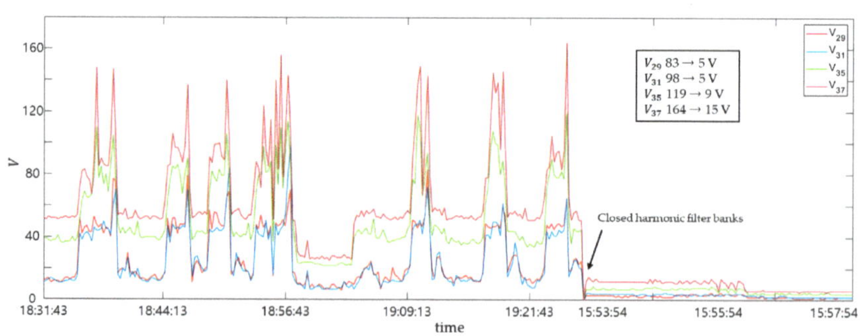

Figure 13. Measured $V_{h(29,\,31,\,35,\,37)}$ on 22 kV system.

4.2.6. Voltage Harmonic Spectrum V_{41}, V_{43}, V_{47}, and V_{49}

Figure 14 shows the measured harmonic voltage spectra V_{41}, V_{43}, V_{47}, and V_{49}. As shown in Figure 14, the largest harmonic voltage V_{49} is improved from 160 to 16 V, and V_{41}, V_{43}, and V_{47} are decreased from 74 to 9 V, 66 to 8 V, and 138 to 14 V, respectively.

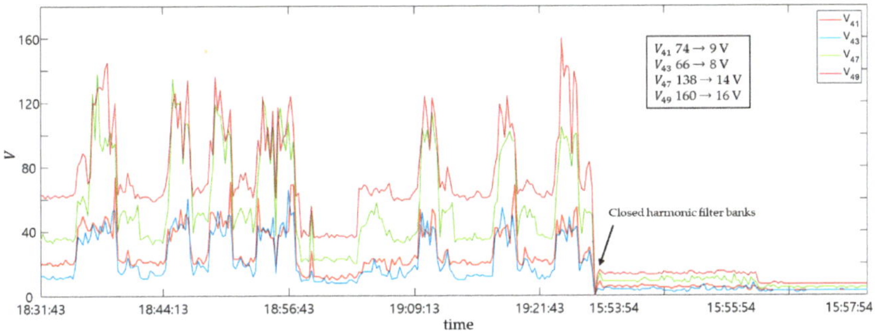

Figure 14. Measured $V_{h(41,\,43,\,47,\,49)}$ on 22 kV system.

5. Discussion

It cannot be done without considering the economics of installing filters for harmonic mitigation in the mill power system. The cost of this project was USD 450,000 including switch gears and cubicles, protection relays, capacitors, series reactors, etc. The amount of energy reduction by the filter depends on the loss characteristics of the transformer, showing a difference of 0.4% before and after application (assume no load loss 0.001 pu, load loss 0.008 pu). The annual energy reduction is calculated as (14) based on the average load (40 MVA). This has the effect of reducing about 84,000$ (US) per year (reflecting Korean electricity bills), and it is estimated that it will take about 5.4 years to recover the investment.

$$0.4\% \times 40,000 \text{ kVA} \times 24 \text{ h} \times 365 \text{ days} = 1,401,600 \text{ kWh} \quad (14)$$

It is also possible to consider applying the active power filter to this system. The active power filter does not cause parallel resonance with the system, and real-time compensation of harmonics and reactive power is possible. For this, a 10 MVA step-up transformer with an impedance of about 1% is required, and it is reasonable to apply the dq transformation control method rather than the FFT control method as it can respond quickly to the load fluctuation characteristics. Except the cost, which is roughly twice that of a passive filter, an active filter is preferred.

6. Conclusions

In this study, it was confirmed that a passive harmonic filter system with an optimal capacity in the mill motor drive system could provide an economical solution that compensated for the reactive power and absorbed the harmonics simultaneously. In addition, the following were discussed: (1) harmonic filter bank configuration parameters, (2) key elements of the filter capacitor and SR design, and (3) attenuation of harmonic voltage by current divider $H_{id}(s)$. To verify the performances of the harmonic filter, voltage fluctuation characteristics and harmonic voltage were also measured.

As a result of the field test, the voltage standard deviation was improved from 153.1 to 83.1 V with reactive power compensation. Excluding V_5, a wide range of harmonic voltages such as V_7, V_{11}, V_{13}, V_{17}, V_{19}, V_{23}, V_{25}, V_{29}, V_{31}, V_{35}, V_{37}, V_{41}, V_{43}, V_{47}, and V_{49}, decreased significantly to a satisfactory level. For V_5, the Q adjustment and tuning frequency of the filter are shown in Figure 11. Finally, this paper is expected to be helpful in the field of harmonic filter applications.

Author Contributions: The main idea was proposed by B.P. and H.Y.; The data and experiment results were collected by B.P. and H.Y.; The simulation results were analyzed by B.P. and J.L.; B.P., J.L. and G.J. wrote the paper. All authors have read and agreed to the published version of the manuscript.

Funding: This research received no external funding.

Institutional Review Board Statement: Not applicable.

Informed Consent Statement: Not applicable.

Data Availability Statement: Not applicable.

Acknowledgments: This research was supported by the Basic Research Program through the National Research Foundation of Korea (NRF) funded by the MSIT (No. 2020R1A4A1019405).

Conflicts of Interest: The authors declare no conflict of interest.

References

1. Mjorning, L. *ABB Drives and Control System for Hot Flat Rolling Mills Improves Yield and Quality*; South East Asia iron and Steel Institute (SEAISI): Shah Alam, Malaysia, 2005.
2. Hosoda, H.; Kodama, S.; Tessendorf, R. Large PWM Inverters for Rolling Mills. *Iron Steel Technol.* **2008**, *5*, 65–73.
3. Emadi, A.; Nasiri, A.; Bekiarov, S.B. *Uninterruptable Power Supplies and Active Filters*; Illinois Institute of Technology: Chicago, IL, USA; CRC Press: Washington, DC, USA, 2005.
4. Yazdani-Asrami, M.; Sadati, S.M.B.; Samadaei, E. Harmonic study for MDF industries: A case study. In Proceedings of the 2011 IEEE Applied Power Electronics Colloquium (IAPEC), Johor Bahru, Malaysia, 18–19 April 2011; pp. 149–154.
5. Arrillaga, J.; Watson, N.R. *Power System Harmonics*, 2nd ed.; John Wiley & Sons: Hoboken, NJ, USA, 2003.
6. Yazdani-Asrami, M.; Song, W.; Zhang, M.; Yuan, W.; Pei, X. AC Transport Loss in Superconductors Carrying Harmonic Current with Different Phase Angles for Large-Scale Power Components. *IEEE Trans. Appl. Supercond.* **2021**, *31*, 1–5. [CrossRef]
7. Yazdani-Asrami, M.; Mirzaie, M.; Akmal, A.A.S. No-load loss calculation of distribution transformers supplied by nonsinusoidal voltage using three-dimensional finite element analysis. *Energy* **2013**, *50*, 205–219. [CrossRef]
8. Yazdani-Asrami, M.; Gholamian, S.A.; Mirimani, S.M.; Adabi, J. Influence of field-dependent critical current on harmonic AC loss analysis in HTS coils for superconducting transformers supplying non-linear loads. *Cryogenics* **2021**, *113*, 103234. [CrossRef]
9. Napoles, J.; Leon, J.I.; Portillo, R.; Franquelo, L.G.; Aguirre, M.A. Selective harmonic mitigation technique for high-power converters. *IEEE Trans. Ind. Electron.* **2009**, *57*, 2315–2323. [CrossRef]
10. Pogaku, N.; Green, T.C. Harmonic mitigation throughout a distribution system: A distributed-generator-based solution. *IEE Proc. Gener. Transm. Distrib.* **2006**, *153*, 350–358. [CrossRef]
11. Sekar, T.C.; Rabi, B.J. A review and study of harmonic mitigation techniques. In Proceedings of the 2012 International Conference on Emerging Trends in Electrical Engineering and Energy Management (ICETEEEM), Tamil Nadu, India, 3–15 December 2012; pp. 93–97.
12. Badrzadeh, B.; Gupta, M. Practical experiences and mitigation methods of harmonics in wind power plants. *IEEE Trans. Ind. Appl.* **2013**, *49*, 2279–2289. [CrossRef]
13. Tischer, H.; Pfeifer, T. Hybrid filter for dynamic harmonics filtering and reduction of commutation notches-a case study. In Proceedings of the 2016 17th International Conference on Harmonics and Quality of Power (ICHQP), Belo Horizonte, Brazil, 16–19 October 2016; pp. 261–265.
14. Kihwele, S. Modelling of Shunt Active Power Filter for Harmonics Case Study of Steel Industry. In Proceedings of the 2019 International Conference on Electronics, Information, and Communication (ICEIC), Auckland, New Zealand, 22–25 January 2019; pp. 1–2.
15. de Miranda, G.C.; Primo, H.F.; Calenzani, H.T.; Camargos, I.A. Harmonic mitigation techniques applied to industrial power systems: Real case study with measurements. In Proceedings of the 2018 Simposio Brasileiro de Sistemas Eletricos (SBSE), Niteroi, Brazil, 12–16 May 2018; pp. 1–6.
16. Gonzalez, D.A.; McCall, J.C. Design of filters to reduce harmonic distortion in industrial power systems. *IEEE Trans. Ind. Appl.* **1987**, *3*, 504–511. [CrossRef]
17. Ludbrook, A. Harmonic filters for notch reduction. *IEEE Trans. Ind. Appl.* **1988**, *24*, 947–954. [CrossRef]
18. Phipps, J.K. A transfer function approach to harmonic filter design. *IEEE Ind. Appl. Mag.* **1997**, *3*, 68–82. [CrossRef]
19. Das, J.C. Passive filters-potentialities and limitations. *IEEE Trans. Ind. Appl.* **2004**, *40*, 232–241. [CrossRef]
20. Jain, S.K.; Agrawal, P.; Gupta, H.O. Fuzzy logic controlled shunt active power filter for power quality improvement. *IEE Proc. Electr. Power Appl.* **2002**, *149*, 317–328. [CrossRef]
21. IEEE Standards Association. *IEEE Std 519-2014. Recommended Practice and Requirements for Harmonic Control in Electric Power Systems*; IEEE Power and Energy Society: New York, NY, USA, 2014.
22. Pérez-Donsión, M.; Pereira, S.J.; Oliveira, F.T. Harmonics and Flicker in an Iron and Steel Industry with AC arc furnaces. In Proceedings of the International Conference on Renewable Energies and Power Quality, Tenerife, Spain, 10–12 April 2019.
23. Harmonic Working Group (IEEE PES T&D Committee). *Guide for Applying Harmonic Limits on Power Systems, P519A/D6*; IEEE: Piscataway, NJ, USA, 1999.
24. Aye, T.M.; Naing, S. Analysis of Harmonic Reduction by Using Passive Harmonic Filters. *IJSETR* **2014**, *3*, 9142–9147.
25. Kapoor, S.R.; Lalwani, M.K.; Tiwari, S. Harmonic Analysis and Reduction of Separately Excited DC Motor. In Proceedings of the 4th International Conference Advance Trend in Engineering, Technology and Research (ICATETR-2015), Kota, India, 19–20 June 2015; pp. 179–185.

26. Almutairi, M.S.; Hadjiloucas, S. Harmonics Mitigation Based on the Minimization of Nonlinearity Current in a Power System. *Designs* **2019**, *3*, 29. [CrossRef]
27. IEEE Std 1531-2003. *IEEE Guide for Application and Specification of Harmonic Filters*; IEEE: Piscataway, NJ, USA, 2003; pp. 1–66.
28. IEEE Std C37.99-2000. *IEEE Guide for the Protection of Shunt Capacitor Banks*; IEEE: Piscataway, NJ, USA, 2000; pp. 1–108.
29. IEEE Std C57.110™-2018. *IEEE Recommended Practice for Establishing Liquid-Immersed and Dry-Type Power and Distribution Transformer Capability when Supplying Nonsinusoidal Load Currents*; IEEE: Piscataway, NJ, USA, 2018; pp. 1–68.
30. Tao, S.; Xiao, X.N. Comparing transformer derating computed using the harmonic loss factor FHL and K-factor. In Proceedings of the 2008 Third International Conference on Electric Utility Deregulation and Restructuring and Power Technologies, Nanjing, China, 6–9 April 2008; pp. 1631–1634.
31. Meng, J.; Jiang, L.; Wang, Y. Study on the Influence of Harmonics on Load Loss of Transformer. In *IOP Conference Series: Earth and Environmental Science, Proceedings of the 2020 International Conference on Green Energy, Environment and Sustainable Development, Wuhan, China, 24–25 April 2020*; IOP Publishing: Bristol, UK, 2020; p. 012126.
32. IEEE Std 3002.8-2018. *IEEE Recommended Practice for Conducting Harmonic Studies and Analysis of Industrial and Commercial Power Systems*; IEEE: Piscataway, NJ, USA, 2018; pp. 1–79.
33. Orcajo, G.A.; Rodriguez, D.J.; Ardura, G.P.; Cano, J.M.; Norniella, J.G.; Llera, T.R.; Cifrian, R.D. Dynamic Estimation of Electrical Demand in Hot Rolling Mills. *IEEE Trans. Ind. Appl.* **2016**, *52*, 2714–2723. [CrossRef]

energies

MDPI

Article

Behavior of Residual Current Devices at Frequencies up to 50 kHz

Stanislaw Czapp * and Hanan Tariq

Faculty of Electrical and Control Engineering, Gdańsk University of Technology, Narutowicza 11/12, PL-80-233 Gdańsk, Poland; hanan.tariq@pg.edu.pl
* Correspondence: stanislaw.czapp@pg.edu.pl

Abstract: The use of residual current devices (RCDs) is obligatory in many types of low-voltage circuits. They are devices that ensure protection against electric shock in the case of indirect contact and may ensure additional protection in the case of direct contact. For the latter purpose of protection, only RCDs of a rated residual operating current not exceeding 30 mA are suitable. Unfortunately, modem current-using equipment supplied via electronic converters with a pulse width modulation produces earth fault currents composed of high-frequency components. Frequency of these components may have even several dozen kHz. Such components negatively influence the RCDs' tripping level and, hence, protection against electric shock may be ineffective. This paper presents the results of the RCDs' tripping test for frequencies up to 50 kHz. The results of the test have shown that many RCDs offered on the market are not able to trip for such frequencies. Such behavior was also noted for F-type and B-type RCDs which are recommended for the circuits of high-frequency components. Results of the test have been related to the requirements of the standards concerning RCDs operation. The conclusion is that these requirements are not sufficient nowadays and should be modified. Proposals for their modification are presented.

Keywords: protection against electric shock; residual current devices; earth current; high-frequency currents; harmonics; testing

Citation: Czapp, S.; Tariq, H. Behavior of Residual Current Devices at Frequencies up to 50 kHz. *Energies* **2021**, *14*, 1785. https://doi.org/10.3390/en14061785

Academic Editor: Gabriel Nicolae Popa

Received: 25 February 2021
Accepted: 19 March 2021
Published: 23 March 2021

1. Introduction

Effective protection against electric shock in modern low-voltage electrical installations depends a lot on the proper selection, application, and operation of residual current devices (RCDs). Analysis of provisions of the standard HD 60364-4-41 [1] shows that the highly-sensitivity RCDs (rated residual operating current not exceeding 30 mA) are obligatory in socket-outlets circuits up to 32 A intended for general use, mobile equipment circuits up to 32 A for using in outdoors, and lighting circuits in premises designed to accommodate a single household. Even wide application of RCDs is required in special installations and locations mentioned in the 700 series of the standard HD (IEC) 60364 "Low-voltage electrical installations". Such widespread use of RCDs as well as utilization of electronic equipment producing various shapes of earth fault currents prompt scientists and engineers from many countries to focus on the operation of RCDs under waveforms different than sinusoidal of the 50/60 Hz.

While the negative influence of the DC component of the residual current on the operation of RCDs has been recognized a long time ago [2,3] and the solutions are widely known [4–7], the influence of high frequencies is still being investigated. Papers [8–14] are focused on the tripping of RCDs for higher frequencies. The conclusion is that high frequency residual current changes the tripping threshold of RCDs and in some cases this threshold can be many times higher than for frequency 50 Hz. Analyses and tests presented in these papers were conducted within the relatively low frequency range—up to 1 kHz. A remedy for negative impact of the frequency up to 1 kHz is presented in [15]. The modification of the RCD's structure, giving stable tripping current within the range

50 Hz–1 kHz, is described. The effect of frequencies lower than 50 Hz is conducted in the paper [16]. Such frequencies change the tripping threshold of RCDs—for very low frequency (e.g., 1 Hz) some RCDs may not trip at all. The impact of mixed-frequency residual currents on RCDs tripping is examined in papers [17–19]. They conclude that high-order harmonics may increase the tripping threshold of RCDs which can be quite dangerous for human life. Detection and analysis of the advanced signals, including distorted residual current generated in variable-speed drive circuits, are considered in [20,21]. It is mentioned that a pulse width modulated residual waveform may not be detected by some types of RCDs. A mathematical approach to the detection of distorted currents, including mixed-frequency waveforms, is presented in [22]. However, it is only a simulative study, without laboratory tests. Analysis of the provisions of the international standards referred to RCDs [23–25] as well as the guide [26] show that these standards provide RCDs for higher frequencies but only up to 1 kHz. Admittedly, the national German standard [27] provides RCDs, which are able to react within the extended frequency range—up to 20 kHz but this type of RCDs is not met widely. Therefore, taking into account the requirements of the international standards as well as former research, this paper presents results of the RCDs' tripping test at frequencies up to 50 kHz. The response of the selected RCDs for such a wide frequency range is commented. Contrary to the previous tripping tests (usually up to 1 kHz) based on slowly raised residual current, the authors performed a test with suddenly applied residual current of the predetermined value. Such a test reflects a more real-life situation where exposure of the person is usually due to sudden touching of live part or exposed-conductive part. Based on the results of the test and insufficient requirements of the relevant international standards, modification of these requirements is proposed.

2. RCDs Construction and Normative Requirements Related to Their Tripping

The role of RCDs is the detection of the residual current which occurs in the case of the earth fault or direct contact of a person with live parts. Commonly used RCDs are voltage-independent and their structure is presented in Figure 1a. Elements responsible for the detection of the residual (earth) current i_Δ and the tripping are a current transformer (CT) and an electromechanical relay (RY).

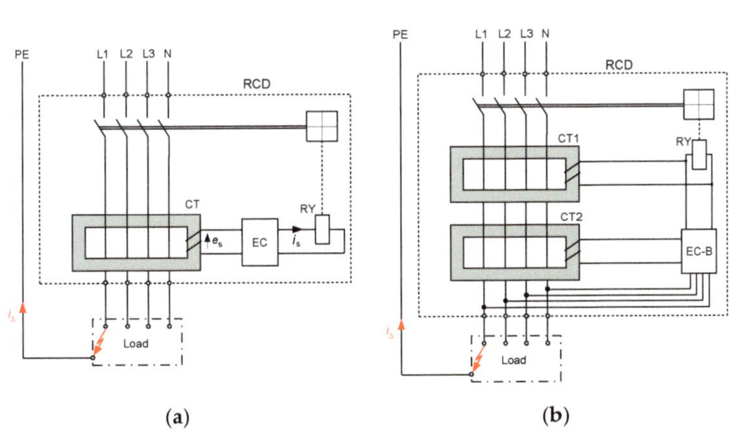

Figure 1. Structure of residual current devices (RCDs): (**a**) voltage-independent (AC-type, A-type, F-type); (**b**) voltage-dependent (B-type); RCD—residual current device; CT, CT1, CT2—current transformers; EC—electronic matching system (e.g., to increase the sensitivity to the DC pulsating waveform); EC-B—electronic system which ensures tripping especially in the case of the smooth DC residual current; RY—relay.

RCDs may also contain an electronic matching system (EC) which is used to increase the sensitivity of the RCD to the DC waveform or to ensure delay in tripping. If an

RCD is dedicated specially to detect a smooth DC residual current, its construction is more complicated (Figure 1b). Moreover, such an RCD requires an auxiliary voltage (see Figure 1b: the EC-B unit is supplied from all live conductors).

The equivalent circuit of the example voltage-independent RCD is presented in Figure 2. In the case of the occurrence of the residual current i_Δ, which is also the primary current of the CT, the secondary current i_s flows through the relay RY. If the secondary current i_s reaches a high enough value, the tripping of the RCD occurs (a detailed description of the RY operation can be found in [11,17]).

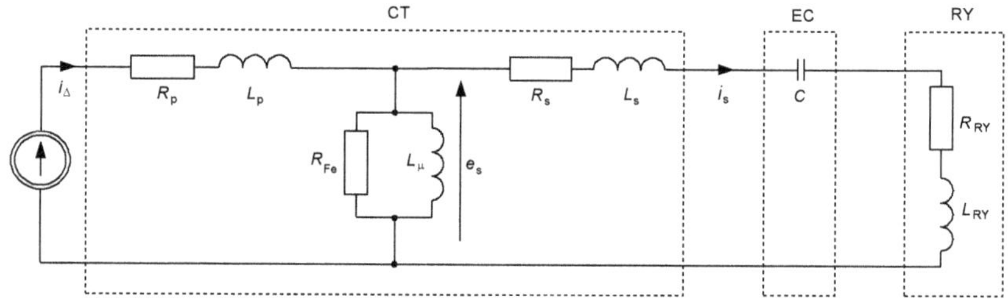

Figure 2. Equivalent circuit of the RCD: e_s—induced secondary voltage; i_Δ—residual (primary) current; i_s—secondary current; C—matching capacitor; R_{Fe}—resistance representing excitation losses; L_μ—inductance with magnetic hysteresis; R_p, L_p,—parameters (resistance and inductance) of the CT primary winding; R_s, L_s—parameters (resistance and inductance) of the CT secondary winding; R_{RY}, L_{RY}—parameters (resistance and inductance) of the relay.

From the point of view of the ability of the waveform shape detection, residual current devices are divided into the following types: AC-type, A-type, F-type and B-type (B+ type). Standards [23–25] indicate normalized tests in order to verify whether a particular type of RCD has relevant sensitivity to a specified type of residual current.

AC-type RCDs ensure tripping for residual sinusoidal alternating currents (suddenly applied or slowly rising). Sinusoidal testing current should have network frequency, usually 50 or 60 Hz.

A-type RCDs ensure tripping for:

- waveform the same as the AC-type;
- residual pulsating direct currents (suddenly applied or slowly rising) having the following current delay angles: 0°, 90° and 135°;
- residual pulsating direct current (current delay angle: 0°) superimposed by smooth direct component of 6 mA;

F-type RCDs ensure tripping for:

- waveforms the same as the A-type;
- residual pulsating direct currents superimposed by smooth direct component of 10 mA;
- mixed-frequency residual current (suddenly applied or slowly rising) intended for circuit supplied between phase and neutral or phase and earthed middle conductor;

B-type RCDs ensure tripping for:

- waveforms the same as the F-type;
- residual sinusoidal alternating currents up to 1 kHz;
- residual alternating currents superimposed by a smooth direct current of 0.4 times the rated residual current;
- residual pulsating direct currents superimposed by a smooth direct current of 0.4 times the rated residual current or 10 mA, whichever has a higher value;
- residual direct currents obtained from rectifying circuits as: two-pulse bridge connection line-to-line for 2-, 3-, and 4-pole RCDs, three-pulse star connection or six-pulse bridge connection for 3- and 4-pole RCDs;

- residual smooth direct currents.

Residual testing currents specified for B-type RCDs may be suddenly applied or slowly increased, independent of polarity.

RCDs of type B+ have extended ability with reference to B-type RCDs—they are suitable for detection of residual sinusoidal alternating currents up to 20 kHz.

It is worth mentioning the definition of the mixed-frequency residual current specified in the standard [25], provided for testing of F-type and B-type RCDs. Table 1 presents the components (also their normative contents) of the mixed-frequency testing waveform. These components reflect:

- I_{fund}—network rated frequency (usually 50 or 60 Hz);
- I_{1kHz}—a power electronics converter switching frequency;
- I_{10Hz}—a power electronics converter output frequency.

Table 1. Components of the mixed-frequency testing waveform used for F-type and B-type RCDs.

Components of the Waveform		
I_{fund}	I_{1kHz}	I_{10Hz}
$0.138 \cdot I_{\Delta n}$	$0.138 \cdot I_{\Delta n}$	$0.035 \cdot I_{\Delta n}$

The mixed-frequency waveform having parameters from Table 1 is presented in Figure 3. One can see that the shape of the waveform practically does not depend on the phase angle of the fundamental component (50 Hz).

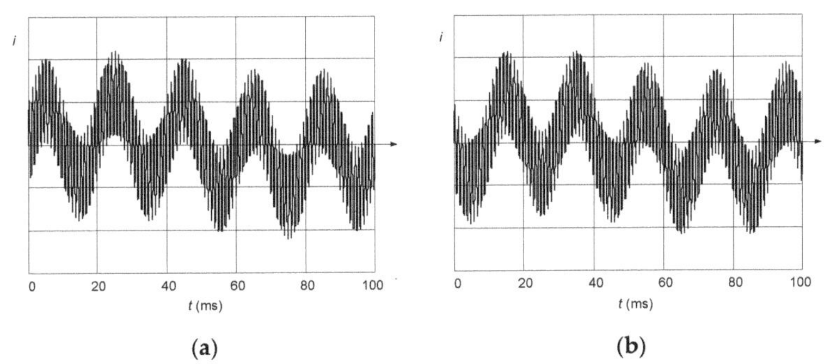

Figure 3. Mixed-frequency waveform composed of components presented in Table 1; phase angle of the I_{1kHz} and I_{10Hz} is 0°; phase angle of the I_{fund} (50 Hz) is: (**a**) 0°; (**b**) 180°.

The intention of the standard related to F-type RCDs is to take into account high-frequency components which may occur in circuits containing power electronic converters to control the power level of current-using equipment or speed of motors in variable-speed drive circuits. In the case of the normative waveform of F-type RCDs, switching frequency on the level only equal to 1 kHz is considered. A similar conclusion is referred to B-type RCDs—only frequencies up to 1 kHz are considered. An analysis of the real circuits equipped with power electronic converters indicates that the level of frequency included in the earth fault current may be significantly higher than 1 kHz. Figure 4 depicts structure of the example circuit (variable-speed drive circuit) producing residual currents having high-frequency components, whereas Figure 5 presents oscillograms of the earth fault current recorded in such a 3-phase circuit. A pulse width modulation (PWM) frequency is equal to 3 kHz. The content of the 3 kHz component depends on the fundamental (main) output frequency which supplies a motor to obtain the desired motor speed. If the fundamental frequency is equal to 50 Hz (upper oscillogram in Figure 5), the 50 Hz component is the highest. However, if the motor is supplied by the fundamental output frequency of

10 Hz (lower oscillogram in Figure 5), the component of 3 kHz is the dominating one. Moreover, the earth current waveform contains multiples of the PWM frequency—even around 20 kHz. The problem is even worse when the PWM frequency is very high and the motor speed is very low. In the case presented in Figure 6, the PWM frequency is equal to 6.67 kHz and the fundamental frequency is 1 Hz. The 6.67 kHz component has become the main component and the multiples of the PWM frequency reach almost 50 kHz.

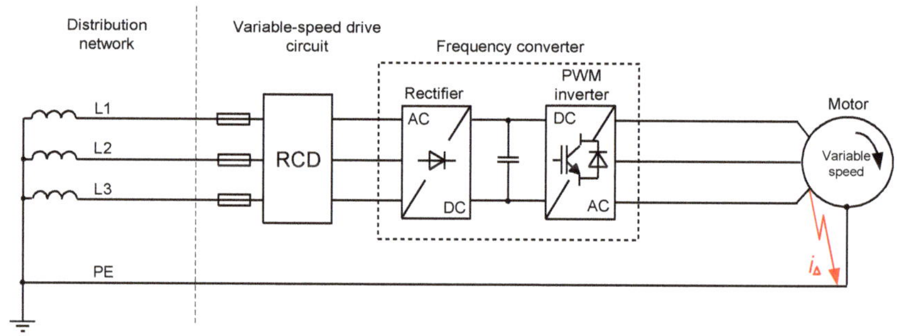

Figure 4. A variable-speed drive circuit producing residual currents of high-frequency components; RCD—residual current device; i_Δ—residual current; PWM—pulse width modulation.

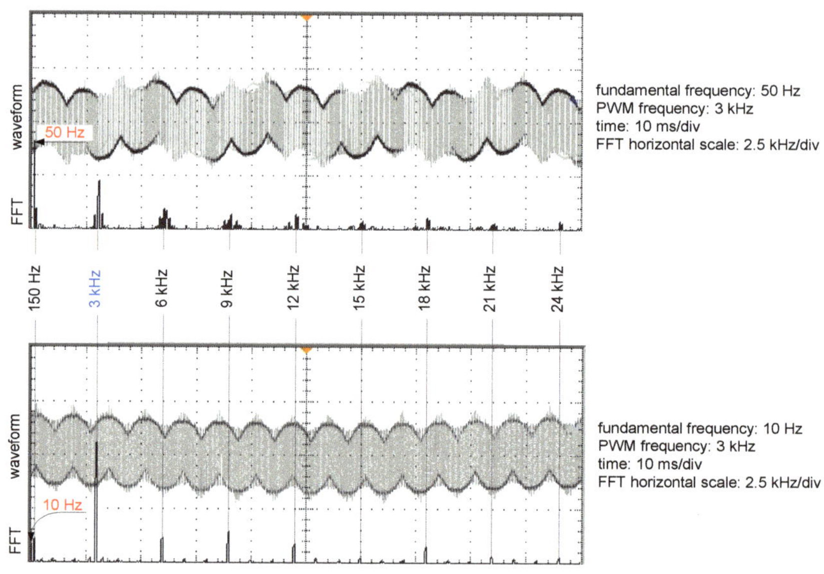

Figure 5. Oscillograms of the earth fault current in a variable-speed drive circuit (earth fault at the motor terminals). Upper oscillogram: the motor is supplied by frequency 50 Hz (rated motor speed); lower oscillogram: the motor is supplied by frequency 10 Hz (slow motor speed); pulse width modulation (PWM) frequency: 3 kHz.

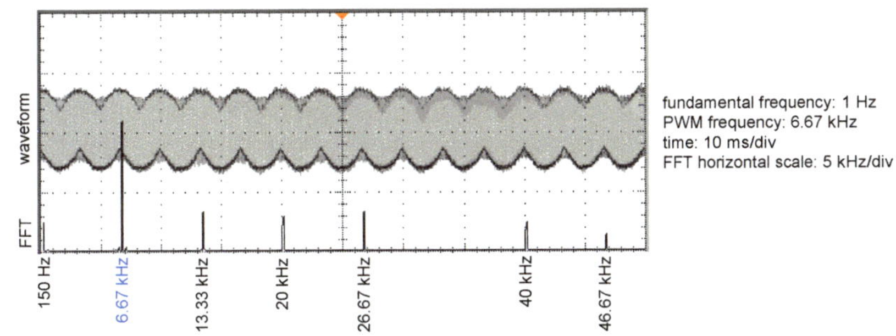

Figure 6. Oscillogram of the earth fault current in a variable-speed drive circuit (earth fault at the motor terminals). The motor is supplied by frequency 1 Hz (very slow motor speed); pulse width modulation (PWM) frequency: 6.67 kHz.

Taking into account the current requirements of standards related to waveforms to be detected by RCDs (mainly that RCDs are tested for frequencies only up to 1 kHz), it seems to be reasonable to perform verification of RCD's ability for detection of residual currents having frequency even several dozen kHz (see real waveforms in Figure 5). Therefore, the latter part of this paper presents results of the tripping test of RCDs for frequencies up to 50 kHz as well as important practical conclusions resulting from this test.

3. Laboratory Test of RCDs

3.1. Laboratory Stand

The behavior of RCDs (AC-type, A-type, B-type, and F-type) has been verified in the laboratory using a laboratory stand. Its generalized diagram is presented in Figure 7. The laboratory stand is comprised of:

- a power supply of 230 V, 50 Hz responsible for powering up the generator (mixed-frequency waveform generator); the generator can create a mixed-frequency signal (residual current content) up to 50 kHz;
- an ammeter for the measurement of the current's true rms value across the circuit during the testing stage;
- a rheostat to achieve the exact value of residual current necessary to perform the test under the specified condition (suddenly applied residual current);
- an RCD, to be tested.

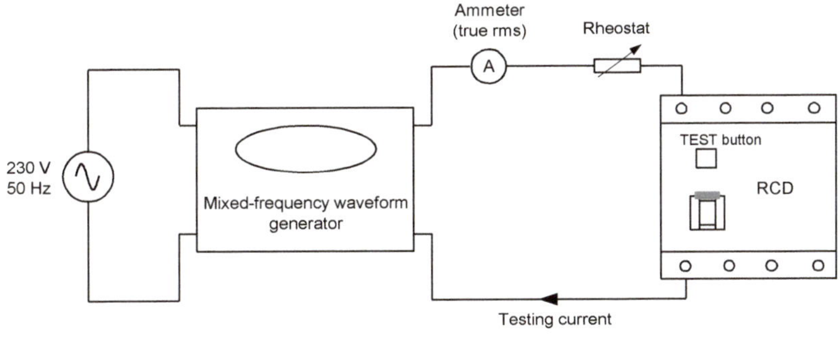

Figure 7. Laboratory stand for RCDs testing.

Figure 8 visualizes the waveforms composed of a mix of two different frequency contents (as an example: 50 Hz + 1000 Hz) generated by the mixed-frequency waveform generator. All of the waveform samples were accessed with the help of dedicated software. During the RCDs test under mixed-frequency waveforms, such contents of fundamental component (50 Hz) and high-frequency component (500 Hz or 1000 Hz or 2000 Hz) were applied. When the content of both components (50 Hz and 1000 Hz) is equal to 50% (Figure 8c) the waveform shape is similar to those presented in Figure 3 (normative mixed-frequency waveform for F-type RCDs testing). The laboratory generator was also used to produce a pure sine waveform of frequency from 50 Hz to 50 kHz (for details see Section 3.2). The laboratory test aims to verify the RCDs behavior within the range much wider than the normative.

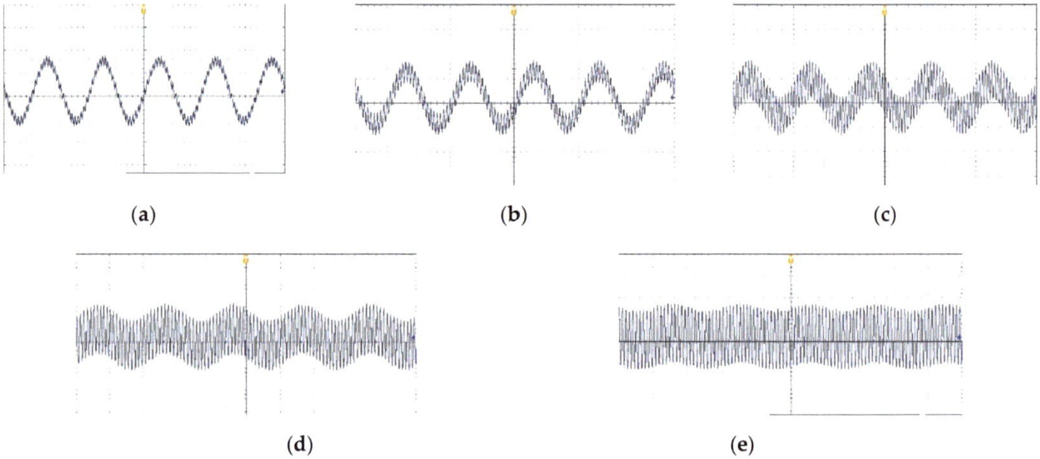

Figure 8. Example waveforms (time: 10 ms/div) composed of two frequencies, generated by the laboratory generator: (**a**) 50 Hz (90%) and 1000 Hz (10%); (**b**) 50 Hz (75%) and 1000 Hz (25%); (**c**) 50 Hz (50%) and 1000 Hz (50%); (**d**) 50 Hz (25%) and 1000 Hz (75%); (**e**) 50 Hz (10%) and 1000 Hz (90%).

3.2. Results of the Laboratory Test

RCDs having rated residual operating current $I_{\Delta n}$ = 30 mA were tested under the suddenly applied residual current of predetermined values: $I_{\Delta n}$; $2I_{\Delta n}$; $5I_{\Delta n}$; $8I_{\Delta n}$; $10I_{\Delta n}$ and $15I_{\Delta n}$ (i.e., 30 mA; 60 mA; 150 mA; 240 mA; 300 mA and 450 mA). Admittedly, this type of test is less restrictive than slowly raised current but it reflects the real-life situation, where a person touches a live part or exposed-conductive part after an insulation fault. The same phenomenon occurs in the case of the earth fault—the earth fault current rises suddenly. For presenting the results of the test, 10 RCDs of $I_{\Delta n}$ = 30 mA were selected as specified in Table 2. These RCDs were selected as a representative group from 14 tested RCDs (from 7 manufacturers).

Table 2. List of the selected RCDs ($I_{\Delta n}$ = 30 mA).

Consecutive RCD	Symbol/No. of the RCD	Type	Manufacturer (Symbol)
1	RCD_1AC	AC	Man_1
2	RCD_2AC	AC	Man_2
3	RCD_3AC	AC	Man_3
4	RCD_1A	A	Man_1
5	RCD_2A	A	Man_3
6	RCD_3A	A	Man_4
7	RCD_1F	F	Man_5
8	RCD_2F	F	Man_6
9	RCD_1B	B	Man_7
10	RCD_2B	B	Man_6

The first part (initial part) of the laboratory test relied on the verification of the RCDs' tripping under the residual current composed of the fundamental frequency (50 Hz) and one high-frequency component, consecutively 500 Hz, 1000 Hz, and 2000 Hz. This type of waveform reflects, with some approximation, dominating components included in the waveform specified in Table 1 (testing of F-type RCDs). However, for a broader look at the problem of sensitivity of RCDs, the content of both the low-frequency component (50 Hz) and the high-frequency component was changed. It is also important to underline that for a set content of the aforementioned components, e.g., 50 Hz (10%) and 1000 Hz (90%) the laboratory generator keeps the ratio of these components constant (here 10%/90%), regardless of the value of the testing current $I_{\Delta n}$, $2I_{\Delta n}$, $5I_{\Delta n}$, $8I_{\Delta n}$ $10I_{\Delta n}$, or $15I_{\Delta n}$. For each tested RCD, the threshold of RCDs' sensitivity for the 50 Hz (reference value, in milliamps) was verified. According to [23–25], for the 50 Hz sinusoidal waveform, the normative range of the tripping threshold is $(0.5–1.0)I_{\Delta n}$ whereas for the mixed-frequency testing waveform (Table 2) is $(0.5–1.4)I_{\Delta n}$. Results of the initial test are presented in Figures 9–12.

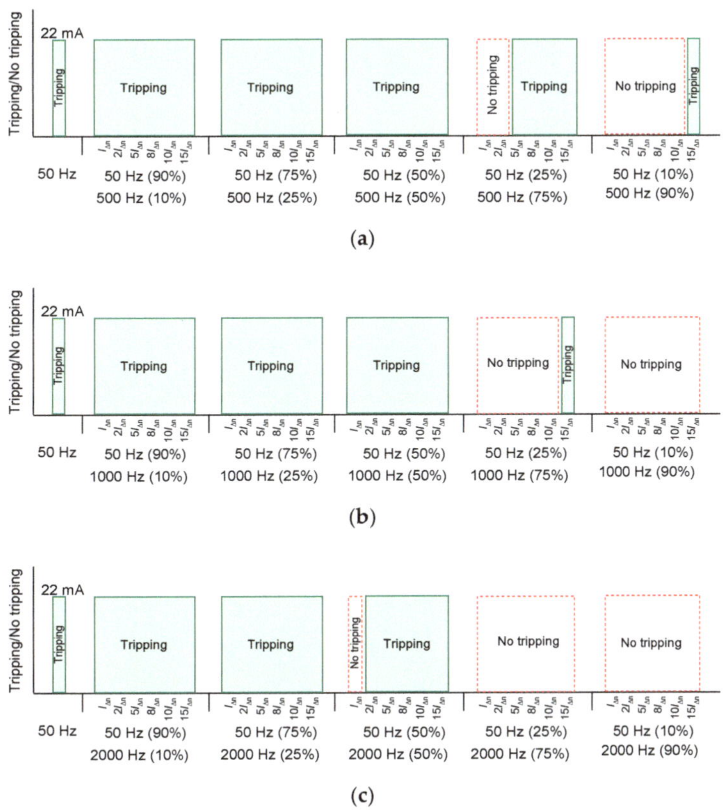

Figure 9. Tripping of the 30 mA AC-type RCD (no. RCD_1AC—symbol defined in Table 2) for waveform composed of the fundamental frequency (50 Hz) and high-frequency component: (**a**) 500 Hz; (**b**) 1000 Hz; (**c**) 2000 Hz. Values in brackets indicate the content of the particular component; 22 mA—real tripping current for a pure sinusoidal signal of 50 Hz.

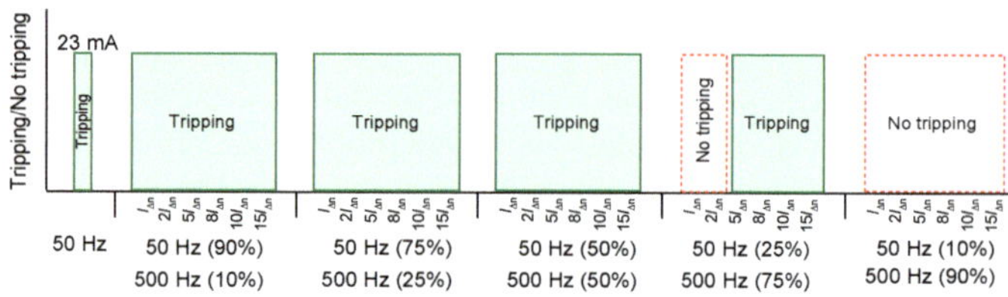

(a)

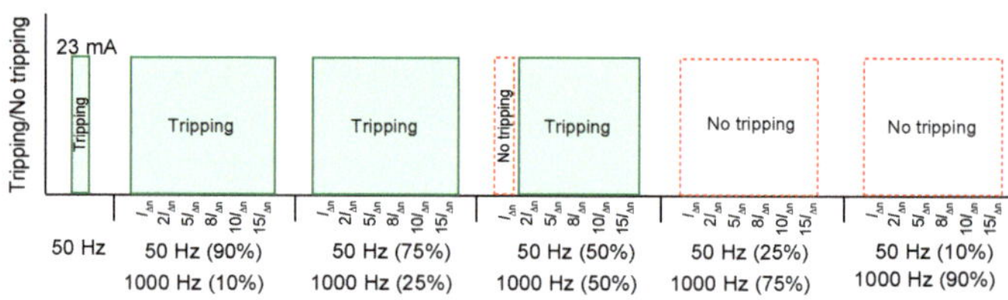

(b)

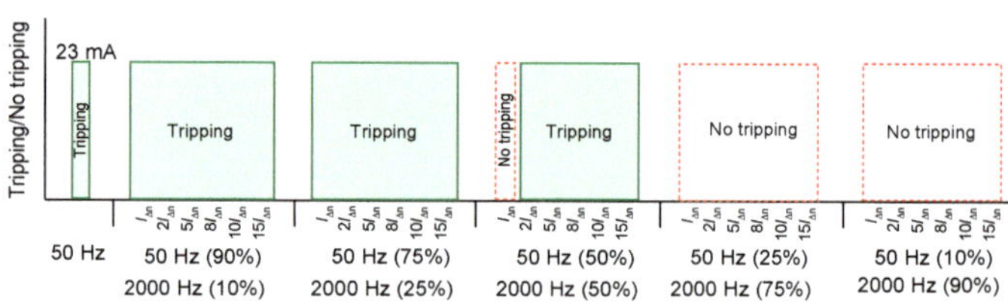

(c)

Figure 10. Tripping of the 30 mA A-type RCD (no. RCD_1A—symbol defined in Table 2) for waveform composed of the fundamental frequency (50 Hz) and high-frequency component: (**a**) 500 Hz; (**b**) 1000 Hz; (**c**) 2000 Hz. Values in brackets indicate the content of the particular component; 23 mA—real tripping current for a pure sinusoidal signal of 50 Hz.

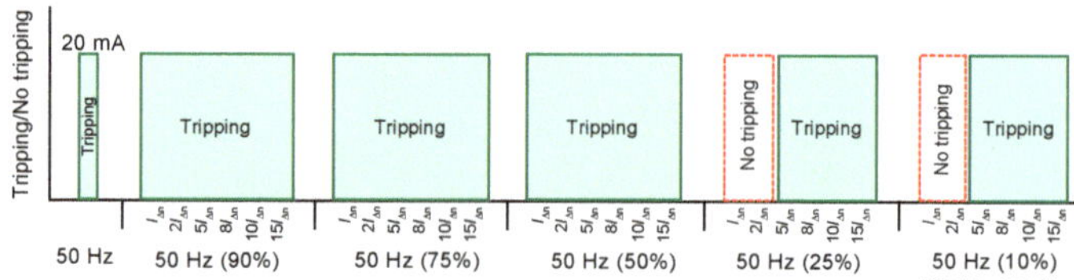

(a)

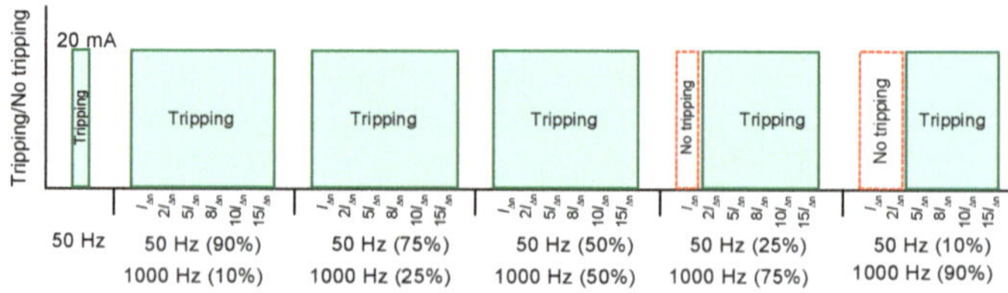

(b)

Figure 11. *Cont.*

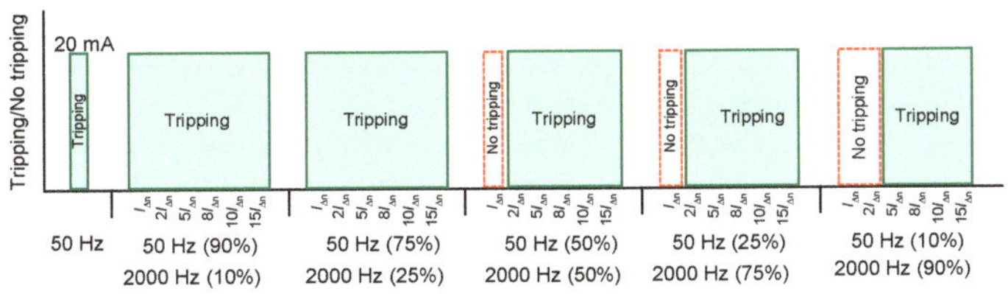

(c)

Figure 11. Tripping of the 30 mA F-type RCD (no. RCD_1F—symbol defined in Table 2) for waveform composed of the fundamental frequency (50 Hz) and high-frequency component: (**a**) 500 Hz; (**b**) 1000 Hz; (**c**) 2000 Hz. Values in brackets indicate the content of the particular component; 20 mA—real tripping current for a pure sinusoidal signal of 50 Hz.

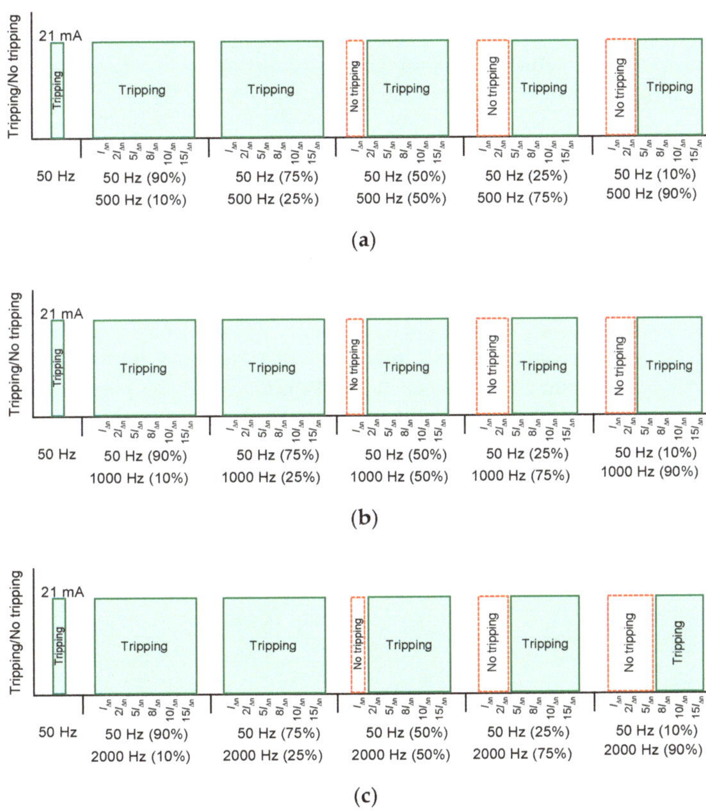

Figure 12. Tripping of the 30 mA B-type RCD (no. RCD_2B—symbol defined in Table 2) for waveform composed of the fundamental frequency (50 Hz) and high-frequency component: (**a**) 500 Hz; (**b**) 1000 Hz; (**c**) 2000 Hz. Values in brackets indicate the content of the particular component; 21 mA—real tripping current for a pure sinusoidal signal of 50 Hz.

When comparing results for the AC-type RCD (Figure 9) and the A-type RCD (Figure 10) one can say that their behavior (AC-type vs. A-type) is similar. Tripping of these RCDs is possible for relatively low content of the high-frequency component (10% and 25%). In such cases, the analyzed two RCDs (AC-type and A-type) tripped even for the testing current equal to $I_{\Delta n}$ (30 mA), regardless of the aforementioned share of the high-frequency component. The content of the high-frequency component equal to 50% made the problem for the AC-type RCD (Figure 9c) when the frequency was 2000 Hz. There was no tripping for the testing current equal to $I_{\Delta n}$.

For the A-type RCD, the same problem occurred in the case of frequencies 1000 Hz and 2000 Hz (Figure 10b,c). The worst condition for RCDs tripping occurred for the highest content of the high-frequency component (90%). In the case of a high-frequency component of 500 Hz, the AC-type RCD (Figure 9a) tripped only for the testing current $15I_{\Delta n}$ (450 mA). However, for high-frequency component equal to 1000 Hz and 2000 Hz, none of the two aforementioned RCDs reacted, even for the testing current $15I_{\Delta n}$.

Afterwards, while concluding the results of F-type RCD (Figure 11) and B-type RCD (Figure 12), a moderate similarity is observed in their performance, i.e., F-type vs. B-type. While considering the facts, it can be seen that both types of RCDs (F-type and B-type) behaved satisfactorily during the exposure of low content of the high-frequency component (500 Hz, 1000 Hz, 2000 Hz) i.e., 10% and 25% share. In this condition, both RCD types reacted very well to a current of $I_{\Delta n}$ (30 mA), irrespective of the share of the high-frequency component. However, subject to the 50% share of the high-frequency component, F-type RCD did not trip in the case of 2000 Hz (Figure 11c) for the residual current value of $I_{\Delta n}$ (30mA). Similarly, B-type RCD depicted the same reaction not only for 2000 Hz (Figure 12c) but also for the 500 Hz (Figure 12a) and 1000 Hz (Figure 12b). For the next contents (75% and 90%), these RCDs (F-type and B-type) behaved slightly better than the previously discussed RCDs (A-type and AC-type) but still not as suspected. During 75% high-frequency component share, F-type RCD did not trip on the current value of $I_{\Delta n}$ (30 mA) in the case of 1000 Hz (Figure 11b) and 2000 Hz (Figure 11c). On the other hand, both RCDs (F-type and B-type) remained untripped on the current value of $I_{\Delta n}$ (30 mA) and $2I_{\Delta n}$ (60 mA) for the rest of the circumstances having a 75% share of the high-frequency component (Figures 11a and 12a–c).

The scenario is similar in the case of 90% high-frequency component share, i.e., no tripping for the current of $I_{\Delta n}$ (30 mA) and $2I_{\Delta n}$ (60 mA) (Figures 11a–c and 12a,b) except for 2000 Hz where B-type RCD (Figure 12c) tripped only in the case of $8I_{\Delta n}$ (240 mA), $10I_{\Delta n}$ (300 mA), and $15I_{\Delta n}$ (450 mA).

The second part (main part) of the laboratory test was devoted to the verification of the RCDs tripping under the sinusoidal residual current of the following higher frequencies (consecutively): 500 Hz; 1000 Hz; 2000 Hz; 5000 Hz; 10,000 Hz; 20,000 Hz; and 50,000 Hz. Similar to the previous test, the residual current was suddenly applied and had predetermined values: $I_{\Delta n}$; $2I_{\Delta n}$; $5I_{\Delta n}$; $8I_{\Delta n}$; $10I_{\Delta n}$; and $15I_{\Delta n}$. Results of the test are presented in Figures 13–16.

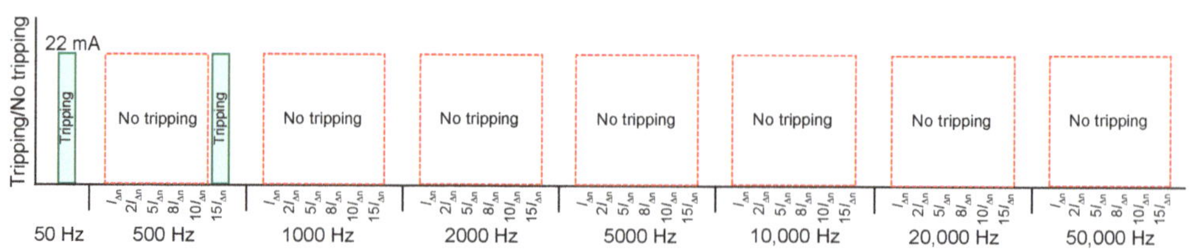

(a)

Figure 13. *Cont.*

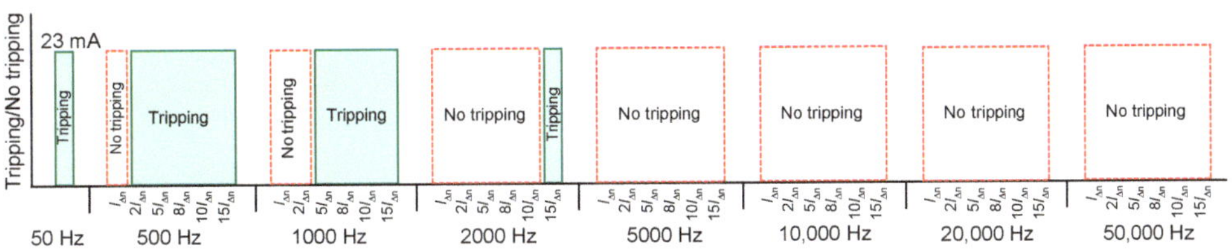

(**b**)

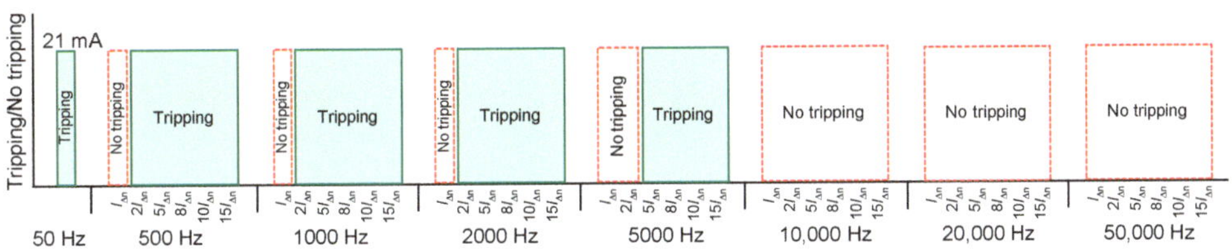

(**c**)

Figure 13. Tripping of the 30 mA AC-type RCDs for a sine waveform of the specified frequency from 50 Hz to 50 kHz: (**a**) RCD_1AC; (**b**) RCD_2AC; (**c**) RCD_3AC; symbols defined in Table 2.

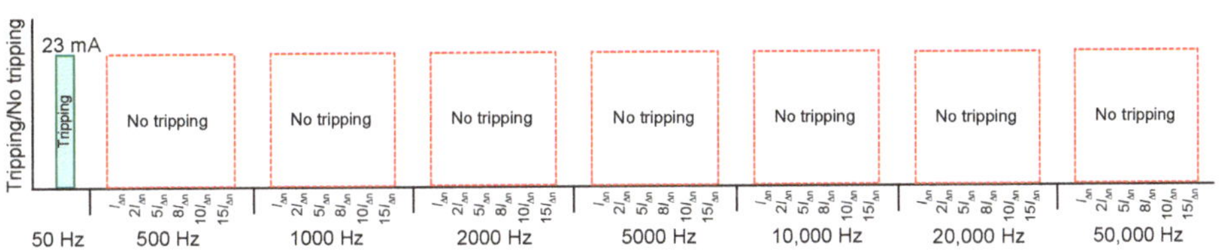

(**a**)

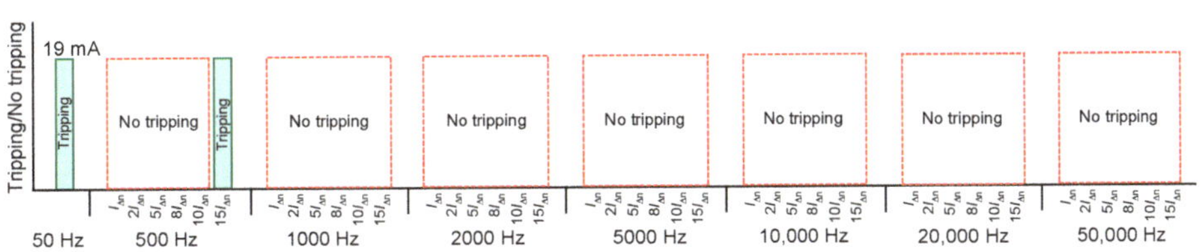

(**b**)

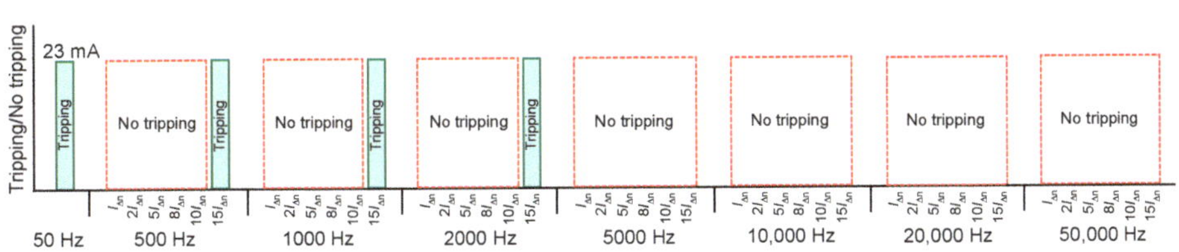

(**c**)

Figure 14. Tripping of the 30 mA A-type RCDs for a sine waveform of the specified frequency from 50 Hz to 50 kHz: (**a**) RCD_1A; (**b**) RCD_2A; (**c**) RCD_3A; symbols defined in Table 2.

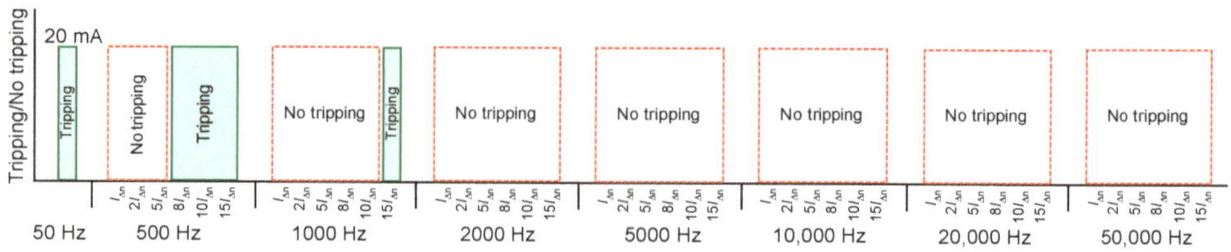

(a)

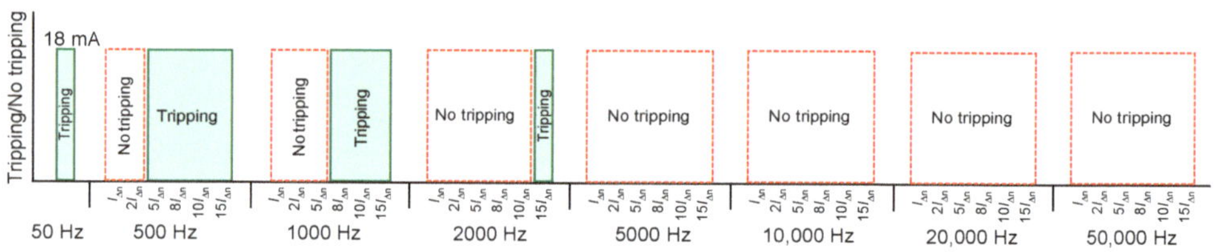

(b)

Figure 15. Tripping of the 30 mA F-type RCDs for a sine waveform of the specified frequency from 50 Hz to 50 kHz: (**a**) RCD_1F; (**b**) RCD_2F; symbols defined in Table 2.

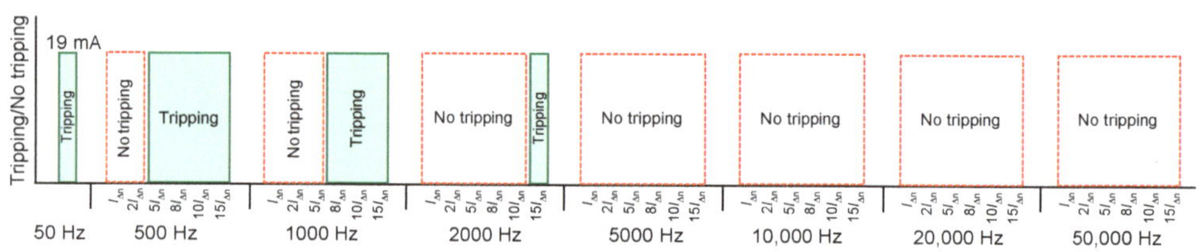

(a)

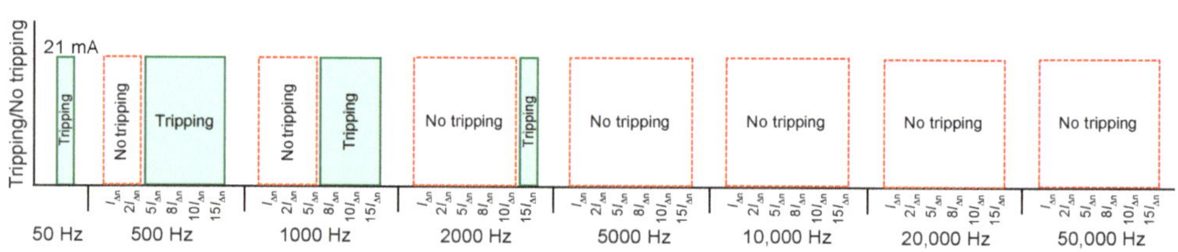

(b)

Figure 16. Tripping of the 30 mA B-type RCDs for a sine waveform of the specified frequency from 50 Hz to 50 kHz: (**a**) RCD_1B; (**b**) RCD_2B; symbols defined in Table 2.

Analysis of the results from Figure 13 enables one to conclude that AC-type RCDs may have various sensitivity to high-frequency residual currents. Very unfavorable behavior was observed for RCD_1AC (Figure 13a). Its tripping was noted only for frequency 500 Hz and value of the residual current equal to $15I_{\Delta n}$ = 450 mA. Clearly better behavior was noted for RCD_3AC (Figure 13c). Tripping of this RCD occurred even for 5 kHz but the value of the residual current had to be higher than $2I_{\Delta n}$ = 60 mA. Unfortunately, frequencies of the residual current 10 kHz, 20 kHz, and 50 kHz were too high to make tripping of the tested AC-type RCDs, even for the value of the current 15 times higher than the rated residual operating current of the RCD.

While going through the results achieved from the testing of A-type RCDs (Figure 14), their outcomes are not very promising. It is evident that all three A-type RCDs reacted well on the nominal frequency, i.e., 50 Hz. Afterwards, RCD_1A (Figure 14a) did not show any reaction on the rest of the frequencies (500 Hz, 1 kHz, 2 kHz, 5 kHz, 10 kHz, 20 kHz, and 50 kHz). A similar reaction was witnessed in the case of RCD_2A (Figure 14b), where the A-type RCD could not trip at all except for the frequency of 500 Hz but only for the highest residual current value, i.e., $15I_{\Delta n}$ (450 mA). Likewise, an identical outcome was experienced for the RCD_3A (Figure 14c), where RCD_3A tripped for the frequency values of 500 Hz, 1 kHz, and 2 kHz but only at the highest residual current value which is $15I_{\Delta n}$ (450 mA). Apart from that, unfortunately, RCD_3A failed to trip on the rest of the higher frequencies.

Figure 15 states the results achieved from the testing of F-type RCDs. The results again are not very favorable. Although the F-type RCDs are supposed to perform well on the higher frequencies, contrary to this, RCD_1F (Figure 15a) only tripped typically on the nominal frequency (50 Hz). During testing of higher frequencies, for 500 Hz, this RCD failed to trip at the residual current values of $I_{\Delta n}$, $2I_{\Delta n,}$ $5I_{\Delta n}$, and for 1 kHz it only reacted for the residual current value equal to $15I_{\Delta n}$. Beyond this point, RCD_1F remained untripped even for the highest residual current of $15I_{\Delta n}$. For the RCD_2F (Figure 15b), the reaction was identical for nominal frequency but for 500 Hz the RCD_2F did not react for the residual current values of $I_{\Delta n}$, $2I_{\Delta n}$, whereas, in the case of 1 kHz, RCD_2F tripped only beyond the residual current value of $5I_{\Delta n}$. For the higher frequencies (5 kHz, 10 kHz, 20 kHz, and 50 kHz), RCD_2F showed negative results and did not trip at any of the residual current value (Figure 15b).

Figure 16 depicts the results for the most advanced RCDs—B-type RCDs. They are considered typically for higher frequency purpose but only up to 1 kHz, according to [25]. Starting from the nominal frequency (50 Hz), both RCD_1B (Figure 16a) and RCD_2B (Figure 16b) performed well. At the threshold of 500 Hz both RCDs did not react at the residual current value of $I_{\Delta n}$ and $2I_{\Delta n}$. For 1 kHz, both RCDs (RCD_1B and RCD_2B) tripped only above the residual current value of $5I_{\Delta n}$. Moving to the higher frequency of 2 kHz, unfortunately, both B-type RCDs only showed tripping just for the highest value of the residual current ($15I_{\Delta n}$). Past this point, for frequencies 5 kHz, 10 kHz, 20 kHz, and 50 kHz, RCD_1B as well as RCD_2B did not react to the testing residual currents.

4. Discussion: Summary of the Test; Proposed Changes in Standards

Residual current devices' usage has been made obligatory not only from the modern domestic point of view but as well as for industrial purposes. The principal objective of this device is protection against electric shock in the case of either direct or indirect contact. The most challenging situation for such devices is when they are exposed to residual currents containing high-frequency contents. Under such circumstances this device may not be able to trip at the expected threshold and, therefore protection against electric shock may not be ensured. RCDs in this research were subjected to two different test categories. The first stage of the test included a mixed-frequency signal (nominal frequency + high-frequency component) and the reaction of the RCDs was quite unsatisfactory. In this initial testing stage, all types of RCDs (AC-type, A-type, F-type, and B-type) demonstrated tripping for the lowest testing current ($I_{\Delta n}$) only when the high-frequency content share was low, which means 10% or 25%. Once the high-frequency content was raised to 50% and beyond (75% or 90%), the RCD's tripping threshold moved to higher values or no tripping occurred (Figures 9–12). Furthermore, in the second testing stage, RCDs were exposed to high-frequency sinusoidal residual current. Again, the behavior of RCDs was unexpected as one of the AC-type RCD (Figure 13c) performed well enough, although the AC-type RCD is not dedicated for higher frequencies. As far as the other types are concerned, B-type and F-type RCDs functioned unsatisfactorily for higher frequencies and did not even respond to the very high residual current value ($15I_{\Delta n}$).

Regarding the risk of harmful effects of the electric shock, it should be commented that, according to IEC 60479-2 [28], the threshold of perception, the threshold of let-go and the threshold of ventricular fibrillation move to higher values when a high-frequency current flows. The most important is the last threshold (fibrillation effect) and it is the most-dependent on frequency. Analyzing provisions of the [25,28], it can be indicated that for the frequency equal to 400 Hz, the threshold of ventricular fibrillation is around 6 times higher than at the 50 Hz (i.e., one may assume 30 mA for 50 Hz but 6 × 30 mA = 180 mA for 400 Hz). For the frequency equal to 1000 Hz, this threshold changes 14 times (analogically: 14 × 30 mA = 420 mA). What is more, the standard [28] informs that for a mixed-frequency signal, the ventricular fibrillation hazard may be estimated (a rough approximation) as equivalent to the hazard caused by a pure sinusoidal current $I_{\text{ev-sin}}$ having the fundamental frequency with an amplitude equivalent to the quadratic summation of all component amplitudes I_{h} individually affected by the appropriate frequency factor F_{factor} (e.g., $F_{\text{factor}} = 6$ for 400 Hz and $F_{\text{factor}} = 14$ for 1000 Hz):

$$I_{\text{ev-sin}} = \sqrt{\sum_{h=1}^{n} \left(\frac{I_{\text{h}}}{F_{\text{factor,h}}} \right)^2} \quad (1)$$

Therefore, in terms of the ventricular fibrillation, high-frequency currents are less dangerous for persons than a current having frequency equal to 50 Hz. This phenomenon (tripping characteristic vs. the threshold of ventricular fibrillation) is utilized by some manufacturers of 30 mA B-type RCDs. The standard [25], in provisions dedicated only to B-type RCDs, specifies residual operating current $I_{\Delta\text{nf}}$ for two higher frequencies: 400 Hz ($I_{\Delta\text{nf}} = 6I_{\Delta\text{n}}$) and 1000 Hz ($I_{\Delta\text{nf}} = 14I_{\Delta\text{n}}$), where $I_{\Delta\text{n}}$ is a rated residual operating current for the nominal frequency (50 Hz). It is confirmation of the aforementioned comment related to the effect of high-frequency current on persons—the values of the residual operating current correspond to the threshold of ventricular fibrillation. In practice, such an increase of the tripping threshold is acceptable only for RCDs having $I_{\Delta\text{n}} \leq 30$ mA for 50 Hz. Unfortunately, the standard [25] indicates the increase of the residual operating current only up to 1 kHz, what is insufficient nowadays. Moreover, standards do not differentiate thermal effect (the dissipated power in the human body) of the 50 Hz current vs. high-frequency currents—it is assumed to be approximately constant.

Results obtained within the frame of this research indicate a need to extend provisions of standards related to RCDs performance and tests. Normative test of B-type RCDs up to only 1 kHz is insufficient nowadays. Similarly, a mixed-frequency testing waveform having the high-frequency component 1 kHz (normative testing of F-type and B-type RCDs) is also insufficient. Contemporary installations comprising power electronics converters may produce harmonics of a level equal to several dozen kHz (see Figures 5 and 6). It is proposed to move the value of the aforementioned normative high-frequency component from 1 kHz to at least 10 kHz. It is also recommended to introduce a new type of RCDs, which could ensure stable tripping up to 50 kHz (alternatively 150 kHz; upper limit of supraharmonics—taking into account fast development and wide use of power electronics converters). These proposed modifications must be respected by RCDs' manufacturers.

With reference to the improvement of the construction of RCDs (from the point of view of the high frequency), special attention should be given to the current transformer CT of the RCD, its relay RY, and the optional electronic matching system EC (Figure 2). Parameters of the aforementioned elements should be selected, coordinated, and verified for the expected operating frequency range.

5. Conclusions

Research conducted by the authors and its results presented in this paper show that there is a strong effect of frequency on the tripping threshold of RCDs. While up to 1 kHz the tripping of RCDs was noted, for frequencies 5 kHz, 10 kHz, 20 kHz, and 50 kHz there were no RCDs reactions to the test currents (except one AC-type RCD reacting to the 5 kHz), even 15 times higher than the rated residual current of the RCD. What is worse, unfavorable behavior was noted also for F-type and B-type RCDs, which are dedicated to circuits having earth fault current with harmonics. No tripping of F-type and B-type RCDs for frequencies at levels of a few or several dozen kHz is due to the current state of the normative requirements. International standards require tests for frequencies not higher than 1 kHz—RCDs have to react only to this level of frequency. In the light of the switching frequency used in modern power electronics converters, such a level of testing frequency (1 kHz) seems to be insufficient. Therefore, it is proposed to raise the threshold of the normative testing current from 1 kHz to at least 10 kHz. For special applications, a separate type of RCDs is recommended to be provided. RCDs of the special type (frequency-proof) should be able to trip for frequencies up to 50 kHz.

Author Contributions: Conceptualization, S.C.; methodology, S.C.; validation, S.C.; formal analysis, S.C.; investigation, S.C. and H.T.; resources, S.C.; writing—original draft preparation, S.C. and H.T.; writing—review and editing, S.C. and H.T.; visualization, S.C. and H.T.; supervision, S.C. All authors have read and agreed to the published version of the manuscript.

Funding: This research was supported by Gdańsk University of Technology.

Institutional Review Board Statement: Not applicable.

Informed Consent Statement: Not applicable.

Data Availability Statement: Data is contained within the article.

Conflicts of Interest: The authors declare no conflict of interest.

References

1. CENELEC. *HD 60364-4-41:2017. Low-Voltage Electrical Installations—Part 4-41: Protection for Safety—Protection Against Electric Shock*; European Committee for Electrotechnical Standardization: Brussels, Belgium, 2017.
2. Rösch, H. Current-operated ELCBs for AC. and pulsating DC fault currents. *Siemens Power Eng.* **1981**, *8–9*, 252–255.
3. Rösch, H. Fehlerstrom-Schutzschalter zum Schutz gegen gefährliche Körperströme. *ETZ* **1989**, *12*, 580–584.
4. Solleder, R. Allstromsensitive Fehlerstrom-Schutzeinrichtung fur Industrieanwendung. *ETZ* **1994**, *115*, 896–901.
5. Han, Y.; Ding, C.; Shou, X. Design & implementation of an A-type residual current circuit breaker IC. In Proceedings of the 2012 IEEE International Symposium on Industrial Electronics, Hangzhou, China, 28–31 May 2012. [CrossRef]
6. Guolei, X.; Yan, S.; Juwei, Y.; Chengcong, L.; Binfeng, W.; Feng, J.; Weidong, J. Research on the principle of residual current protection technology based on transient waveform criterion. *E3S Web Conf.* **2020**, *204*, 02009. [CrossRef]
7. Yao, W.; Kui, L.; Can, L.; Zhitao, G.; Feng, N.; Xiujuan, Z. Study on modeling and simulation of AC/DC sensitive residual current transformer. In Proceedings of the 1st International Conference on Electric Power Equipment—Switching Technology, Xi'an, China, 23–27 October 2011. [CrossRef]
8. Czaja, P. Examination of the impact of design of a residual current protective device on the release frequency range. In Proceedings of the 2017 Progress in Applied Electrical Engineering (PAEE), Koscielisko, Poland, 25–30 June 2017. [CrossRef]
9. Czapp, S.; Horiszny, J. Simulation of residual current devices operation under high frequency residual current. *Prz. Elektrotechniczny* **2012**, *2*, 242–247.
10. Erdei, Z.; Horgos, M.; Lung, C.; Pop-Vadean, A.; Muresan, R. Frequency behavior of the residual current devices. *IOP Conf. Ser. Mater. Sci. Eng.* **2017**, *163*, 012053. [CrossRef]
11. Freschi, F. High-frequency behavior of residual current devices. *IEEE Trans. Power Deliv.* **2012**, *27*, 1629–1635. [CrossRef]
12. Horgos, M.; Erdei, Z.; Barz, C.; Birsan, I.; Ilia, M. Contributions to testing residual current devices at different frequency values. In Proceedings of the 6th International Conference on Modern Power Systems (MPS), Cluj-Napoca, Romania, 18–21 May 2015.
13. Shopov, Y.; Filipova-Petrakieva, S.; Boychev, B. Investigation of residual current devices in high frequencies. In Proceedings of the 10th Electrical Engineering Faculty Conference (BulEF), Sozopol, Bulgaria, 11–14 September 2018.
14. Xie, P.; Fang, Z.; Hu, J.; Yang, J.; Zhu, G. Tripping characteristics of residual current devices under different working conditions. In Proceedings of the 3rd IEEE Conference on Energy Internet and Energy System Integration, Changsha, China, 8–10 November 2019.
15. Czapp, S.; Dobrzynski, K.; Klucznik, J.; Lubosny, Z.; Kowalak, R. Improving sensitivity of residual current transformers to high frequency earth fault currents. *Arch. Elect. Eng.* **2017**, *66*, 485–494. [CrossRef]
16. Czapp, S.; Dobrzynski, K.; Klucznik, J.; Lubosny, Z. Low-frequency tripping characteristics of residual current devices. In Proceedings of the 2017 IEEE International Conference on Environmental and Electrical Engineering & 2017 IEEE Industrial and Commercial Power Systems Europe (EEEIC /I&CPS Europe), Milan, Italy, 6–9 June 2017; pp. 298–301. [CrossRef]
17. Czapp, S. The effect of earth fault current harmonics on tripping of residual current devices. In Proceedings of the International School on Nonsinusoidal Currents and Compensation, IX Conference-Seminar (ISNCC), Lagow, Poland, 10–13 June 2008. [CrossRef]

18. Lee, T.M.; Chan, T.W. The effects of harmonics on the operational characteristics of residual current circuit breakers. In Proceedings of the International Conference on Energy Management and Power Delivery, Singapore, 21–23 November 1995; pp. 715–719. [CrossRef]
19. Wieland, T.; Aigner, M.; Schmautzer, E.; Pasker, J.; Fickert, L. Influences on safety issues for inverter supplied grid structures. In Proceedings of the 2012 Electric Power Quality and Supply Reliability, Tartu, Estonia, 11–13 June 2012. [CrossRef]
20. Rabcan, J.; Levashenko, V.; Zaitseva, E.; Kvassay, M.; Subbotin, S. Application of fuzzy decision tree for signal classification. *IEEE Trans. Ind. Inf.* **2019**, *15*, 5425–5434. [CrossRef]
21. Gruhn, T.; Glenney, J.; Savostianik, M. Type B ground-fault protection on adjustable frequency drives. *IEEE Trans. Ind. Appl.* **2018**, *54*, 934–939. [CrossRef]
22. Panda, R.K.; Veeramalla, J. Behavior modeling of a type B RCD. In Proceedings of the 1st IEEE International Conference on Power Electronics, Intelligent Control and Energy Systems (ICPEICES), Delhi, India, 4–6 July 2016.
23. IEC. *IEC 60755:2017. General Safety Requirements for Residual Current Operated Protective Devices*; International Electrotechnical Commission: Geneva, Switzerland, 2017.
24. IEC. *IEC 61008-1:2010. Residual Current Operated Circuit-Breakers without Integral Overcurrent Protection for Household and Similar Uses (RCCB)—Part 1: General Rules*; International Electrotechnical Commission: Geneva, Switzerland, 2010.
25. IEC. *IEC 62423:2009. Type F and Type B Residual Current Operated Circuit-Breakers with and without Integral Overcurrent Protection for Household and Similar Uses*; International Electrotechnical Commission: Geneva, Switzerland, 2009.
26. Eaton. Residual Current Devices. In *Application Guide*; Eaton: Vienna, Austria, 2017.
27. VDE. *DIN VDE 0664-400:2020-03. Residual Current Operated Circuit-Breakers Type B without Integral Overcurrent Protection to Operate at Residual Alternating and Residual Direct Currents for Advanced Preventative Protection against Fire–Part 400: RCCB Type B+*; VDE: Berlin, Germany, 2020.
28. IEC. *IEC 60479-2:2007(2019). Effects of Current on Human Beings and Livestock–Part 2: Special Aspects*; International Electrotechnical Commission: Geneva, Switzerland, 2019.

energies

Article

Application of Decision Trees for Optimal Allocation of Harmonic Filters in Medium-Voltage Networks

Maciej Klimas, Dariusz Grabowski and Dawid Buła *

Electrical Engineering and Computer Science Department, Faculty of Electrical Engineering, Silesian University of Technology, 44-100 Gliwice, Poland; maciej.klimas@polsl.pl (M.K.); dariusz.grabowski@polsl.pl (D.G.)
* Correspondence: dawid.bula@polsl.pl

Abstract: The paper proposes a solution for the problem of optimizing medium voltage power systems which supply, among others, nonlinear loads. It is focused on decision tree (DT) application for the sizing and allocation of active power filters (APFs), which are the most effective means of power quality improvement. Propositions of some DT strategies followed by the results have been described in the paper. On the basis of an example of a medium-voltage network, an analysis of the selection of the number and allocation of active power filters was carried out in terms of minimizing losses and costs keeping under control voltage total harmonic distortion (THD) coefficients in the network nodes. The presented example shows that decision trees allow for the selection of the optimal solution, depending on assumed limitations, expected effects, and costs.

Keywords: power quality; active power filters; decision trees; power losses; optimization; modeling and simulation; frequency domain

Citation: Klimas, M.; Grabowski, D.; Buła, D. Application of Decision Trees for Optimal Allocation of Harmonic Filters in Medium-Voltage Networks. *Energies* **2021**, *14*, 1173. https://doi.org/10.3390/en14041173

Academic Editor: Andrea Mariscotti

Received: 27 January 2021
Accepted: 18 February 2021
Published: 22 February 2021

1. Introduction

These days, when we observe rapid technological progress and constantly increasing energy consumption, it is necessary to pay attention to responsible energy utilization. Electrical energy supply, especially in highly developed regions, is achieved through energy transmission. The concept of alternating current (AC) power systems, which was assumed many years ago, makes the monitoring of some power quality indices below limits necessary. It is more and more challenging as the number of loads having a negative influence on these indices increases. The distortion of voltage and current waveforms, which is caused by nonlinear loads and manifests itself through higher harmonics, belongs to the most important power quality parameters, and it is expressed by the total harmonic distortion (THD) coefficient. The higher harmonic elimination and, thereby, THD minimization is a basic task given to power quality improvement systems.

The placement of compensation devices, including active power filters (APFs), in power networks is one of the key factors in successful power quality improvement not only from the technical but also from the economical point of view. In many cases, filters have to be applied in complex and large power networks with several nonlinear loads and a significant number of nodes. Among these nodes there are numerous ones to which APFs or other devices ensuring high power quality can be connected. Thus, a system designer has to be able to solve an optimization problem that consists in location of the devices ensuring, among others, effective higher harmonic elimination while keeping the solution costs under control.

The optimization of passive and active power filter parameters and location in power systems is a common problem among researchers all over the world due to its crucial role in the attainment of power quality goals. There are many propositions of solutions based on diverse optimization methods that implement classical algorithms or try to develop problem specific ones. The proposed solutions differ in complexity, universality, and most of all the degree to which practical aspects of implementation of power quality systems have been taken into account. The simplest approach to filter location is just the iterative examination of power quality improvement obtained when installing it in successive nodes one by one [1]. The best solution is chosen by comparison of the results. Such an approach can be extended to more sophisticated forms based on the adaptive step presented recently in [2]. The proposed algorithm uses a decision tree and achieves an excellent harmonic suppression effect for an exemplary circuit using less number and capacity of APFs when compared with a solution calculated by a traditional method [2]. This paper also follows this direction in the research aimed at optimal APF location and sizing. In the past, many approaches have been proposed for solution of the problem under consideration. The advanced solutions have been based on complex optimization algorithms, e.g., ant colony system [3], modified harmony search algorithm [4], whale optimization algorithm [5], gray wolf optimizer [6], and bacterial foraging optimization algorithm [7]. Nevertheless, the most popular approaches are based on genetic algorithms [8–11] and particle swarm optimization [12–14].

Among optimization goals, one can find a minimization of the power losses [15–17], which has been used also in this paper along with an economic criterion expressed by the relative cost of the solution. The distinctive feature of this paper is minimization of the power losses caused by the higher harmonics in all system elements (mainly electric cables and transformers). Such an approach is especially efficient in relatively small networks working with little power margin and, thus, exposed to overloads or replacements of main and expensive components, e.g., transformers. The size of the network encouraged us to check the performance of decision trees as a tool for APF location and sizing. So far, decision trees in the field of power systems have been applied to:

- optimal phasor measurement unit (PMU) placement for voltage security assessment [18,19], including power system islanding identification [20] and line outage detection [21–23]—fast and direct measurement results by PMUs combined with decision trees gives more time for corrective or preventive actions;
- classification and detection of power line faults [24–26];
- detection of power system problems, including power quality disturbances [27], active power imbalance [28], voltage stability margin [29], and nontechnical losses [30,31];
- fault location in power distribution systems [32–34];
- optimal planning of storage in power systems integrated with wind power generation [35];
- decision-making in single-device cases [36] as well as in global power plant operation [37].

Thus, decision trees are widely used in power systems—they belong to the 10 major machine learning models frequently used in power systems [38]. However, literature overview has revealed no evidence regarding the application of decision trees to solve the problem under consideration, i.e., APF location. The only similar research includes application of decision trees for the determination of optimal location and rate of series compensation to increase power system loading margin [39]. The paper [39] is focused on series compensation of transmission lines, which is one of the most effective means to increase the loading margin of an interconnected power system. It includes a proposition of methodology for the identification of the critical transmission lines and their proper compensation rate with respect to voltage stability using decision trees. It must be stressed that the application of decision trees in various areas resulted in several descendant methods, for example, random forest [40] and gradient boosted trees [41]. The random forest, as a technique consisting in the aggregation of a large number of decision trees, is usually regarded as a black-box algorithm due to the large number of trees. Gradient-boosted trees represent another approach, which belongs to the so called "ensemble methods" based on more than one decision tree. Such methods are especially useful in solving large-scale problems. They have also been applied in power systems, for example, in the building load model [42]. In this paper, due to the scale of the problem and its type, which is optimization rather than classification or prediction, a classical form of the decision tree algorithm was used.

In this paper, the authors proposed decision trees as an algorithm to solve the problem of APF location in medium voltage networks. For a given test system and two goal functions, the results have been compared with the global optimum obtained by the brute force algorithm. The paper consists of three sections with an introduction. Section 2 is devoted to APF optimum sizing and location, and it includes the definitions of goal functions as well as the description of the brute force and decision tree algorithms. Section 3 includes the description of the test system and optimization results for brute force and decision tree algorithms. Sections 4 and 5 present the result discussion and conclusions.

2. Optimization Problem Definitions and Solution Algorithms

2.1. Goal Functions

This paper proposes optimization of APF placement in power systems based on two different goal functions. The general denotation, R_k, has been introduced in order to formalize goal function descriptions, where R_k is k-th element of a set of all possible APF configurations. It specifies the placement of each APF in the discussed power system in relation to the list of all APFs denoted by K.

$$R_k = [s_1 \; s_2 \; \cdots \; s_i], \quad (1)$$

where

s_i—state of the APF in i-th node (1—APF placed, 0—no APF),
for all APF list $K = [n_1 \; n_2 \; \cdots \; n_i]$, where n_i—denotation of i-th APF.

Along with the goal functions, all optimization processes take into account the maximum value of voltage total harmonic distortion coefficient (THDV) factor occurring in the system. A maximum allowed THDV level of 5% has been assumed for all optimization methods presented in this paper.

2.1.1. Power Losses Criterion

The first implemented goal function is related to criterion of active power losses in system elements:

$$\min_x F_1(x) : \min_{R_k} \sum_{m=1}^{M} P_m, \quad (2)$$

where

m—number of power system elements in which power losses occur (transformers, lines, coils, etc.), $m = 1, 2, \ldots, M$.
P_m—power losses of m-th element.
x—independent variable of the goal function that is subjected to minimization.

The minimization of F_1 function allows reduction of power losses related not only with harmonics but also with reactive power in all system elements excluding loads. Due to the fact that even with full compensation, loads require power transfer, F_1 cannot be reduced completely to 0.

2.1.2. Cost Criterion

Application of the dispersed power quality improvement system is related with the necessity of considering many possible solutions that rely on the placement of at least several APFs. For such an approach, economic criterion is of high importance. A fundamental factor influencing the costs of a solution consists of a set of rated parameters of used APFs and results from conditions imposed by the power system. On the grounds of market recon, public information provided by APF producers and literature overview, a nonlinear function of APF price has been presented below. It was scaled in order to indicate price relative to the most expensive APF. The bar graph of APF relative cost depending on the range of rated current has been shown in Figure 1.

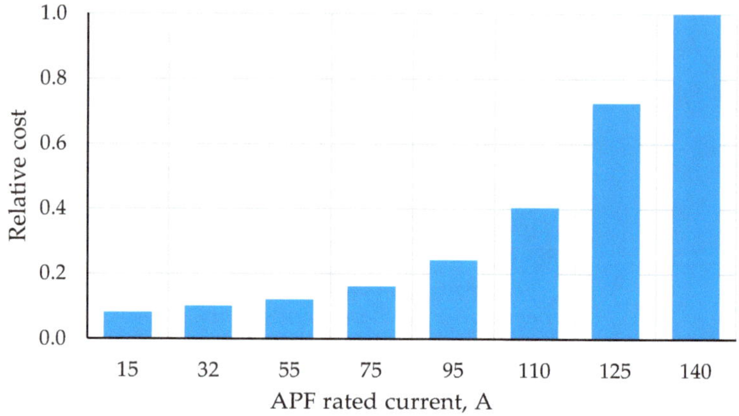

Figure 1. Bar graph of exemplary relative costs of active power filters (APFs) depending on their rated current.

The cost criterion has been described formally through F_2 goal function, formulated as follows:

$$\min_{x} F_2(x) : \min_{R_k} \sum_{i=1}^{N} p\left(\left|I_i^k\right|\right), \tag{3}$$

where

$p(\cdot)$—function assigning the cost to i-th APF depending on its RMS current value,
i—APF number, $i = 1, 2, \ldots, N$,
$\left|I_i^k\right|$—RMS current of i-th APF, calculated as:

$$\left|I_i^k\right| = \sqrt{\sum_{h=1}^{H} |I_{ih}^k|^2}, \tag{4}$$

where
$\left|I_{ih}^k\right|$—RMS current of h-th harmonic of i-th APF,
h—harmonic number, $h = 1, 2, \ldots, H$.

2.2. Brute Force Algorithm

The problem of APF sizing and allocation has been solved using a brute force (BF) algorithm, which due to computation times is especially useful in the case of small-scale problems. In this paper, the BF algorithm allows us to find all solutions, including the global minimum, and evaluate the quality of solutions obtained by the decision tree (DT) algorithm. The block diagram of the BF algorithm has been presented in Figure 2.

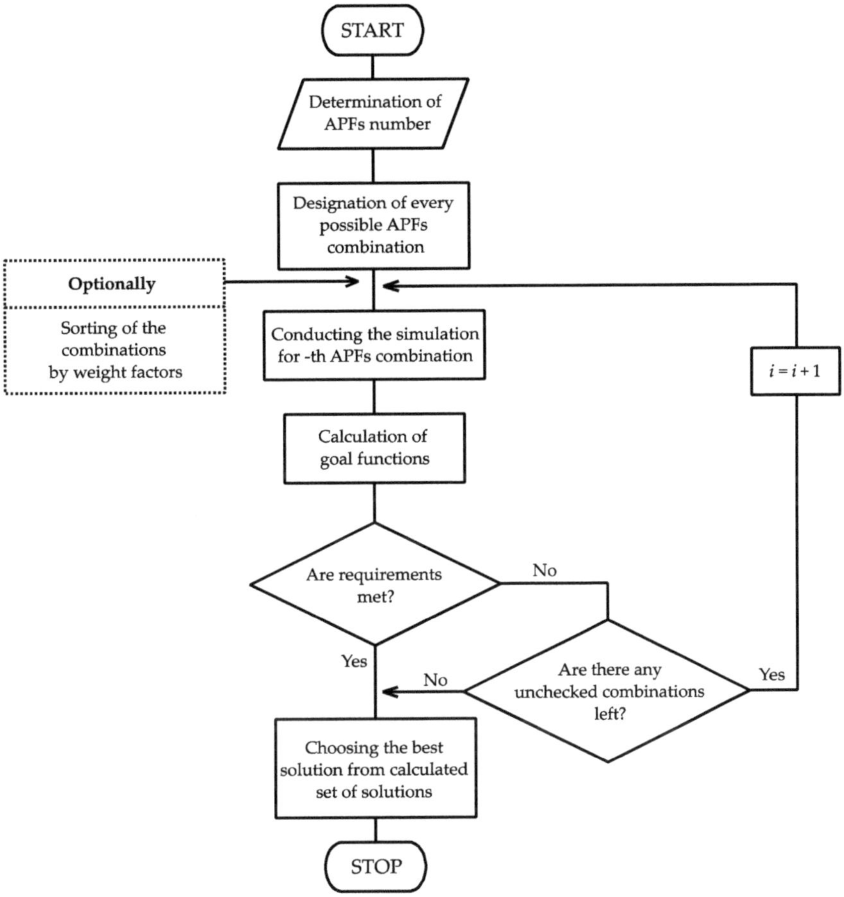

Figure 2. Block diagram of the brute force algorithm.

The BF algorithm is very straightforward and consists in checking every combination representing in our case different allocations of APFs in the power system. If the power system has n nodes (the number of nodes in which APFs can be connected, not the total number of nodes), the number of k-combinations is equal to the binomial coefficient:

$$C_n^k = \binom{n}{k} = \frac{n!}{k!(n-k)!}. \tag{5}$$

In successive steps of the BF algorithm, values of the given goal function (2) or (3) are calculated. The algorithm can be terminated before analyzing all combinations if the goal function value drops below the assumed threshold value—in this case there is no guarantee that the global minimum has been reached. Otherwise, it is terminated if all combinations have been checked. Afterward, the best solution is chosen.

2.3. Optimization with Decision Trees

The application of decision trees is the most popular in classification problems in which an algorithm uses a set of features that can be checked one by one. Such a method is efficient and allows the effective solving of similar issues. However, classification problems are not the only one class of problems that can be addressed with decision trees, and this method can be also applied in general combinatorial optimization. This paper proposes a solution for the application of decision trees in the optimization of APF placement in power systems. Such an issue can be formulated as the problem of deciding where APFs should be placed in order to get the best possible results in terms of previously defined goal functions.

An implemented decision tree algorithm works iteratively by checking which step should be taken next in order to reduce the current value of the minimized goal function. The APF combination through which the algorithm passes is coded in a vector representing the state of each APF (1).

The algorithm operation begins with a state in which only one APF is placed in the power system, on a previously defined starting position. In the next steps, the algorithm makes a decision whether the current APF should be moved to the next or the previous node from R_k vector or whether it should remain in the current position and another one should be placed additionally. There is also a possibility to terminate the decision tree's operation if any of the aforementioned options does not provide a better outcome. The criterion of the decision-making process can be formulated through the difference between the current value of the goal function and a value that could be reached by each possible decision:

$$\min(\Delta f_{+}, \Delta f_{-}, \Delta f_{add}, \Delta f_{0}) \tag{6}$$

where

Δf_{+}—difference between goal function values for the current state and a state after moving the APF to the next node,
Δf_{-}—difference between goal function values for the current state and a state after moving the APF to the previous node,
Δf_{add}—difference between goal function values for the current state and a state after adding a new APF, modified by the correctional coefficient, W_{corr},
Δf_{0}—zero value connected with staying in the current state.

In order to keep the balance between the advantages and disadvantages of adding a new APF, the decision tree algorithm was also fitted with a correctional coefficient, W_{corr}, which allows the regulation of how many times benefits from adding a new APF should be higher than benefits from moving it in order to make such a decision. With appropriate configuration, this coefficient prevents a cost increase, which would occur due to the tendency for adding new APFs in each step while reducing power losses.

The schematic of the recurring element of the decision tree has been presented in Figure 3. The components of this diagram can only be modified in border cases related to reaching the maximum allowed number of APFs, or the lack of movement possibilities—in all other cases, the structure repeats itself until the termination of the algorithm. Such conditions, limiting decision tree's options, include also the THDV value. The whole optimization process is related to improving power quality quantities such as THDV along with the minimized function. In order to guarantee such improvement, only solutions that result in a THDV value below 5% can be accepted. The decision tree can only terminate its action if the THDV is below this level, otherwise it continues its search.

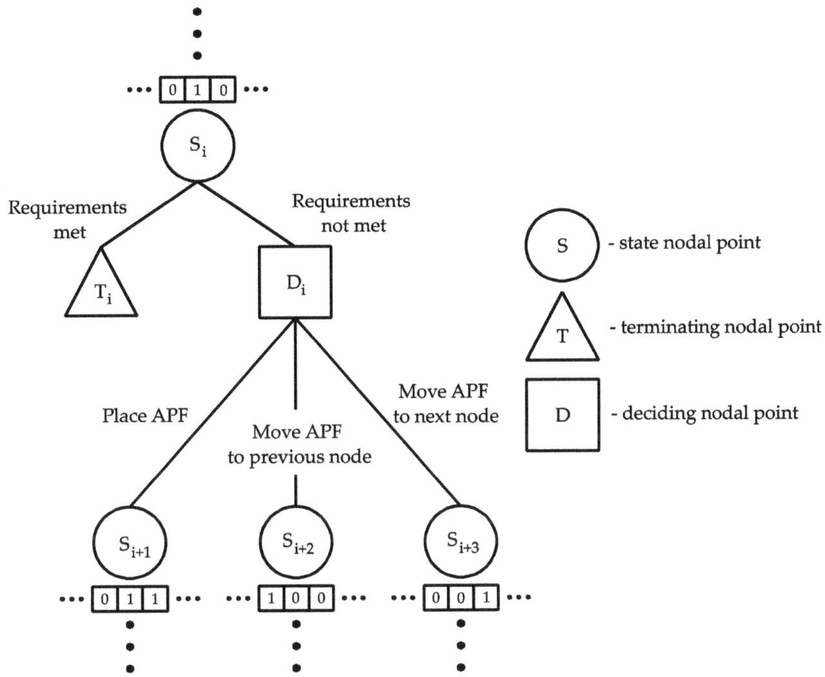

Figure 3. Schematic of recurring part of decision tree used for the optimization of APF placement.

3. Optimization Results

3.1. Test System

In this paper, a test system was adopted based on the power supply scheme of an extended ski station. The data on the system and its topology have been taken from the documentation of the PCFLO simulation software [43] and are often used in other papers [9,44,45]. The advantages of the selected test system are its complexity, the presence of nonlinear loads, access to complete information on its components, and medium voltage for which there are technical solutions of APF systems. The diagram of the system with marked potential connection points of active power filters is presented in Figure 4.

It should be noted that despite the series connection of APFs (F_1–F_{14}) in the test power system, the APF itself is constructed as parallel, based on a current-controlled current source (Figure 5). The nonlinear loads in the presented system are six pulse rectifiers, which are the cause of current and voltage waveform distortions in the entire circuit. The modeling and simulation software based on the iterative method in the frequency domain described in [46] was used for calculations. In this case, a simple, without losses, ideal model of the active power filter was used in the simulations (Figure 5).

Table 1 summarizes the voltage total harmonic distortion coefficients (THDVs) for all nodes and the current total harmonic distortion coefficients (THDIs) for all lines of the analyzed system in the case without APFs.

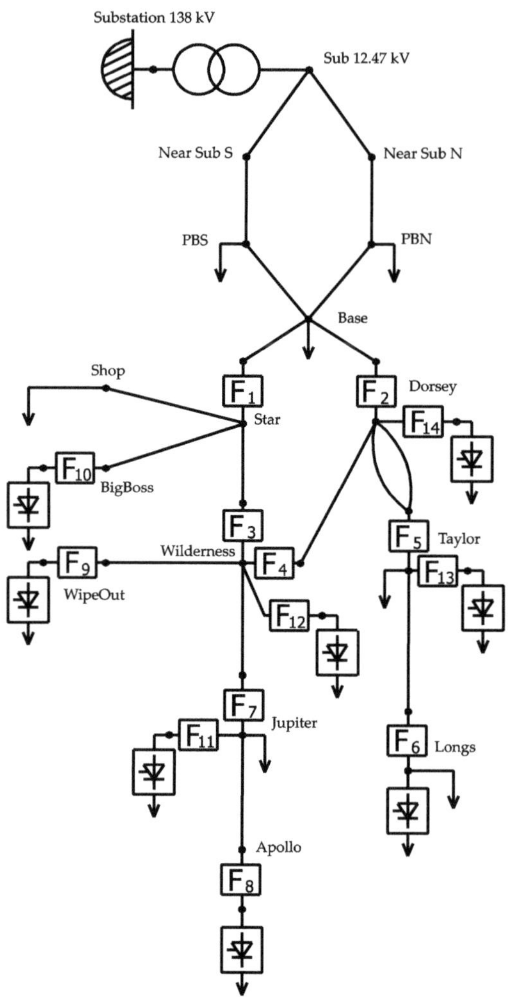

Figure 4. Diagram of the test power system with marked possible locations of APFs (F_1–F_{14}).

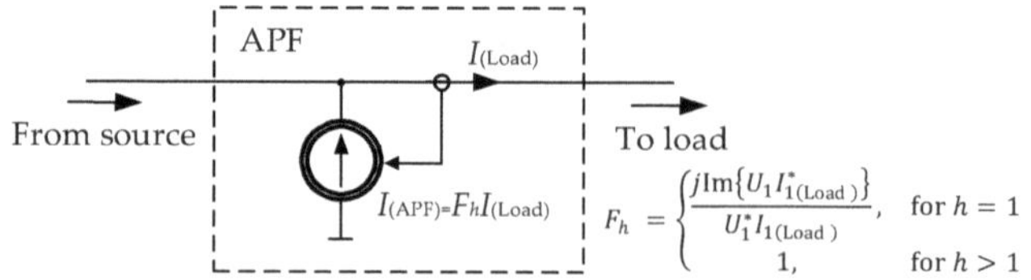

Figure 5. Model of active power filter.

Table 1. Values of voltage total harmonic distortion coefficients (THDVs) for all voltage nodes and current total harmonic distortion coefficients (THDIs) for all line currents.

Node Name	Node Number	THDV, %	Line	THDI, %
Sub 12.47 kV	1	11.3	20-1	11.0
Near Sub S	2	11.4	1-2	11.0
Near Sub N	3	11.4	1-3	11.0
PBS	4	11.7	2-4	10.9
PBN	5	11.7	3-5	10.9
Base	6	11.9	4-6	12.2
Star	7	12.4	5-6	12.2
Wilderness	8	12.5	6-7	22.3
Dorsey	9	12.4	6-9	19.3
Taylor	10	12.4	7-16	3.4
Longs	11	12.5	7-15	30.7
Apollo	12	13.1	7-8	16.7
Jupiter	13	12.8	8-9	40.2
WipeOut	14	12.7	8-14	32.8
BigBoss	15	13.0	8-13	23.0
Shop	16	12.4	13-12	29.7
Sub 138 kV	20	3.3	9-10	6.9
			10-11	6.1

As can be seen in Table 1, the THDV and THDI values are significant and can potentially cause power losses and power quality problems in the analyzed system. Distortion is also visible in the transformer current and voltage waveforms, which have been shown in Figures 6 and 7, respectively. The voltage waveform shows instantaneous voltage drops, which increase the THD coefficient. These drops are due to the high steepness of the transformer output current. The oscillations around these drops result directly from the used model of nonlinear receivers and the simulation method in the frequency domain. The presented waveforms are a combination of a finite number of harmonics obtained as a result of simulations and not measured waveforms.

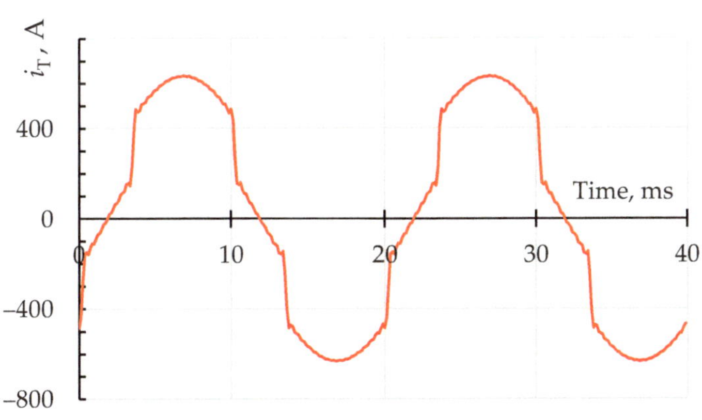

Figure 6. Current waveform of the secondary side of the transformer in the analyzed system.

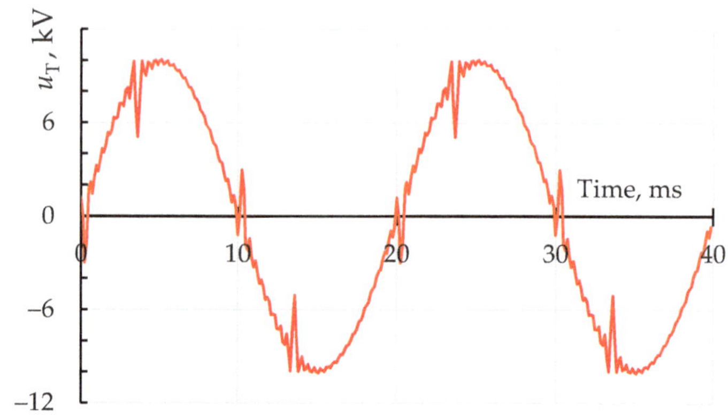

Figure 7. Voltage waveform of the secondary side of the transformer in the analyzed system.

3.2. Brute Force Results

As a part of the work, optimization by means of the brute force method was carried out in order to compare results with those of the proposed method based on decision trees. The obtained results have been analyzed and presented below. The first aspect related to the operation of the algorithm is the calculation time, which in the case of the complete solution set search method may be long enough to prevent the efficient use of the software. Along with the increase in the number of possible cases according to Equation (5), the time needed to analyze all the combinations increased. Figure 8 presents both the aforementioned quantities as a function of the number of APFs, the arrangement of which was optimized. The obtained results confirm that such an approach, when applied to complex problems, despite its high effectiveness, may not be the optimal choice.

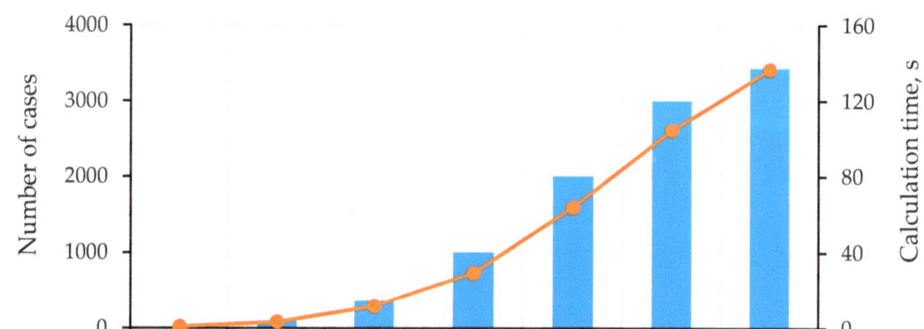

	1	2	3	4	5	6	7
Number of cases	14	91	364	1001	2002	3003	3432
Calculation time	0.8	3.4	11.9	29.3	63.8	104.5	136.6

APFs number

Figure 8. Dependence of the number of cases and the calculation time of the brute force algorithm on the APF number.

Table 2 contains the APF connection cases, which were selected by the software as the best in terms of minimizing each of the considered goal functions with the limitation related to the maximum THDV coefficient. From these data, it can be seen that the best solutions for a small number of APFs are in most cases the solutions for a larger number of APFs.

Table 2. List of cases meeting the minimum criterion of the analyzed goal functions, taking into account the assumed maximum of THDVs.

APF Numbers	Number of APFs						
	For a Minimum of F_1						
1	2						
2	3	10					
3	2	8	10				
4	3	5	8	10			
5	2	6	8	10	11		
6	2	3	4	8	10	13	
7	1	2	4	6	8	10	11
	For a Minimum of F_2						
1	1						
2	4	10					
3	4	10	13				
4	4	8	10	12			
5	4	8	9	10	14		
6	2	3	4	8	10	13	
7	4	7	8	9	10	12	13

The impact of placing subsequent APFs on the values achieved by the individual goal functions in the cases selected as optimal was also analyzed. The results were presented for the group with the limitation resulting from the maximum allowable THDV level. The results of this analysis have been shown in Figure 9. For the goal function F_1, the increase in the number of APFs results in a clear decrease in power losses, which is natural and results directly from the concept of an APF, i.e., harmonic reduction and reactive power compensation. In the case of the goal function F_2, the relative value (related to the maximum APF cost) was used. With the assumption of limiting the THDV factor, the minimum costs for the case with two APFs is visible. Reaching the required maximum THDV in the system is, therefore, the most cost effective when using two APFs. Any other solution is more expensive.

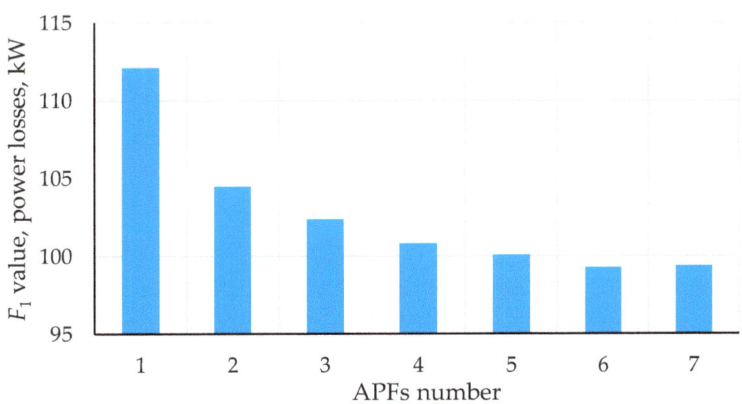

(a)

Figure 9. *Cont.*

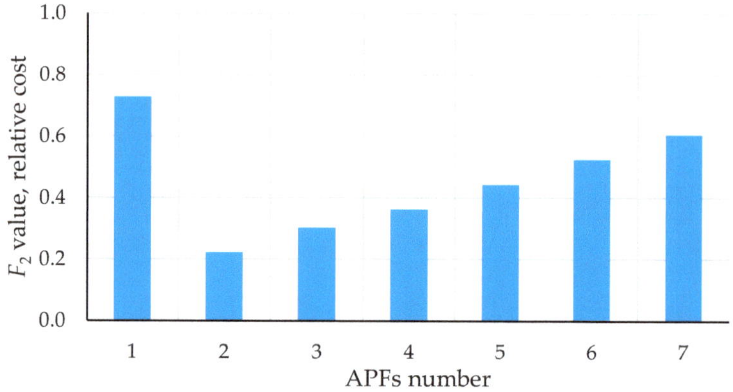

(b)

Figure 9. Values of the goal functions depending on the number of optimally placed APFs; (**a**) for F_1; (**b**) for F_2.

All the results obtained by applying the brute force algorithm, even before the analysis in terms of finding the minimum for the individual goal functions, are in the form of a set of values of these functions associated with each possible solution. For a larger number of APFs, the number of cases is so large that the graphical representation becomes unclear, therefore an example bar graph of the maximum THDV value in the system for selecting the position of only two APFs has been presented below (Figure 10). This graph shows how large the variability of the maximum THDV is for the analyzed cases.

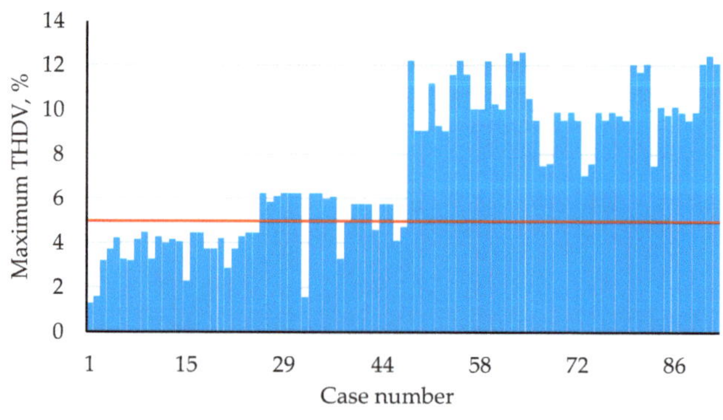

Figure 10. Bar chart of the maximum THDV value among all network nodes for the allocation of two APFs.

In order to illustrate the results from a wider perspective, for the case of three APFs in the system, all feasible solutions have been presented in Figure 11. As can be read, for 191 solutions out of 364 (see Figure 8) the constraint on the THDV coefficient (less or equal 5%) is fulfilled. Moreover, the Pareto frontier includes five solutions for which the objective functions given by (2) and (3) take approximately the values between 102 and 112 kW for F_1 and between 0.30 and 0.60 for F_2. These solutions are optimal in the Pareto sense and the choice of the one to be implemented depends on other aspects, e.g., easier APF allocation in some nodes or long-term financial analysis. Therefore, selection of the optimum solution can vary among decision-makers because it is based on their preferences and criteria.

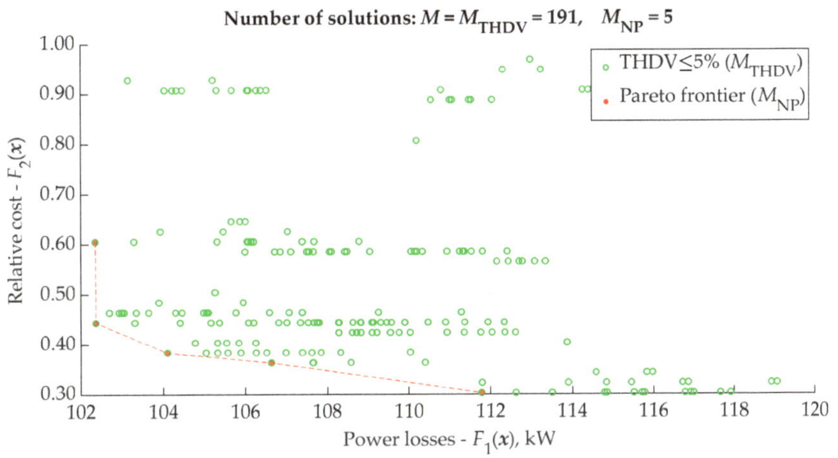

Figure 11. Objective function values for all feasible solutions in the case of three APFs installed in the system (M_{THDV}—number of feasible solutions, M_{NP}—number of noninferior points).

All solutions for three APFs presented in Figure 12 allow us to have a deeper insight into this example. The Pareto frontier in such a case consists of eight points representing the optimum solutions for the unconstrained problem. The feasible points presented in Figure 11, for which THDV $\leq$ 5%, have been also marked out in Figure 12. The same data have been presented in 3D space in Figure 13.

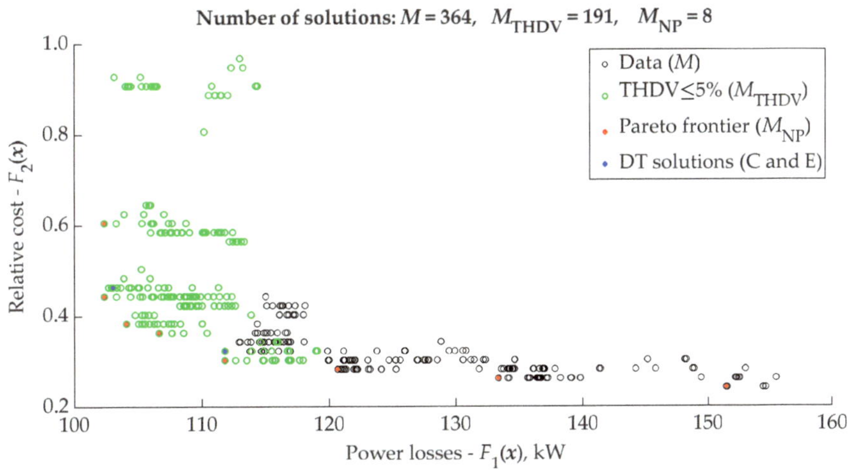

Figure 12. Objective function values for all solutions in the case of three APFs installed in the system (M—total number of solutions, M_{THDV}—number of feasible solutions, M_{NP}—number of noninferior points).

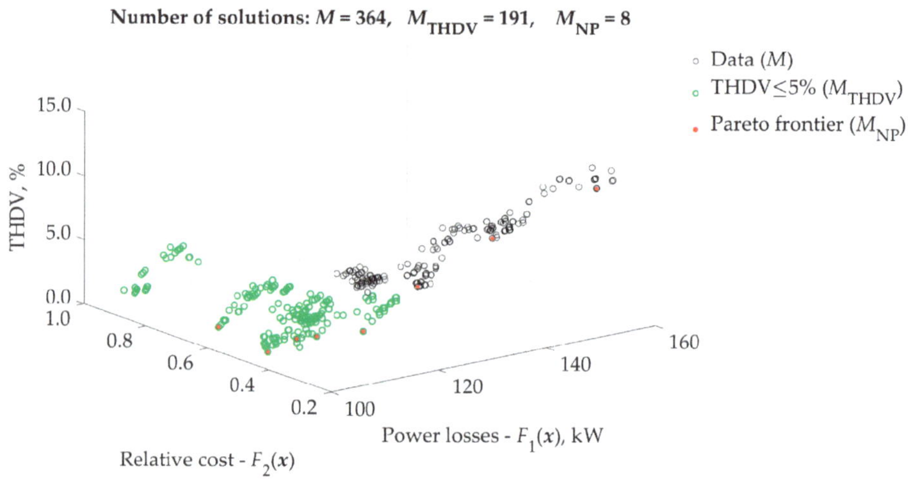

Figure 13. Objective function and THDV values for all solutions in the case of three APFs installed in the system (M—total number of solutions, M_{THDV}—number of feasible solutions, M_{NP}—number of noninferior points).

3.3. Decision Tree Results

Optimization with the decision tree algorithm is challenging considering how complex the hierarchical structure of the power system is. The first factor affecting the decision-making process is the algorithm's starting point. The decision tree is highly sensitive to the choice of the starting point because it determines the position of the first APFs that are going to be checked. Consequently, different parts of the simulated power system can be examined at the beginning. The mentioned problem is only a specific case of the more general issue of arrangement of APF combination in a single solution vector, R_k. In a particular case, it only results in the change of the starting point. Considering the high importance of APF arrangement, and therefore the algorithm's route, different strategies of its setup have been examined. The optimization has been conducted for both F_1 and F_2 goal functions with differences between routes that arise from different methods of sorting the APFs in R_k vector. The research assumes three sorting strategies: by node THDV value (descending), by APF current value (ascending), and by APF number (descending). Another factor differentiating tested routes is the W_{corr} coefficient, which has been included and excluded in turns, for the consecutive algorithm workflows. The results for all tested approaches have been presented in Table 3.

Table 3. List of route parameters and results of the decision tree algorithm applied for F_1 and F_2 minimization.

Index	Minimized Function	Route Sorted by	W_{corr}	F_1, kW	F_2	Calculation Time, s	APF Number	Number of Installed APFs
A	F_1	THDV	Yes	111.8	0.32	1.4	3	10, 7, 11
B	F_1	THDV	No	100.4	0.69	1.3	7	8, 10, 7, 11, 9, 6, 3
C	F_1	Number of APF	Yes	103.0	0.46	1.4	3	10, 7, 3
D	F_1	Number of APF	No	99.4	1.23	4.3	14	14, 13, 12, 11, 10, 9, 8, 7, 6, 5, 4, 3, 2, 1
E	F_2	THDV	Both	111.8	0.32	0.7	3	10, 7, 11
F	F_2	Number of APF	Both	105.6	0.54	2.1	6	13, 12, 11, 10, 9, 8
G	F_1	Current	Yes	108.7	0.40	1.3	4	13, 10, 11, 7
H	F_1	Current	No	99.4	1.15	3.9	13	12, 14, 9, 13, 6, 8, 10, 11, 7, 5, 4, 1, 2
I	F_2	Current	Both	110.1	0.54	2.2	6	14, 9, 13, 6, 8, 10

As presented, different routes result in a different number of placed APFs and optimization efficiency. The application of this method highly reduces computation time in comparison to the brute force algorithm. The calculation time mostly depends on the number of steps taken by the decision tree, because in each step, before the decision, the algorithm repeats simulation the same number of times. The only exceptions occur if the algorithm encounters a position where the number of possible choices is limited, and in that case, there is no need for repeating the simulation for every option.

Figures 14 and 15 present THDV, F_1, and F_2 values for each step of the decision tree algorithm. Results were grouped in such a way that each axis contains a representation of routes differing from each other only by presence of the W_{corr} coefficient. Figure 14 shows routes for F_1 minimization and Figure 15 for F_2 minimization. In the case of F_1, every solution shows a clear tendency of reducing power losses in the system for each step taken. This effect is connected with placing additional APFs or with moving APFs to better positions. The requirement of reducing the THDV coefficient below 5% level enforces placing more than one APF, which consequently raises the cost of the solution. It is worth stressing that the THDV limitation requirement also causes the algorithm to increase the cost of the solution by placing another APF even during F_2 minimization. The reduction of costs can only be achieved and is visible in cases when the THDV requirements are met.

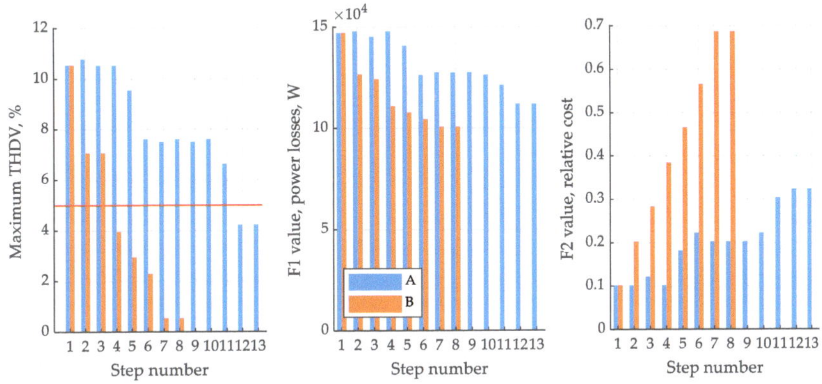

(a)

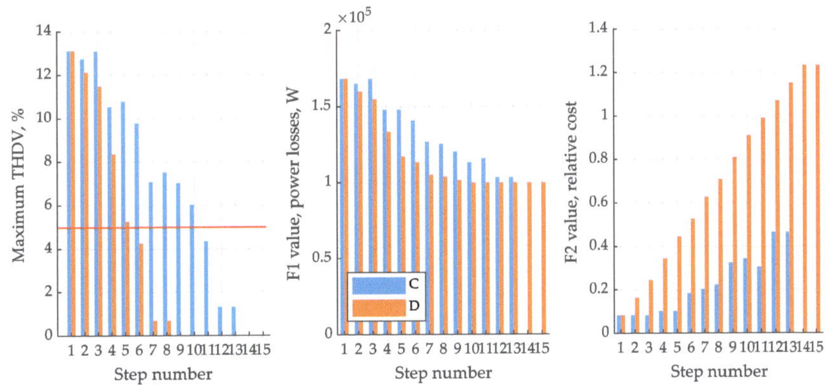

(b)

Figure 14. *Cont.*

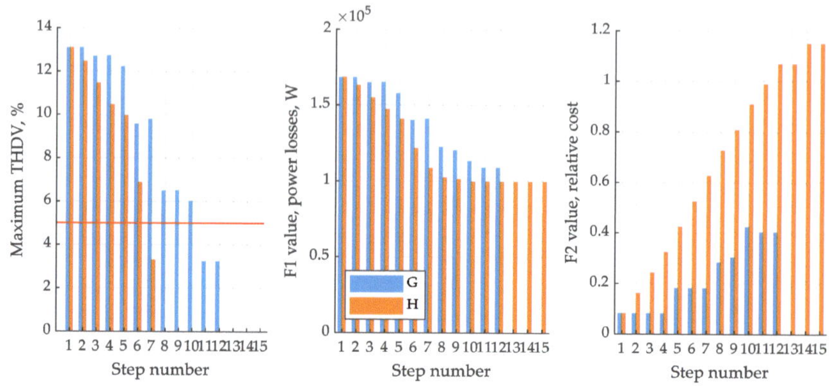

(**c**)

Figure 14. Maximum THDV, F_1, and F_2 values for each decision tree algorithm step for F_1 minimization; (**a**) A and B routes; (**b**) C and D routes; (**c**) G and H routes.

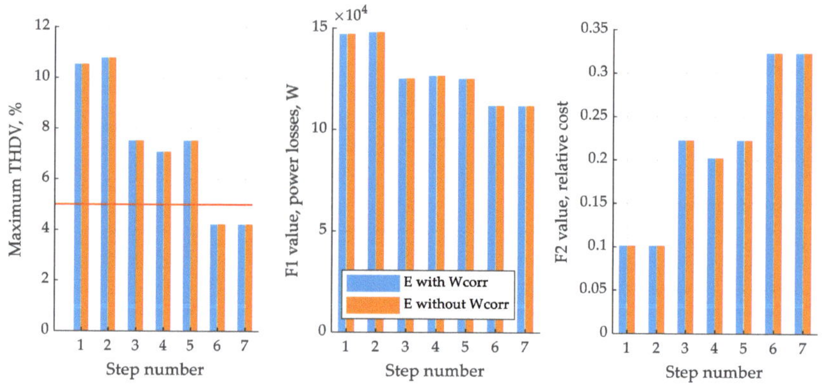

(**a**)

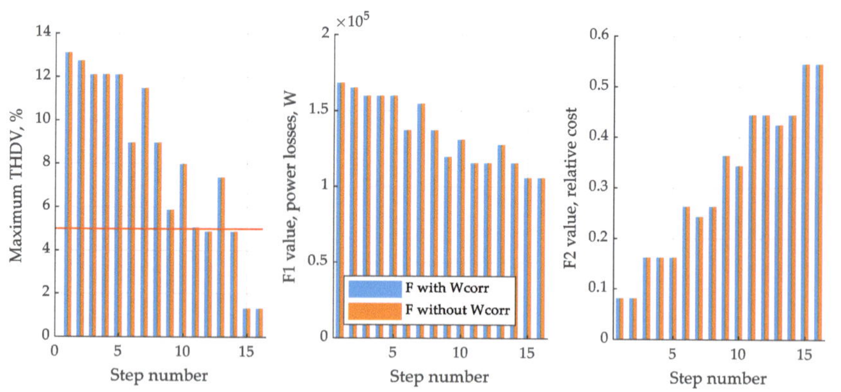

(**b**)

Figure 15. *Cont.*

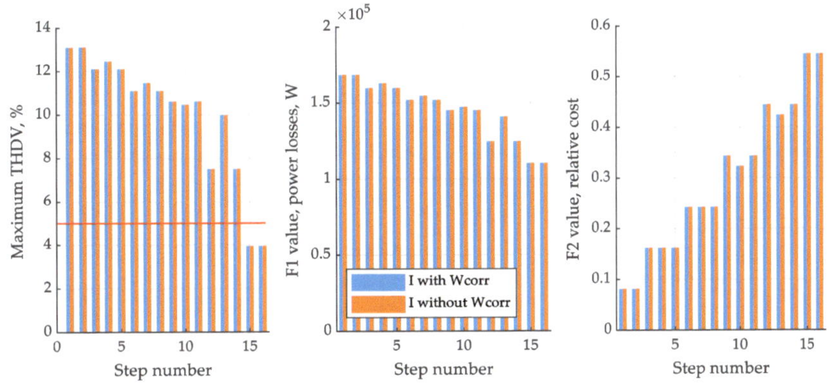

(c)

Figure 15. Maximum THDV, F_1, and F_2 values for each decision tree algorithm step for F_2 minimization; (**a**) E route; (**b**) F route; (**c**) I route.

The decision tree algorithm is much faster compared to the brute force algorithm, especially for complex systems, but computation speed is achieved at the price of reduced quality of optimization. Although the decision tree provides results close to globally optimal (see DT solutions in Figure 12), it does not guarantee finding them exactly. Final optimized values of F_1 and F_2 functions for each route, along with the number of placed APFs required for obtaining them, have been presented in Figure 16. As shown, there are solutions that reduce the goal functions highly, but in order to accomplish that effect, a large number of APFs have to be placed in the system. On the other hand, solutions that minimize the goal functions similarly can be found. In this case, the number of required APFs is significantly smaller.

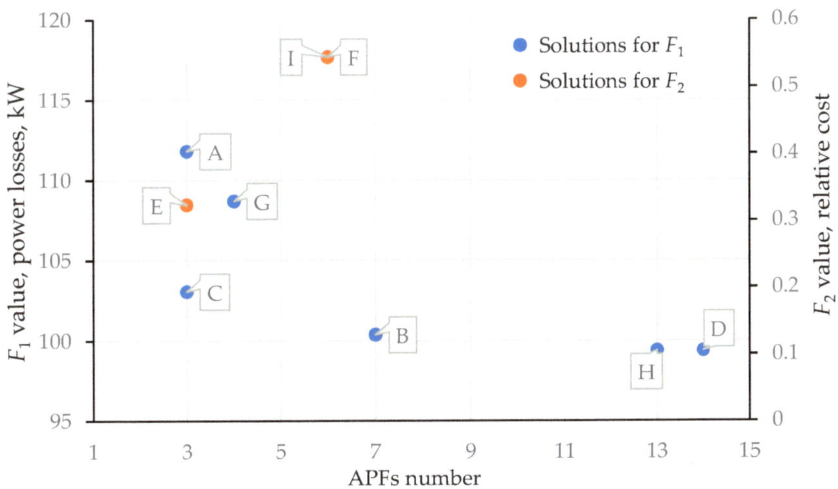

Figure 16. Optimized values of F_1 and F_2 functions for each route A–H (Table 3) along with the number of placed APFs required for obtaining them.

From the set of all examined routes, two examples were chosen in order to provide a more detailed analysis of the decision tree algorithm workflow. The chosen C and E routes are the best compromise between the number of APFs and the goal function values. Each step and the decision made by the algorithm for those routes have been presented in Table 4. Additionally, all steps taken by the decision tree have been also superimposed graphically on the power system schematic in Figure 17. Both ways of presenting the data provide a clear view on the capabilities of the decision tree algorithm implementation. Although C and E routes result in placing three APFs, they vary by APF sorting method and final results. The different sorting also causes E route to turn around at some point in order to examine a previously checked position but with the first APF already placed. Such behavior was not necessary in C route. Those examples represent different strategies that can be applied by the algorithm during the optimization process.

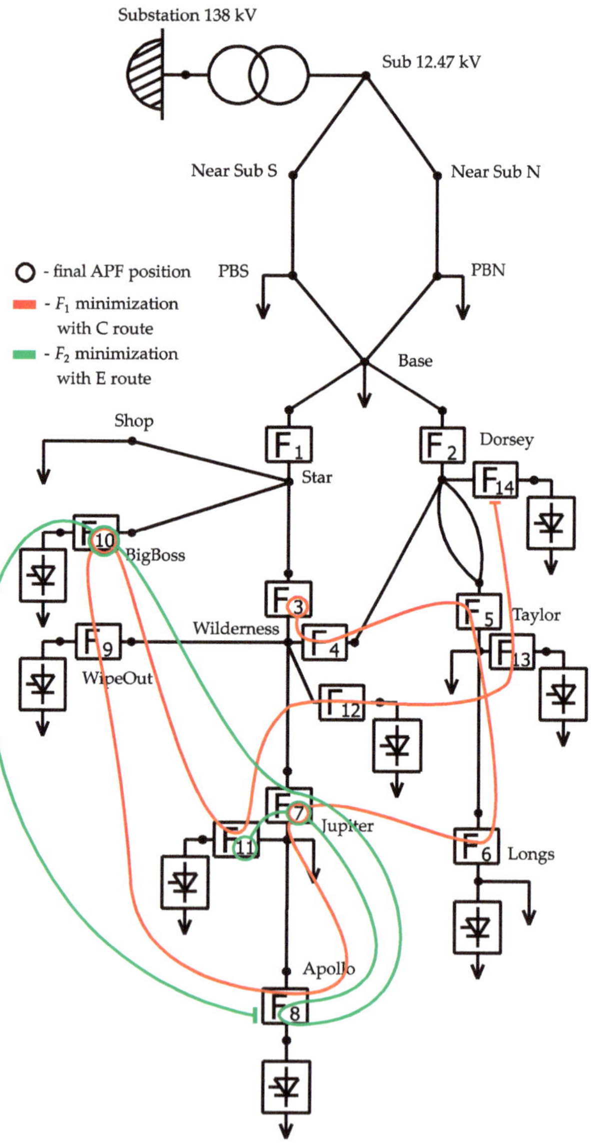

Figure 17. Graphical representation of C and E routes of the decision tree algorithm with final positions of APFs.

Table 4. Details of decisions taken by the decision tree algorithm within C and E routes.

	F_1 Minimization with C Route		F_2 Minimization with E Route	
Step	**Number of APF**	**Decision**	**Number of APF**	**Decision**
1	14	Next	8	Next
2	13	Next	10	Place APF
3	12	Next	7	Previous
4	11	Next	8	Next
5	10	Place APF	7	Place APF
6	9	Next	11	Place APF
7	8	Next	Terminate algorithm	
8	7	Place APF		
9	6	Next		
10	5	Next		
11	4	Next		
12	3	Place APF		
13	Terminate algorithm			

Table 5 summarizes the voltage total harmonic distortion coefficients (THDVs) for all nodes and current total harmonic distortion coefficients (THDIs) for all lines of the analyzed system in the case of optimization using the decision tree algorithm within C and E routes. A clear improvement of THDVs for all nodes as compared to results without APF (Table 1) can be noticed. However, the results for the optimization within the E road are slightly worse because it was associated with cost minimization (F_2).

Table 5. Values of THDVs for all node voltages and THDIs for all line currents for C and E route.

Node	THDV, % C Route	THDV, % E Route	Line	THDI, % C Route	THDI, % E Route
1	0.6	3.3	20-1	0.7	11.0
2	0.6	3.3	1-2	0.7	11.0
3	0.6	3.3	1-3	0.7	11.0
4	0.6	3.4	2-4	0.7	10.9
5	0.6	3.4	3-5	0.7	10.9
6	0.6	3.5	4-6	0.8	12.2
7	0.6	3.7	5-6	0.8	12.2
8	0.6	3.8	6-7	0.0	22.3
9	0.7	3.8	6-9	2.3	19.3
10	0.7	3.8	7-16	0.6	3.4
11	0.9	4.0	7-15	0.0	30.7
12	1.4	4.2	7-8	0.0	16.7
13	0.9	3.8	8-9	12.6	40.2
14	0.8	4.0	8-14	32.8	32.8
15	0.6	3.7	8-13	13.9	23.0
16	0.6	3.7	13-12	29.7	29.7
20	0.1	0.8	9-10	6.9	6.9
			10-11	6.1	6.1

Figures 18–21 show the current (sum of currents of line 1-2 and 1-3) and voltage (node no. 1) waveforms of the transformer secondary side. These waveforms can be compared with the waveforms in Figures 6 and 7.

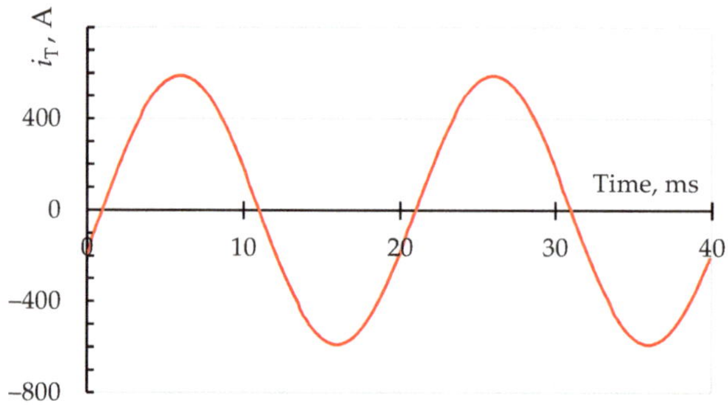

Figure 18. Current waveform of the secondary side of the transformer in the analyzed system in the case of optimization using the decision tree algorithm within C route.

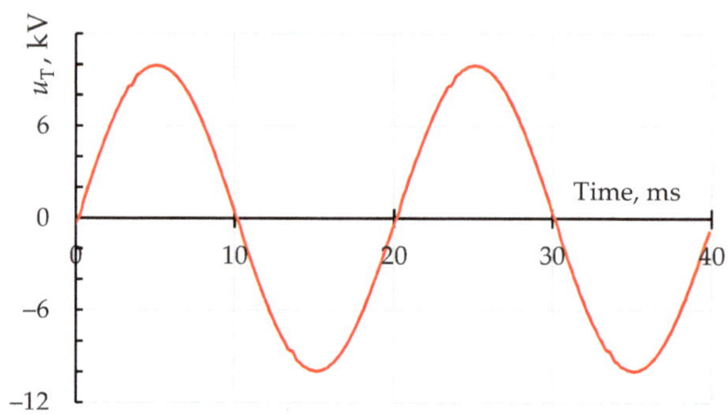

Figure 19. Voltage waveform of the secondary side of the transformer in the analyzed system in the case of optimization using the decision tree algorithm within C route.

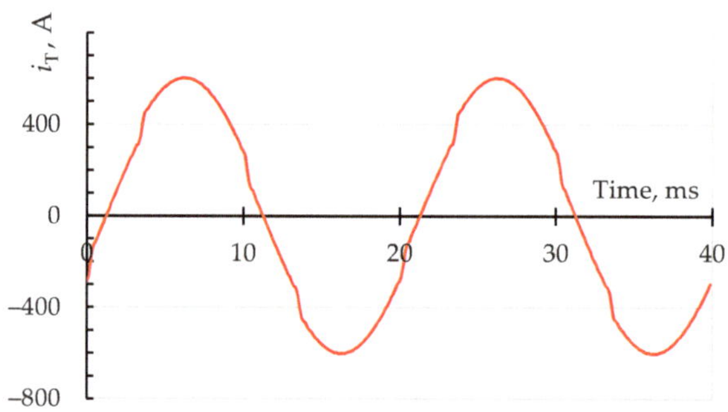

Figure 20. Current waveform of the secondary side of the transformer in the analyzed system in the case of optimization using the decision tree algorithm within E route.

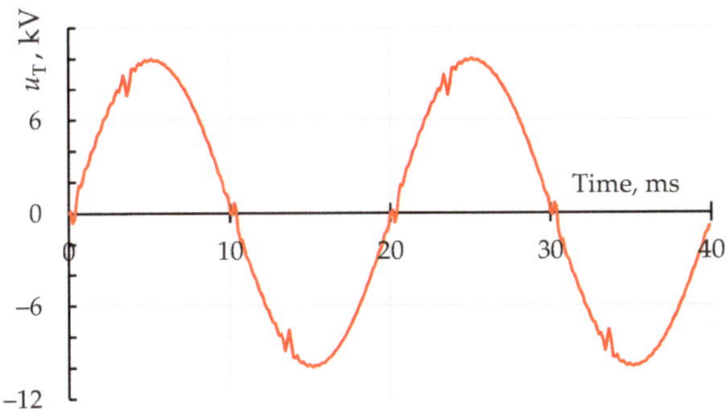

Figure 21. Voltage waveform of the secondary side of the transformer in the analyzed system in the case of optimization using the decision tree algorithm within E route.

4. Discussion

This paper addresses the problem of optimization of APF placement in an exemplary power system. However, the presented solutions are universal and can be applied to any power system in which the necessity of improvement of power quality occurs due to nonlinear loads or reactive power issues. The research concerned various strategies of the best way of designing a dispersed power quality improvement system. The test circuit used in this paper fulfilled its role in terms of sufficient structural complexity and presence of current and voltage distortions.

The validity of used goal functions was verified. The minimization of power losses has a direct positive effect on power quality in modeled systems. The reduction of power losses through elimination of higher harmonics and reactive power compensation leads to a decrease in THDV levels. This fact contributes to an improvement in economic factors related to excessive power consumption, viability, or restrictions imposed by energy providers. The minimization of power losses along with THDV limitation is connected with the need to install several APFs, which should be placed optimally in order to prevent an unnecessary increase in costs. The minimization of costs is in turn related with the pricing of APFs available in the market. Total costs of a solution depend on rated currents of every APF included in the system, which on the other hand depend on the harmonic distortions and amount of reactive power. Those parameters are strictly connected with placement of APFs in the circuit, and for some solutions, total costs can be the factor that prejudge the final decisions about the power quality improvement system. However, in order to find the best possible outcome, a multi-criterion analysis is necessary. As presented in this paper, there are solutions located in the Pareto frontier that represent a compromise between high price and proper power loss reduction.

The optimization of APF placement described in this paper was conducted using two different methods. The first one is a method consisting in a complete search through all possible solutions (brute force algorithm). Such an approach is the least complex one and has exemplary effectiveness—it always leads to finding the best possible outcome. Despite its efficacy, it is also linked to the necessity of simulating the power system for every possible combination of APF placement, which requires large computation power and time, especially for complex circuits. The second used method was implemented with a premise of reduction of time and algorithm steps leading to a final conclusion. It uses a decision tree algorithm with sorting of the algorithm's planned route. This approach is highly sensitive for its starting point and configuration of following steps in general. Due to its sensitivity, this paper shows results of analysis for different route sorting strategies. In case of good conditioning of the planned route, the decision tree finds a solution very quickly. The reason is that in comparison to brute force, there are very few simulations that need to be conducted in order to find the best outcome. However, such an outcome may not be the best globally, but as presented in previous chapters, decision tree results are placed close to the Pareto frontier of all solutions in multi-criterion analysis. This paper presents the application of three different approaches to route design for the decision tree. The first is based on indexes assigned to APFs on the schematic and can be captious if the system is designed in an unstructured way. The second one relates to THDV values of nodes where APFs can be installed and sorts them in descending order. The final sorting method relies on the ascending order of APFs currents. As presented, each approach can lead to good results of optimization. Implementing additional conditions regulating the balance between the decision of adding a new APF, instead of moving the current one into better placement, further improves decision tree results. Due to that fact, there is no risk of obtaining trivial solutions such as placing APFs in every possible location in order to minimize power losses as much as possible.

5. Conclusions

The problem of the optimization of harmonic filter allocation in terms of reducing power losses and costs of APFs has been presented in this paper. The emphasis was put on the decision tree algorithm, which is widely known in different research areas, although up to now it has not been commonly applied for power filter allocation in medium-voltage networks. For comparison purposes, a simple brute force algorithm was also implemented. Results indicate that application of the decision tree provide very fast and well-optimized solutions on condition that a route of the algorithm is appropriate. Due to the fact the that decision tree is highly sensitive to its route design, three different approaches of its arrangement have been examined. They consist in sequential analysis of APFs sorted by node THDV value (descending order), APF current value (ascending order), and random manner, which in this case was APF numbers on the schematic (ascending order). The best results obtained from the decision tree were located closely to the Pareto frontier of both power losses and solution cost reduction.

Author Contributions: Conceptualization, M.K., D.G., and D.B.; methodology, M.K., D.G., and D.B.; software, M.K. and D.B.; validation, M.K., D.G., and D.B.; formal analysis, D.G.; writing—original draft preparation, M.K., D.G., and D.B.; writing—review and editing, M.K., D.G., and D.B.; supervision, D.G. All authors have read and agreed to the published version of the manuscript.

Funding: This research was partly financed by the EU funding for 2014–2020 within the Smart Growth Operational Program.

Institutional Review Board Statement: Not applicable.

Informed Consent Statement: Not applicable.

Data Availability Statement: The data presented in this study are available on request from the corresponding author. The data are not publicly available due to at the date of publication, no data repository has been established yet.

Conflicts of Interest: The authors declare no conflict of interest.

References

1. Ziari, I.; Jalilian, A. Optimal placement of an active filter in a power system. In Proceedings of the 2007 42nd International Universities Power Engineering Conference, Brighton, UK, 4–6 September 2007; pp. 1150–1154.
2. Guo, M.; Jin, Q.; Yao, Z. Implementation of Adaptive Step-Size Algorithm in Distribution Network for Optimal Location and Sizing of SAPFs. In Proceedings of the 2020 IEEE/IAS Industrial and Commercial Power System Asia (I&CPS Asia), Weihai, China, 13–16 July 2020; pp. 1562–1566.
3. Alhaddad, F.M.; El-Hawary, M. Optimal Filter Placement and Sizing Using Ant Colony Optimization in Electrical Distribution System. In Proceedings of the 2014 IEEE Electrical Power and Energy Conference, Calgary, AB, Canada, 12–14 November 2014; pp. 128–133.
4. Shivaie, M.; Salemnia, A.; Ameli, M.T. A multi-objective approach to optimal placement and sizing of multiple active power filters using a music-inspired algorithm. *Appl. Soft Comput.* **2014**, *22*, 189–204. [CrossRef]
5. Rosyadi, A.; Penangsang, O.; Soeprijanto, A. Optimal filter placement and sizing in radial distribution system using whale optimization algorithm. In Proceedings of the 2017 International Seminar on Intelligent Technology and Its Applications (ISITIA), Surabaya, Indonesia, 28–29 August 2017; pp. 87–92. [CrossRef]

6. Lakum, A.; Mahajan, V. Optimal placement and sizing of multiple active power filters for radial distribution system using grey wolf optimizer. In Proceedings of the 2017 7th International Conference on Power Systems (ICPS), Pune, India, 21–23 December 2017; pp. 562–567.
7. Sindhu, M.R.; Jisma, M.; Maya, P.; Krishnapriya, P.; Vivek, M.M. Optimal Placement and Sizing of Harmonic and Reactive Compensators in Interconnected Systems. In Proceedings of the 15th IEEE India Council International Conference, Coimbatore, India, 16–18 December 2018; pp. 1–6.
8. Belchior, F.N.; De Lima, L.R.; Ribeiro, P.F.; Castro, J.F.C. A novel approach towards passive filter placement. In Proceedings of the 2015 IEEE Power & Energy Society General Meeting, Denver, CO, USA, 26–30 July 2015; pp. 1–5.
9. Maciążek, M.; Grabowski, D.; Pasko, M. Genetic and combinatorial algorithms for optimal sizing and placement of active power filters. *Int. J. Appl. Math. Comput. Sci.* **2015**, *25*, 269–279. [CrossRef]
10. El-Arwash, H.M.; Azmy, A.M.; Rashad, E.M. A GA-based initialization of PSO for optimal APFS allocation in water desali-nation plant. In Proceedings of the Nineteenth International Middle East Power Systems Conference (MEPCON), Cairo, Egypt, 19 December 2017; pp. 1378–1384.
11. Milovanović, M.; Radosavljević, J.; Klimenta, D.; Perović, B. GA-based approach for optimal placement and sizing of passive power filters to reduce harmonics in distorted radial distribution systems. *Electr. Eng.* **2019**, *101*, 787–803. [CrossRef]
12. Moghbel, M.; Masoum, M.A.S.; Fereidouni, A.; Deilami, S. Optimal Sizing, Siting and Operation of Custom Power Devices with STATCOM and APLC Functions for Real-Time Reactive Power and Network Voltage Quality Control of Smart Grid. *IEEE Trans. Smart Grid* **2018**, *9*, 5564–5575. [CrossRef]
13. Moghbel, M.; Deilami, S.; Masoum, M.A. Optimal Siting and Sizing of Multiple Active Power Line Conditioners to Minimize Network THD Considering Harmonic Couplings. In Proceedings of the 2019 9th International Conference on Power and Energy Systems (ICPES), Perth, Australia, 10–12 December 2019; pp. 1–6.
14. Tian, S.; Jia, Q.; Xue, S.; Yu, H.; Qu, Z.; Gu, T. Collaborative optimization allocation of VDAPFs and SVGs for simultaneous mitigation of voltage harmonic and deviation in distribution networks. *Int. J. Electr. Power Energy Syst.* **2020**, *120*, 106034. [CrossRef]
15. Subramani, C.; Jimoh, A.-G.A.; Dash, S.S.; Harishkiran, S. PSO Application to Optimal Placement of UPFC for Loss Minimization in Power System. In *Advances in Intelligent Systems and Computing, Proceedings of the Sproceedings of the 2nd International Conference on IntelPubligent Computshing and Applications, Amsterdam, The Netherlands, 22 February 2017*; Springer: Singapore, Singapore, 2017; pp. 223–230.
16. Chakeri, V.; Hagh, M.T. Optimal allocation of the distributed active filters based on total loss reduction. *Int. J. Smart Electr. Eng.* **2017**, *6*, 171–175.
17. Li, D.; Wu, Z.; Zhao, B.; Zhang, X.; Zhang, L. An Adaptive Active Power Optimal Allocation Strategy for Power Loss Minimization in Islanded Microgrids. In Proceedings of the IECON 2019-45th Annual Conference of the IEEE Industrial Electronics Society, Lisbon, Portugal, 14–17 October 2019; Volume 1, pp. 292–297.
18. Lopez, J.A.; Lu, C.-N. A comparison of placement methods for collecting PMU data used in angular stability detection. In Proceedings of the 2017 19th International Conference on Intelligent System Application to Power Systems (ISAP), San Antonio, TX, USA, 17–20 September 2017; pp. 1–6.
19. Alhalaseh, R.; Tokel, H.A.; Chakraborty, S.; Alirezaei, G.; Mathar, R. Feature-Selection based PMU Placement for Detection of Faults in Power Grids. In Proceedings of the 2018 28th International Telecommunication Networks and Applications Conference (ITNAC), Sydney, NSW, Australia, 21–23 November 2018; pp. 1–6.
20. Samudrala, A.N.; Amini, M.H.; Kar, S.; Blum, R.S. Optimal Sensor Placement for Topology Identification in Smart Power Grids. In Proceedings of the 2019 53rd Annual Conference on Information Sciences and Systems (CISS), Baltimore, MD, USA, 20–22 March 2019; pp. 1–6.
21. Sevlian, R.A.; Zhao, Y.; Rajagopal, R.; Goldsmith, A.; Poor, H.V. Outage Detection Using Load and Line Flow Measurements in Power Distribution Systems. *IEEE Trans. Power Syst.* **2018**, *33*, 2053–2069. [CrossRef]
22. Alhalaseh, R.; Tokel, H.A.; Chakraborty, S.; Alirezaei, G.; Mathar, R. PMU Placement with Power Grid Partitioning for Line Outage Detection. In Proceedings of the 2019 4th International Conference on Smart and Sustainable Technologies (SpliTech), Split, Croatia, 18–21 June 2019; pp. 1–6.
23. Samudrala, A.N.; Amini, M.H.; Kar, S.; Blum, R.S. Sensor Placement for Outage Identifiability in Power Distribution Net-works. *IEEE Trans. Smart Grid* **2020**, *11*, 1996–2013. [CrossRef]
24. Sharma, G.; Mahela, O.P.; Kumar, M.; Kumar, N. Detection and Classification of Transmission Line Faults Using Stockwell Transform and Rule Based Decision Tree. In Proceedings of the 2018 International Conference on Smart Electric Drives and Power System (ICSEDPS), Nagpur, India, 12–13 June 2018; pp. 1–5.
25. Godse, R.; Bhat, S. Mathematical Morphology-Based Feature-Extraction Technique for Detection and Classification of Faults on Power Transmission Line. *IEEE Access* **2020**, *8*, 38459–38471. [CrossRef]
26. Gong, C.-S.A.; Su, C.-H.S.; Tseng, K.-H. Implementation of Machine Learning for Fault Classification on Vehicle Power Transmission System. *IEEE Sens. J.* **2020**, *20*, 15163–15176. [CrossRef]
27. Ribeiro, E.G.; Dias, G.L.; Barbosa, B.H.G.; Ferreira, D.D. Real-time system for automatic classification of power quality disturbances. In Proceedings of the 17th International Conference on Harmonics and Quality of Power (ICHQP), Belo Horizonte, Brazil, 16–19 October 2016; pp. 908–913.

28. Shams, N.; Wall, P.; Terzija, V. Active Power Imbalance Detection, Size and Location Estimation Using Limited PMU Measurements. *IEEE Trans. Power Syst.* **2019**, *34*, 1362–1372. [CrossRef]
29. Meng, X.; Zhang, P.; Zhang, D. Decision Tree for Online Voltage Stability Margin Assessment Using C4.5 and Relief-F Algorithms. *Energies* **2020**, *13*, 3824. [CrossRef]
30. Jindal, A.; Dua, A.; Kaur, K.; Singh, M.; Kumar, N.; Mishra, S. Decision Tree and SVM-Based Data Analytics for Theft Detection in Smart Grid. *IEEE Trans. Ind. Inform.* **2016**, *12*, 1005–1016. [CrossRef]
31. Guerrero, J.I.; Monedero, I.; Biscarri, F.; Biscarri, J.; Millan, R.; Leon, C. Non-Technical Losses Reduction by Improving the Inspections Accuracy in a Power Utility. *IEEE Trans. Power Syst.* **2017**, *33*, 1209–1218. [CrossRef]
32. Swetapadma, A.; Yadav, A. A Novel Decision Tree Regression-Based Fault Distance Estimation Scheme for Transmission Lines. *IEEE Trans. Power Deliv.* **2017**, *32*, 234–245. [CrossRef]
33. Pessoa, A.L.D.S.; Oleskovicz, M.; Martins, P.E.T. A multi-stage methodology for fault location in radial distribution systems. In Proceedings of the 2018 18th International Conference on Harmonics and Quality of Power (ICHQP), Ljubljana, Slovenia, 13–16 May 2018; pp. 1–6.
34. Araújo, M.A.; Flauzino, R.A.; Moraes, L.A.; Borges, F.A.S.; Spatti, D.H. Decision Trees Applied to Fault Locations in Distri-bution Systems with Smart Meters. In Proceedings of the IEEE International Conference on Environment and Electrical Engi-neering and 2019 IEEE Industrial and Commercial Power Systems Europe (EEEIC/I&CPS Europe), Genova, Italy, 11–14 June 2019; pp. 1–6.
35. Xiong, P.; Singh, C. Optimal Planning of Storage in Power Systems Integrated with Wind Power Generation. *IEEE Trans. Sustain. Energy* **2016**, *7*, 232–240. [CrossRef]
36. Lee, C.-T.; Horng, S.-C. Abnormality Detection of Cast-Resin Transformers Using the Fuzzy Logic Clustering Decision Tree. *Energies* **2020**, *13*, 2546. [CrossRef]
37. Sai, T.; Kumar, P.S.; Reddy, K. Towards intelligent decision making in power plant operation. In Proceedings of the 2014 Annual IEEE India Conference (INDICON), Puni, India, 11–13 December 2014; pp. 1–6.
38. Mosavi, A.; Salimi, M.; Ardabili, S.F.; Rabczuk, T.; Shamshirband, S.; Varkonyi-Koczy, A.R. State of the Art of Machine Learning Models in Energy Systems, a Systematic Review. *Energies* **2019**, *12*, 1301. [CrossRef]
39. Leonidaki, E.; Georgiadis, D.; Hatziargyriou, N. Decision Trees for Determination of Optimal Location and Rate of Series Compensation to Increase Power System Loading Margin. *IEEE Trans. Power Syst.* **2006**, *21*, 1303–1310. [CrossRef]
40. Couronné, R.; Probst, P.; Boulesteix, A.-L. Random forest versus logistic regression: A large-scale benchmark experiment. *BMC Bioinform.* **2018**, *19*, 1–14. [CrossRef] [PubMed]
41. Ke, G.; Meng, Q.; Finley, T.; Wang, T.; Chen, W.; Ma, W.; Ye, Q.; Liu, T.Y. Lightgbm: A highly efficient gradient boosting decision tree. *Adv. Neural Inf. Process. Syst.* **2017**, *30*, 3146–3154.
42. Tian, W.; Lei, C.; Tian, M. Dynamic Prediction of Building HVAC Energy Consumption by Ensemble Learning Approach. In Proceedings of the 2018 International Conference on Computational Science and Computational Intelligence (CSCI), Las Vegas, Nevada, USA, 13–15 December 2018; pp. 254–257.
43. Grady, W.M. Understanding Power System Harmonics. Available online: http://web.ecs.baylor.edu/faculty/grady/ (accessed on 18 January 2021).
44. Buła, D.; Lewandowski, M. Steady state simulation of a distributed power supplying system using a simple hybrid time-frequency model. *Appl. Math. Comput.* **2018**, *319*, 195–202. [CrossRef]
45. Buła, D.; Grabowski, D.; Lewandowski, M.; Maciążek, M.; Piwowar, A. Software Solution for Modeling, Sizing, and Allocation of Active Power Filters in Distribution Networks. *Energies* **2020**, *14*, 133. [CrossRef]
46. Buła, D.; Lewandowski, M. Comparison of frequency domain and time domain model of a distributed power supplying system with active power filters (APFs). *Appl. Math. Comput.* **2015**, *267*, 771–779. [CrossRef]

energies

MDPI

Article

Active Power Filter Pre-Selection Tool to Enhance the Power Quality in Oil and Gas Platforms†

Lais Abrantes Vitoi [1,*,‡], Danilo Brandao [1,‡] and Elisabetta Tedeschi [2,3,‡]

1 Department of Electrical Engineering, Federal University of Minas Gerais (UFMG), Belo Horizonte 31270-901, Brazil; dibrandao@ufmg.br
2 Department of Electric Power Engineering, Norwegian University of Science and Technology (NTNU), 7030 Trondheim, Norway; elisabetta.tedeschi@ntnu.no
3 Department of Industrial Engineering, University of Trento, 38123 Trento, Italy
* Correspondence: laisvitoi@gmail.com

† This paper is an extended version of our paper "Power quality enhancement by SiC Active Power Filters in Oil and Gas Platforms" published in 2019 IEEE Energy Conversion Congress and Exposition (ECCE), Baltimore, MD, USA, 29 September–3 October 2019; pp. 4299–4304, doi:10.1109/ECCE.2019.8913034.
‡ These authors contributed equally to this work.

Abstract: This paper proposes a preliminary design tool for active power filters' (APFs) solutions to be applied in offshore oil and gas platforms, where power quality indices are typically low, and reactive power compensation and current harmonic mitigation are often desired. The proposed approach considers that APF selection and rating is a trade-off between performance and size, and that both component and system aspects need to be optimized to achieve a well-tailored solution. As size and weight are critical constraints in offshore applications, possible benefits of using Silicon Carbide (SiC) switches for the APF implementation are investigated. Moreover, different compensation strategies are compared, varying the connection point of the APF between two different voltage levels and assigning the APFs different compensation goals. Improvements in power quality indices, as well as APFs rating, number of components, power losses, and filter size, have been considered for both SiC and Silicon-based solutions to identify the best trade-offs suitable for the considered, energy intensive industrial application.

Keywords: active power filter; O&G platform; power quality; wide band-gap semiconductors

Citation: Vitoi, L.A.; Brandao, D.; Tedeschi, E. Active Power Filter Pre-Selection Tool to Enhance the Power Quality in Oil and Gas Platforms. *Energies* **2021**, *14*, 1024. https://doi.org/10.3390/en14041024

Academic Editor: Gabriel Nicolae Popa

Received: 8 January 2021
Accepted: 1 February 2021
Published: 16 February 2021

1. Introduction

In spite of the development of the renewable energy sector, gas and oil are determined to remain the two head energy assets until 2050 and beyond. In the past few decades, offshore oil and gas exploration and drilling have increased significantly. Currently, offshore oil and gas (O&G) exploration accounts for 27% and 30%, respectively [1,2]. Still, fossil fuel combustion in power plants, oil refineries, and large industrial facilities [3] (including O&G platforms) is the main source of the anthropocentric CO_2 emissions, and the environmental problem is concurrent with the technical challenge of O&G platforms powering, in indicating the need for more electric and more efficient platforms. In most cases, the high distance between the platform and the mainland, and the high-power needs of local processing equipment (5–200 MW) prevent the cable-connection to the land-based power system. In this case, electricity is generated locally by gas turbines or diesel generators, and the grid operates as an isolated power system, which is characterized by a weak grid.

Figure 1 represents a typical power system of an O&G platform. The main elements are gas turbines coupled with synchronous generators, power transformers, power converters, and loads. Although DC-based power distribution for O&G drilling applications has recently been under investigation [4], AC systems represent the only industrially applied solution. Nonetheless, the utilization of different loads with high power demand (e.g., pumps and compressors), and the developing use of electric drives, combined with generator impedances (12–25%) considerably higher than those of grid-connected power systems contribute to deteriorate the local power quality (PQ). A significant disadvantage of these AC power distribution systems is the inherent presence of reactive power (with measured power factors, PF, that in extreme cases can be as low as 0.36 [5]), which leads to increased current and losses. In addition, the connection of power electronic converters, typically for AC and DC drives, results in non-linear loads, and therefore harmonic pollution with reported Total Harmonic Voltage Distortion (THDv) and Total Harmonic Current Distortion (THDi) as high as 12% and 27%, respectively [5].

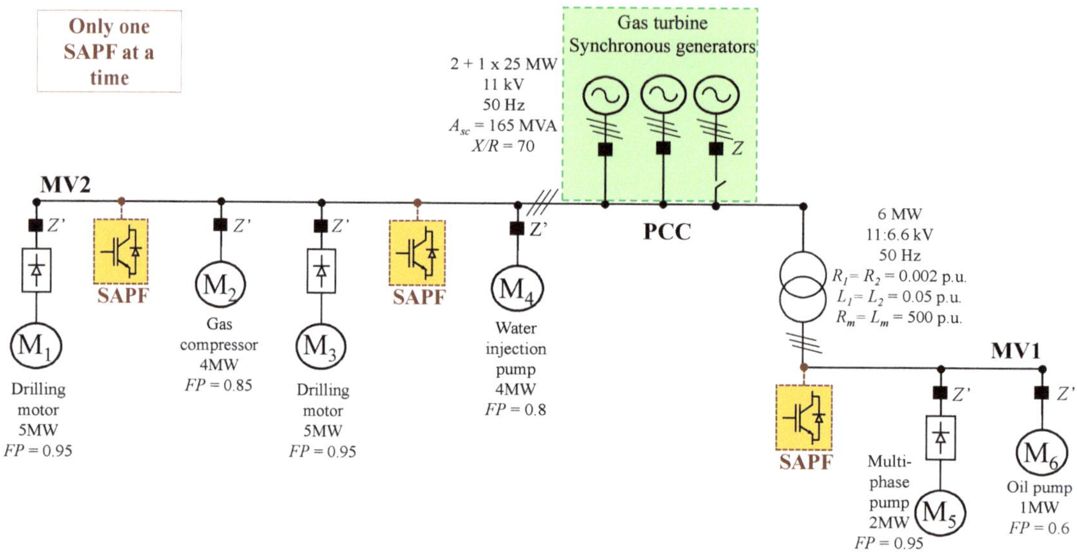

Figure 1. Diagram of a typical power grid of an offshore oil and gas platform.

As space and weight are critical factors in offshore O&G platforms, the deployment of any additional equipment occupying deck-space is preferably avoided, or carefully weighed versus its added value, as proved by the recent trend to place more processing equipment sub-sea [4]. On the other hand, the cost of a single power-quality incident offshore can be up to 750,000 EUR per day [5]. In order to avoid such incidents and guarantee that the system and the equipment work correctly, it is necessary to comply with relevant standards.

That situation is not different from that of other offshore systems, such as maritime microgrids, for which specific standards are also available [6–8] and power quality issues have been more widely investigated [9].

Several methods have been proposed in the literature and also applied in practice to compensate reactive power and harmonic components generated by industrial [10,11] and maritime loads [9,12], but just a few contributions specifically targeted O&G drilling rigs [13–15]. In [13], an analysis for passive filters selection in O&G industry is proposed. The authors of [14] present the design and tuning of passive filters for offshore applications using genetic algorithm technique. The authors of [15] compare passive and active filters for oil rig power systems, and conclude that active filters are more suitable where space constrain is an issue. Additionally, due to the typical load cycle of the oil drilling rig, the source impedance seen by the filter varies and, in the passive filter, this will shift the resonant peak.

On the contrary, Shunt Active Power Filters (SAPFs) based on three-phase Voltage Source Converters (VSCs) are proper for this application, as they use the ability of the converter to generate reactive power without using bulky energy storage components, in addition to flexibly compensating reactive power, they can also compensate for multiple harmonics. Hence, they are faster, lighter, smaller, and with better performances at reduced voltages compared to other solutions. In [16], a chain circuit active power filter for high voltage and power applications is proposed, a prototype of 10 kV/1 MVA is built. However, although a preliminary analysis has been presented in [17], a holistic and detailed study bench-marking APF design and PQ performance for this specific application is not available in the literature.

Over the past decades, there has been a breakthrough in devices based on wide-bandgap materials, such as silicon-carbide (SiC). With the increasing demand for high efficiency, voltages, and switching frequencies, the traditional switches made of silicon (Si) may not be able to satisfy all the requirements. The emerging of silicon carbide (SiC) devices has brought new design possibilities for high-voltage high-power converters. In contrast to the Si technology, the SiC exhibits superior material properties. The higher thermal conductivity, dielectric breakdown field strength, and wide band-gap allow an increase in the operational switching frequency and voltage without increasing the losses [18]. The higher the switching frequency, the smaller are the passive filtering components, and then the cooling requirements can be reduced. Therefore, the overall system (converter and filter) volume, area and weight are decreased [19–23]. In [24], a dynamic voltage restorer based on SiC Mosfet is studied to mitigate voltage sag in the O&G industry. References [25,26] present the design, performance analysis, and experimental results of high-speed motor drives that are usually required in the O&G industries that use SiC Mosfet technology.

Based on these trends and needs of the O&G sector, the core contribution of the paper is to provide a SAPF pre-selection tool to orient the O&G platform designer. The tool encompasses both component and system-level analyses tailored to such industrial applications. More precisely, the paper considers: (1) the impact of using wide-band-gap semiconductors and reduced output filter in the SAPF, for equipment size and loss reduction; (2) the PQ performance achievable by connecting the SAPF to different buses/voltage levels, as well as assigning it different compensation targets; and (3) the suitability of two compensation strategies, i.e., sinusoidal current synthesis and resistive load synthesis, and respective control implementations, which are analyzed in detail both by theoretical investigations and dynamic simulations.

With respect to [17], this paper details the description of the SAPF control scheme, with a thoroughly analytical analysis of the advantages of each compensation strategy, supported by additional results. Finally, the paper concludes by presenting an integrated SAPF pre-selection tool that allows to compare the different alternatives to identify the most suitable one. It is worth noting that the proposed tool can have a wider validity than for the proposed O&G application, if decision criteria are properly adapted and weighted.

The paper is organized as follows: Section 2 describes the main industrial processes that take place on an O&G platform and the required electrical power components; Section 3 shows the theoretical and mathematical analysis of SAPFs; in Section 4 the case studies and results are presented and discussed. Finally, Section 5 concludes.

2. Processing of O&G on Offshore Platforms and Related Power Sources

On an O&G platform, the transformation process to convert the fluid extracted from the well into marketable products and clean the waste products, such as produced water, requires several stages and large equipment with high power consumption [27,28]. Each O&G platform has generally multiple wells, divided into injection and production wells. While the former is used for the production of O&G, the latter are drilled to inject gas or water into the reservoir to increase its pressure and push the fluid towards the production well in a process called "enhanced oil recovery". This process needs large pumps or compressors with high power demand ranging from a few MW up to more than 25 MW [29].

Modern installations include also electrical submerged pumps, with power consumption of a few MW, into the well.

In addition, as the well-stream often consists of crude oil, gas, water, condensates, and contaminants, a separator is used to divide the different components. Owing to the low pressure of the gas flowing from the separators, it must be recompressed before transport. Several types of compressors can be used for this purpose, with the largest centrifugal compressors having a power in the 80 MW range [2]. The production cycle ends with the metering, storage, and export process, in addition to the treatment of chemicals and wastewater.

With a few exceptions, power generation on O&G platforms is usually supplied by local gas turbines (GTs) coupled to synchronous generators (SGs), as shown in Figure 1. Their capacity range is normally between a few MW and 40 MW per turbine [30]. The number of turbines is usually limited to three or four, with one used as a back-up for reliability purposes. SGs and GTs are connected to the highest voltage bus on the platform (i.e., Europe—11 kV, 50 Hz, USA—11 kV, 60 Hz, or South America—13.8 kV, 60 Hz).

For the platform power distribution system, two main AC voltage levels, i.e., 6.6 kV and 11 kV, hereafter specified as medium voltage 1 (MV1) and medium voltage 2 (MV2), respectively, are often used. Large drilling equipment, pumps, and compressors are the main loads powered by electric motors. They usually reflect 75–80% of the overall electric load on the platform and, as their individual power consumption is in the multi-MW range (up to several dozens MW), they are typically connected to the MV2 bus. A low-voltage bus (LV, i.e., 400 V) is also present to allow the interconnection of several small loads (e.g., lightning and living-quarter loads).

Due to the differences in load types and voltage levels, transformers and power electronic converters are needed. Power converters, particularly large 6- or 12-pulse rectifiers [31] coupled to fully controlled inverters, are increasingly connected to electric motors for drilling, pumping, etc., to allow variable speed operation for improved efficiency. This, however, leads to harmonic generation and power quality deterioration [31].

The considered electric power system of an O&G platform is shown in Figure 1, and further described in Section 4. Such an electric grid supplies different load types based on induction motors. The loads M1 to M4 are connected to the MV2 bus (11 kV), while loads M5 and M6 are connected to the MV1 bus (6 kV). The points where the SAPFs are connected are called Points of Connection (PoCs), and the Point of Common Coupling (PCC) is at the output of generators, as shown in Figure 1.

3. Theoretical and Mathematical Analysis of Shunt Active Power Filter

Herein, the SAPF has been adopted as a power quality conditioner for reactive power and harmonic pollution mitigation in an offshore O&G platform. The selected SAPF topology is the three-phase two-level VSC, as shown in Figure 2. It consists of three arms, each one comprising two half-bridge modules. Therefore, six sets of power switches are required. Each set of power switches can be implemented with multiple power semiconductor devices connected either in series or in parallel, depending on the voltage and current rating. There are several methods to optimize the switching device number [32], but this analysis is out of the scope of this work. To determine the number of power semiconductors, the method presented in Section 3.1 is used. In addition to the switching stage, the SAPF requires passive components: DC side capacitor and AC output filter. The following subsections detail the SAPF.

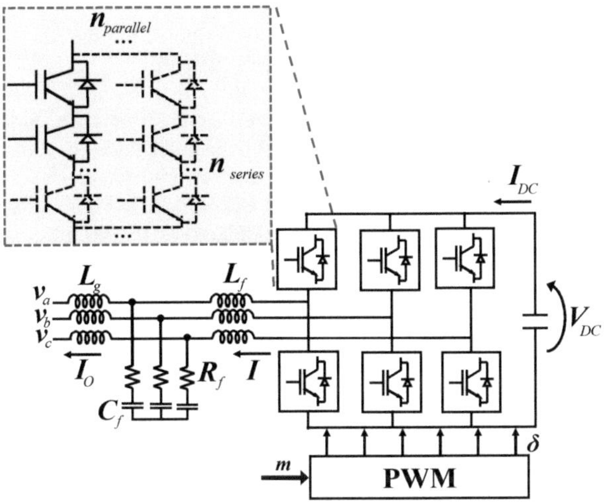

Figure 2. Power electronics structure of the shunt active power filter—SAPF [17].

3.1. Power Semiconductor Devices

An combination of series and parallel switches is necessary to comply with the requirements of the circuit current and voltage [33]. The number of devices in parallel ($n_{parallel}$) and series (n_{series}) for two-level converters is calculated by (1) and (2), respectively. Where I_{device} is the device datasheet direct current, V_{device} is the datasheet blocking voltage, and SF is a voltage safety factor. The SF must be chosen according to typical values used and values reported in the datasheet. A thermal runaway can happen with too high a repetitive voltage peak, even if this value is below the avalanche break-down limit. Therefore, it is necessary to consider a SF for the voltage. More details about this procedure can be found on [34].

$$n_{parallel} = \frac{I_{DC}}{I_{device}} \tag{1}$$

$$n_{serie} = \frac{V_{DC}}{V_{device}SF} \tag{2}$$

The converter losses calculation follows the methodology used in [22,35,36] that includes both conduction and switching losses. The MOSFET and IGBT average conduction losses are given in (3) and (4), respectively. V_{ce} is the collector-emitter voltage, R_0 is the slope resistance, and I_{av} and I_{rms} are the average and root mean square currents, respectively.

$$P_{cond-mosfet} = R_0 I_{rms}{}^2 \tag{3}$$

$$P_{cond-igbt} = V_{ce} I_{av} \tag{4}$$

The switching losses for a two-level converter are calculated by means of (5) where n_{device} is the total number of devices, f_{sw} is the switching frequency, and E_{on} and E_{off} are the switching loss energy obtained from the datasheet and testing materials. Such calculation is generic for both devices. Since the datasheet switching loss energy is measured for a specific V_{ref}, it is necessary to correct the losses for the actual voltage across each device (V_{cc}).

$$P_{sw,2L} = n_{device} f_{sw} (E_{on} + E_{off}) \frac{V_{cc}}{V_{ref}} \tag{5}$$

3.2. Passive Components

The proper design of the output filter is of great importance for performance and size of SAPF [19,20]. Literature reports different methods for designing passive components. Herein the LCL filter configuration shown in Figure 2 has been considered. LCL is inductor-capacitor-inductor output filter at the converter's output to reduce high frequency ripples on the current waveform caused by the pulse width modulation (PWM) technique. The LCL filter attenuates more the high order harmonics with lower cost and reduced overall weight and size compared to the L and LC filter [37]. The filter design follows the methodology of [37]; C_f, L_f, L_g, and R_f are calculated by (6)–(9), respectively.

$$C_f = 0.05C_b \tag{6}$$

$$L_f = \frac{V_{DC}}{6f_{sw}\Delta I_{Lmax}} \tag{7}$$

$$L_g = \frac{\sqrt{\frac{1}{k_a{}^2}} + 1}{C_f w_{sw}{}^2} \tag{8}$$

$$R_f = \frac{1}{3w_{res}C_f} \tag{9}$$

where ΔI_{Lmax} is the maximum current ripple at the inverter output, k_a is the desired harmonic attenuation, C_b is the base capacitance, and w_{res} is the resonant frequency.

3.3. Control Scheme and Compensation Strategies

The block diagram of SAPF control scheme is shown in Figure 3. The control scheme consists of a fast inner loop to regulate the current, and a slow outer loop to control the DC-link voltage. V_{DC}^* is the DC-link voltage reference, i^* is the AC current reference, and s^* is the synthesis signal that can be the normalized voltage at the Point of Connection (PoC) or can come from a phase-locked loop (PLL) algorithm, depending on the compensation strategy adopted [38]. C_v and C_i are the controllers of voltage and current loops, respectively. The switches are controlled by PWM (Pulse Width Modulation) control where m is the reference signal, and δ is the control signal to the converter.

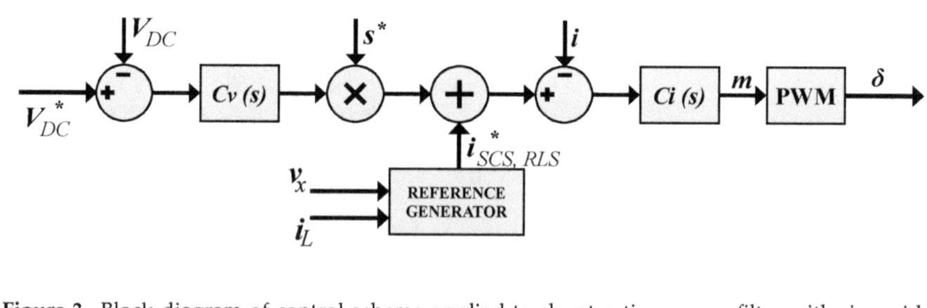

Figure 3. Block diagram of control scheme applied to shunt active power filter with sinusoidal current synthesis (SCS) or resistive load synthesis (RLS).

Two compensation strategies can be implemented: (i) resistive load synthesis (RLS) and (ii) sinusoidal current synthesis (SCS) [39]. The reference generator block is responsible for synthesizing both compensation strategies and creating the current reference signal (i^*). In short, v_x is equal to the measured SAPF PoC voltage in each phase (v_m) for RLS, resulting in i^*_{RLS}, whereas the SCS needs $v_x = v_{m1}$ resulting in i^*_{SCS}, where the subscript *1* stands for the fundamental value of the variable.

Figure 4 represents the equivalent model of the electric grid shown in Figure 1, in which the SAPF compensates three MV2 loads (M1, M2, and M3). Two assumptions can be made for the analysis: (i) the operating generators are represented as a single equivalent voltage source, and (ii) the circuit is represented in per-unit (p.u.); therefore, the transformer does not appear. Figure 5 shows the simplified model of Figure 4, where the current source represents the non-linear load, the constant impedances the linear loads, and Z_{comp} the loads to be compensated (M1, M2, and M3) associated with SAPF. Z_{comp} assumes different features depending on the control strategy (RLS or SCS) applied. The capacitor branch represents the typical shunt capacitor at the output terminals of SAPF.

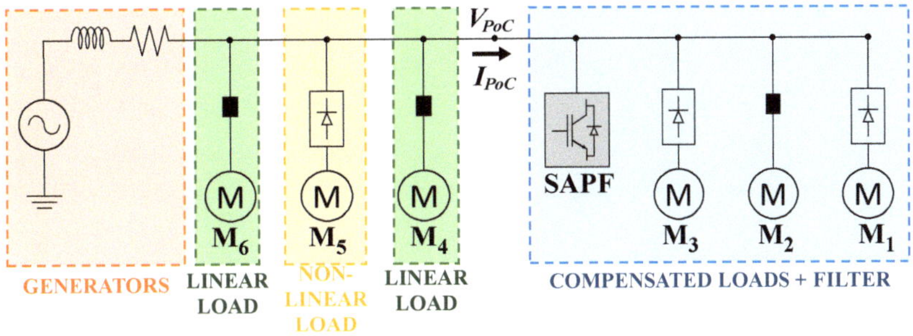

Figure 4. Equivalent model of the electric grid of Figure 1.

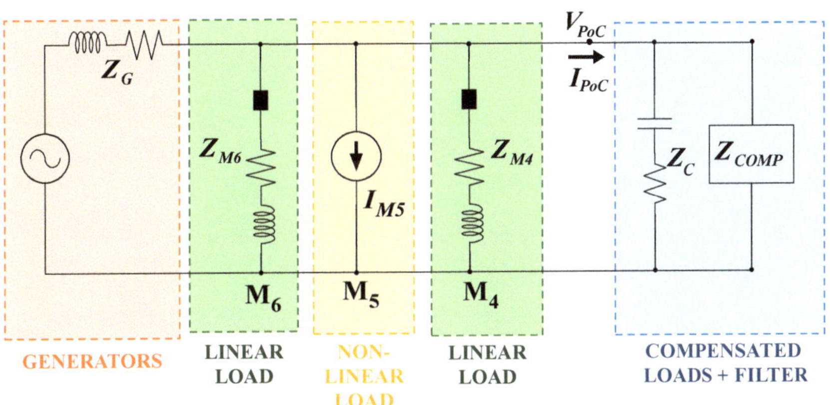

Figure 5. Simplified model of the electric grid of Figure 4.

3.3.1. Resistive Load Synthesis (RLS) Strategy

This strategy aims at emulating a resistive load behavior; therefore, the line current, I_{PoC}, has the same voltage waveform at the point of SAPF connection, V_{PoC}. The current reference i^* is calculated as:

$$i_{RLS}{}^* = i_{Lm} - \frac{P}{V^2} v_m = i^{RLS}_{na_{Lm}} \tag{10}$$

where P is the total active power and i_{Lm} and v_m are the measured load current and SAPF PoC voltage in each phase ($m = a, b, c$), respectively. V is the collective value of the SAPF PoC voltage, $V^2 = V_a^2 + V_b^2 + V_c^2$. Note that the current reference signal is equivalent to the non-active current of the load ($i^{RLS}_{na_{Lm}}$).

In this strategy, the set of compensated loads (M1, M2, and M3) associated with SAPF emulates a resistor behavior for every harmonic order within the SAPF bandwidth, as given in (11), and the Z_{comp} is calculated by (12).

$$I_{PoC}(h) = kV_{PoC}(h), \qquad \forall h \tag{11}$$

$$Z_{comp-RLS}(h) = \frac{V_{PoC}(h)}{I_{PoC}(h)} = \frac{1}{k}, \qquad \forall h \tag{12}$$

Such that k is a constant, I_{PoC} and V_{PoC} are the current and voltage at the SAPF PoC, as shown in Figure 5, and h is the harmonic order.

3.3.2. Sinusoidal Current Synthesis (SCS) Strategy

This strategy produces a sinusoidal line current, I_{PoC}, whatever the waveform of the voltage at the SAPF PoC, V_{PoC}, is. The current reference signal i^* is calculated by (13), such that $\mathbf{V_1}$ is the collective value of voltage considering only the fundamental components of each phase. Note that the current reference signal is equivalent to the non-active current of the load considering a sinusoidal voltage supply ($i_{na_{Lm}}^{SCS}$).

$$i_{scs}{}^* = i_{Lm} - \frac{P}{\mathbf{V_1^2}} v_{m_1} = i_{na_{Lm}}^{SCS} \tag{13}$$

The equivalent circuit model is the same as Figure 5. Differently from the RLS, for the SCS, the ratio of I_{PoC} to V_{PoC} varies with frequency. In terms of fundamental component, the current is proportional to the voltage waveform emulating a resistor; however, for the other harmonic orders, I_{PoC} is equal to zero behavings like an open circuit, as expressed in (14). The Z_{comp} is calculated then by (15).

$$I_{PoC}(h) = \begin{cases} kV_{PoC}(h), & \text{for } h = 1 \\ 0, & \text{for } h \neq 1 \end{cases} \tag{14}$$

$$Z_{comp-SCS}(h) = \frac{V_{PoC}(h)}{I_{PoC}(h)} = \begin{cases} \frac{1}{k}, & \text{for } h = 1 \\ \infty, & \text{for } h \neq 1 \end{cases} \tag{15}$$

3.3.3. Comparison between RLS and SCS

As shown in (12) and (15), both strategies have the same behavior at the fundamental frequency. However, for other harmonic orders, the RLS emulates a resistor, while the SCS emulates an open-circuit. Figure 6 represents the equivalent model for RLS and SCS strategies considering frequency components higher than the fundamental one. It is worth mentioning that the voltage source is short-circuited since the synchronous generators are considered as purely sinusoidal, while the non-linear load, represented by a current source, is still included because it generates harmonic currents.

From Figure 6, the transfer functions between PoC harmonic voltage (V_h) and non-linear load harmonic current (I_h) for SCS and RLS are given by (16) and (17), respectively.

$$\frac{V_{PoC-h_{SCS}}(s)}{I_{M5-h_{SCS}}(s)} = \frac{Z_{M6}(s)Z_{M4}(s)Z_G(s)Z_C(s)}{Z_{M6}(s)Z_{M4}(s)Z_C(s) + Z_G(s)Z_{M4}(s)Z_C(s) + Z_{M6}(s)Z_G(s)Z_C(s) + Z_{M6}(s)Z_G(s)Z_{M4}(s)} \tag{16}$$

$$\frac{V_{PoC-h_{RLS}}(s)}{I_{M5-h_{RLS}}(s)} = \frac{\frac{Z_{M6}(s)Z_{M4}(s)Z_G(s)Z_C(s)}{k}}{\frac{Z_{M6}(s)Z_{M4}(s)Z_C(s)}{k} + \frac{Z_G(s)Z_{M4}(s)Z_C(s)}{k} + \frac{Z_{M6}(s)Z_G(s)Z_C(s)}{k} + \frac{Z_{M6}(s)Z_G(s)Z_{M4}(s)}{k} + Z_{M6}(s)Z_G(s)Z_{M4}(s)Z_C(s)} \tag{17}$$

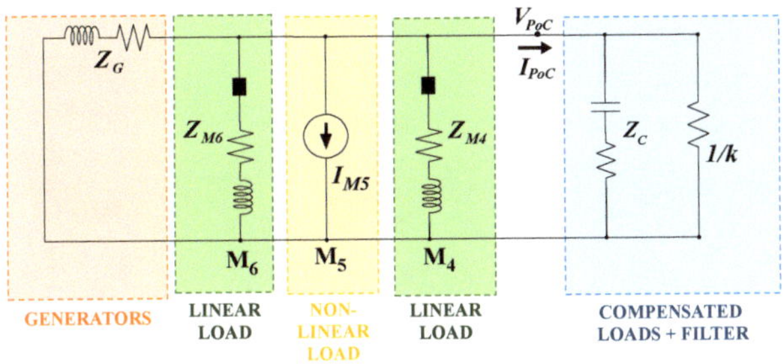

(**a**) RLS

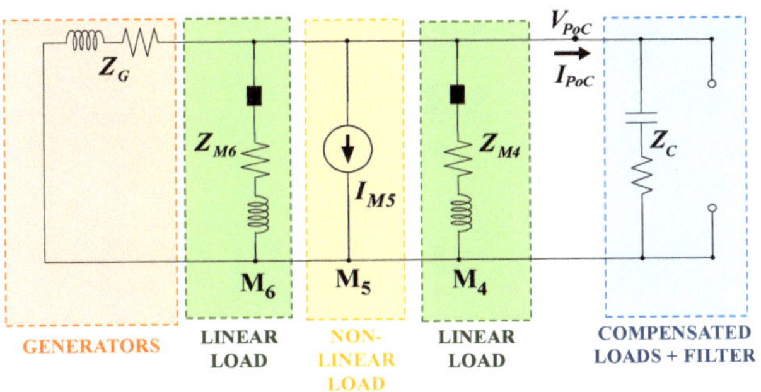

(**b**) SCS

Figure 6. Equivalent circuit model for the frequency components superior to the fundamental one: (**a**) RLS and (**b**) SCS.

4. Simulation Results and Analysis

The power grid of Figure 1 was used as a test-case and was modeled for dynamic simulations in Matlab/Simulink. The Simulink block diagram for Case 3 is shown in Figure 7. It includes the gas turbines—synchronous generators (2×25 MVA), two 5 MW drilling motors controlled by 6-pulse rectifier-based variable speed drive (VSD)—non-linear loads, a 4 MW gas compressor, and a 4 MW water pump-induction motor (IM) directly connected to the MV1 bus. Another induction motor (1 MW) was connected to the MV1 bus, together with a multi-phase pump of 2 MW (non-linear load). Table 1 shows the parameters of the loads, Table 2 highlights the case studies that are addressed in this section, and Table 3 shows the simulation values to plot the Bode diagram of Figure 6. Cases 1, 2 and 3 were related to the position of the SAPF in Figure 1, while *.a* or *.b* sub-indices are related to the semiconductor technology, i.e., Si or SiC.

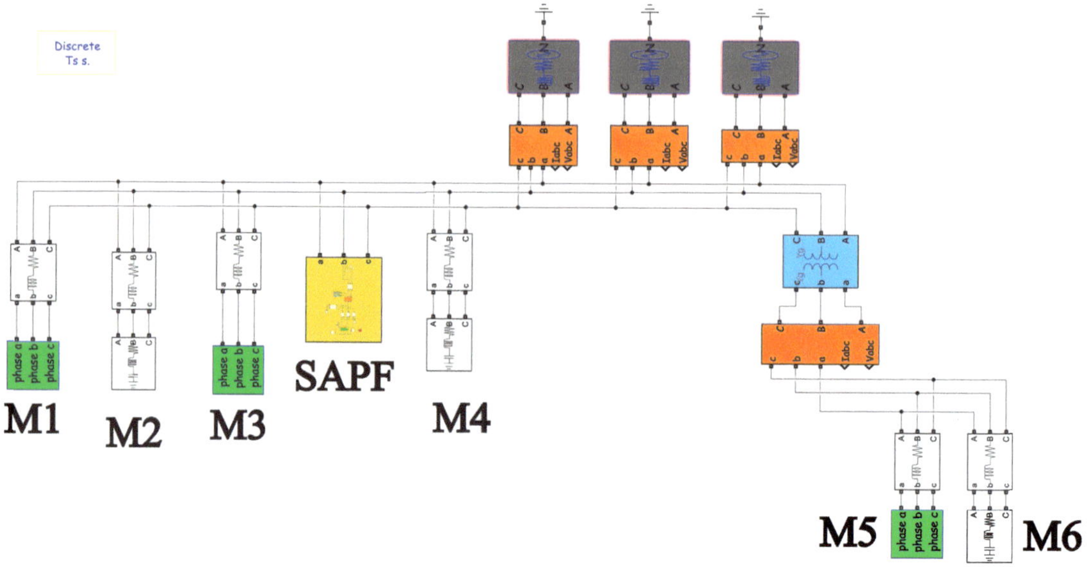

Figure 7. Simulink block diagram for Case 3.

Table 1. Parameters of the loads of Figure 1.

Load	Type	Active Power—P	PF	Apparent Power—A	$\sqrt{A^2-P^2}$
M1	Drilling motor	5 MW	0.95	5.36 MVA	1.93 MVA
M2	Gas compressor	4 MW	0.85	4.71 MVA	2.48 MVA
M3	Drilling motor	5 MW	0.95	5.35 MVA	1.90 MVA
M4	Water injection pump	4 MW	0.8	5.00 MVA	3.00 MVA
M5	Multi-phase pump	2 MW	0.95	2.10 MVA	0.64 MVA
M6	Oil pump	1 MW	0.6	1.67 MVA	1.33 MVA

Table 2. Details of the case studies considered.

Case	Comp. Loads	Comp. Objective	APF Bus	DC Volt.	Semic. Used
1.a	M6 and M5	SAPF at MV1 bus	MV1 (6.6 kV)	12 kV	Si
1.b	M6 and M5	SAPF at MV1 bus	MV1(6.6 kV)	12 kV	SiC
2.a	M1	SAPF at M1	MV2 (11 kV)	18 kV	Si
2.b	M1	SAPF at M1	MV2 (11 kV)	18 kV	SiC
3.a	M1, M2 and M3	SAPF at M1, M2 and M3	MV2 (11 kV)	18 kV	Si
3.b	M1, M2 and M3	SAPF at M1, M2 and M3	MV2 (11 kV)	18 kV	SiC

Table 3. Simulation values to plot the Bode diagram of Figure 8.

Variable	Value
k	0.115
$Z_G = Z + Z_{generator}$	14.34 + j734.2 mΩ
$Z_{M4} = Z' + Z_{motor4}$	6.46 + j4.84 Ω
$Z_{M6} = Z' + Z_{motor6}$	14.56 + j19.37 Ω
Z_C	10 + j318.3 Ω

4.1. Bode Analysis of the Circuit

Figure 8 shows the Bode diagrams for both strategies using the values of Table 3. From Figure 8 one sees that the RLS strategy provided more damping effect for the harmonic currents than SCS, minimizing the voltage distortion [38]. The SCS strategy created non-linearity in the circuit, which may have amplified the THD_v value up to 10 times [39]. Therefore, the RLS showed better performance than the SCS.

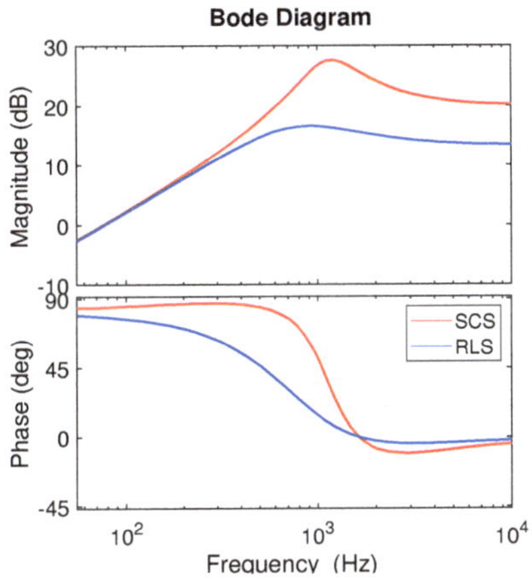

Figure 8. Bode diagram of the $V_{PoC\text{-}h}/I_{M5\text{-}h}$.

4.2. Performance Analysis of Semiconductors

The first analysis carried out was the performance analysis of the semiconductor. Two variants were considered for each case:

(*a*) refers to the implementation of SAPFs using Si-based IGBTs, with two voltage ratings: 3.3 kV Si-IGBT 5SNA 1200E33100 [35] and 6.5 kV Si-IGBT 5SNA 0400J650100 [40] with rated current capacity of 1200 A and 400 A, and maximum blocking voltage of 3.3 kV and 6.5 kV, respectively;

(*b*) refers to the implementation of SAPFs using SiC-based MOSFETs (10 kV SiC MOSFET/SiC-JBS diode [41]) with a rated current capacity of 100 A, and a blocking voltage of 10 kV.

The first analysis was between the two Si-IGBT-based solutions. Following the method described in Section 3, the total losses and number of switching components were calculated, as shown in Table 4. The number of components needed was smaller for the 6.5 kV Si-IGBT, but the overall loss was higher than for the 3.3 kV Si-IGBT. The 3.3 kV Si-IGBT solution (a-cases) was then chosen to compare further with the SiC alternative (b-cases).

Table 4. Comparison results for the Si-IGBT SAPF.

Parameter	Case 1	Case 2	Case 3	Case 1	Case 2	Case 3
Semicondutor	Si					
Switching frequency	2 kHz					
Voltage safety factor (SF)	0.6					
Model	5SNA 1200E330100 [35] 3.3 kV			5SNA 0400J650100 [40] 6.5 kV		
RMS current of SAPF (A)	163.9	91.8	243.7	163.9	91.8	243.7
Number of devices	36	54	54	24	30	30
Switching losses (kW)	61.92	73.48	114.81	97.29	77.32	175.58
Conduction losses (kW)	1.61	0.76	5.35	4.29	1.69	11.88
Total losses (kW)	63.53	74.24	120.16	101.58	79.01	178.49

Table 5 shows the number of components, the losses and efficiency of the power electronics part, and the required LCL filter for the 3.3 kV Si-IGBT and the 10 kV SiC-MOSFET. Note that the high-power IGBT-based converter typically switched with lower frequency (e.g., 2 kHz) than the high-power SiC-MOSFET-based converter that switched at 10 kHz. The results show that the use of SiC considerably decreased the switching and conduction losses and was better applicable for low-current applications, where it provided a count of fewer switches than Si-counterparts. From the perspective of the size of the LCL filter, the most compact solutions were those based on SiC, which provided lighter weight and smaller volume than with Si-IGBT, as expected.

Table 5. Results of the SAPF design.

Parameter		Case 1a	Case 1b	Case 2a	Case 2b	Case 3a	Case 3b
Point of connection		MV1	MV1	MV2	MV2	MV2	MV2
Semiconductor device		Si [35]	SiC [41]	Si [35]	SiC [41]	Si [35]	SiC [41]
Switching frequency (kHz)		2	10	2	10	2	10
RMS current of SAPF (A)		163.9	163.9	91.8	91.8	243.7	243.7
Number of devices		36	24	54	18	54	54
Switching losses (kW)		61.92	42.72	73.48	32.04	114.81	96.12
Conduction losses (kW)		1.61	6.45	0.76	6.07	5.35	14.25
Total losses (kW)		63.53	49.17	74.24	38.11	120.16	110.37
Efficiency (%)		96.61	97.38	95.76	97.82	97.41	97.62
ΔI_{Lmax}		25% of $I_{SAPF-peakvalue}$					
k_a		0.2					
Passive filter requirements	L_f (mH)	17.26	3.45	46.22	9.24	18.88	3.78
	L_g (mH)	5.27	0.21	14.96	0.59	4.58	0.18
	C_f (uF)	7.21	7.21	2.54	2.54	8.30	8.30
	r_f (Ω)	7.89	1.75	22.23	4.96	7.02	1.53
Resonance frequency (kHz)		0.92	4.20	0.94	4.21	0.91	4.18

4.3. *Performance Analysis of Power Quality in O&G Platform*

The following analyses correspond to PQ performance. In order to select the more suitable SAPF, several options considering different connection points (i.e., MV1 or MV2 and loads targeted) and compensation strategies (i.e., RLS or SCS) were analyzed.

4.3.1. Performance Comparison for RLS and SCS

Sinusoidal current is generated by the SCS technique independently of the voltage waveform, and it may cause other non-linearities in the system and increase the system harmonic content, as discussed in Section 3.3.3. The PCC voltage and current waveforms are shown in Figure 9 for RLS and SCS in case 3 (M1, M2, and M3 compensated by the SAPF). As expected, the harmonic content for RLS was lower than for SCS. For the latter, the THD_v was 4.39%, and THD_i was 2.26%, while for RLS, the THD_v was 2.07%, and THD_i was 1.73%. Given this, further analyses will focus on the RLS strategy, which is more appropriate to O&G applications than SCS strategy.

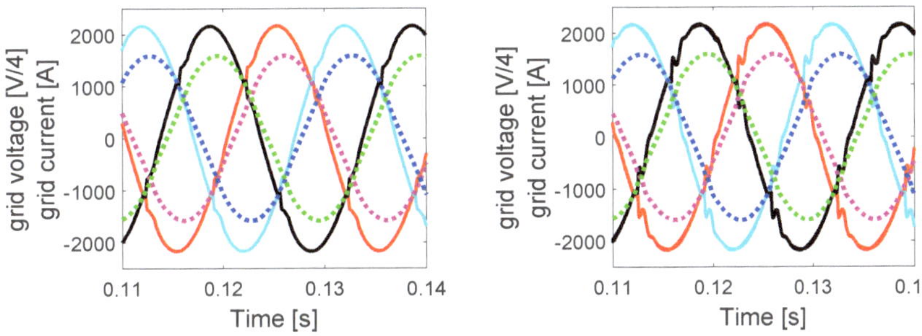

(**a**) RLS (**b**) SCS

Figure 9. Waveforms of V_{PCC} (continuous line-x0.25) and I_{PCC} (dashed line) for (**a**) RLS and (**b**) SCS.

4.3.2. Power Quality Indices for Different Points of SAPF Connection

Table 6 shows the power quality parameters for the three cases and without compensation. With no SAPF, the power factor (defined as the ratio between active and apparent power) measured at the PCC was 0.91 and THD_i = 12.57%. At t = 0.1 s, the SAPF was activated with the RLS mode to achieve a purely active current (i.e., unity power factor) at the chosen bus. The PQ output indices, regardless of the type of semiconductor used, were similar. Therefore, the main reason for the performance difference introduced by different locations of SAPFs was the power rating compensated by the SAPF. Large loads had the most significant impact on PCC's PQ indices.

Table 6. Analysis of the power quality performance.

Variable	Base Case:No comp	Case 1	Case 2	Case 3
SAPF bus voltage (kV)	-	6.6	11	11
SAPF current (A)	-	163.9	91.78	243.7
SAPF rating (MVA)	-	1.87	1.75	4.64
THD_v—@PCC (%)	10.39	9.73	5.92	2.07
THD_i—@PCC (%)	12.57	11.65	7.09	1.73
PF—@PCC	0.91	0.94	0.94	0.98

Case 1 employed the SAPF connected to the MV1 bus and had the smallest power rating among the other case studies. Despite being successful in the local compensation task (it achieved PF at MV1 equal to 1), the improvement of the PQ indices at PCC was limited, as the power rating of the MV1 loads was negligible compared to the MV2 loads. The THD_v at PCC was reduced to below 6% with the SAPF connected at the MV2 bus compensation only one load (M1—Case 2). The PF at PCC was almost unchanged due to the presence of large uncompensated linear and non-linear loads at MV2 in both cases. Finally, in Case 3, the SAPF compensated all the MV2 loads except M4, and the THD_i was decreased to 1.73% and THD_v to 2.07%, with PF = 0.98. However, the goal was achieved at the expense of a significantly high power rating for the SAPF converter.

Figures 10 and 11 show the waveforms corresponding to Case 3. Figure 10 shows the instantaneous (p) and the average active power (P), the instantaneous (q) and the average reactive power (Q), and the PF at PCC. As can be seen, the active power remained unchanged since the filter did not exchange active power with the grid; however, the reactive power at the PCC decreased when the SAPF operated (t = 0.1 s), increasing the PF. Figure 11 shows the grid current and voltage (I_{PCC} and V_{PCC}), and the SAPF current. As expected, when the SAPF operated, I_{PCC} and V_{PCC} became less distorted, which is quantified in Table 6.

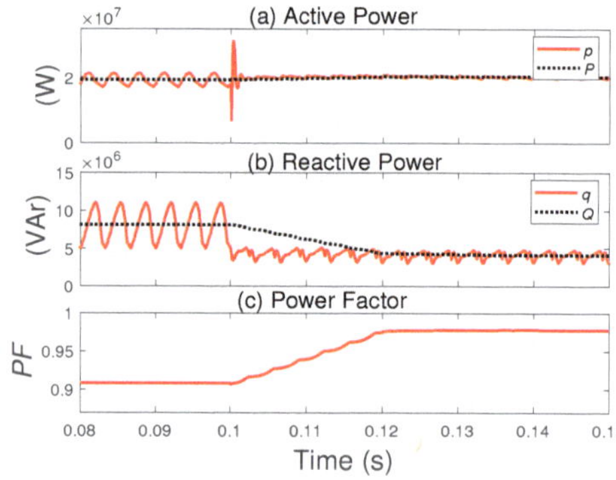

Figure 10. Waveforms of power and PF at the MV2 bus (PCC) in Case 3: (**a**) Active power, (**b**) reactive power and (**c**) Power factor (FP).

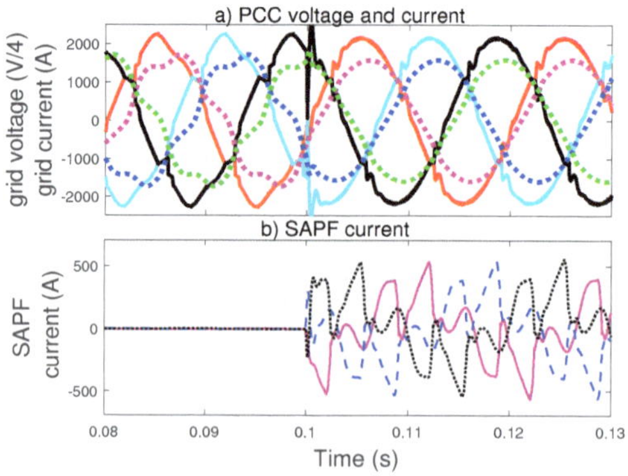

Figure 11. Circuit waveforms corresponding to Case 3: (**a**) Point of Common Coupling (PCC) voltage and current, and (**b**) SAPF current.

The PQ indices should comply with the standards for electrical installations on offshore units. According to ABNT NBR IEC 61892-1:2016 [42] and NEK IEC 61892-1:2019 [43] that are based on IEC 61892-1:2015 and IEC 61892-1:2019 respectively; the voltage harmonic distortion acceptance limits are 3% for each harmonic component, and 5% for the THD (i.e., class 1). However, for certain electrical installations where it is not practical to comply with those requirements, higher values are accepted: no single harmonic shall exceed 6%, and the THD shall not exceed 8% (i.e., class 2). Therefore, this requirement was met only in Case 2 and Case 3.

4.4. Shunt Active Power Filter Pre-Selection Tool

The final choice of the SAPF to be applied to an O&G platform was a complicated decision that emerged as a trade-off between multiple factors, including the APF rating, design (e.g., output filter sizing, number, and type of semiconductor switches), overall PQ performance (i.e., THDv, THDi, and PF), and operation (e.g., losses).

Figure 12 shows a summary of the different aspects considered in this study, proposed as a preliminary selection method to orient the choice of the SAPF. The parameters were normalized over their maximum value. Assuming they were weighted similarly, the smaller area delimited by the line corresponded to the case that was considered indicative of the better SAPF alternative. Table 7 shows the areas of Figure 12 for each case, in which 100% was the largest area.

It is worth highlighting that this was not the final tool to select the SAPF's best choice. A complete analysis involving several factors e.g., reliability, cost, user choice, and physical size, must be provided to determine the final selection. Although, the technique presented in this paper is useful as a pre-selection tool to orient the designer in the first analyses and immediately disregard options clearly resulting in poor performances.

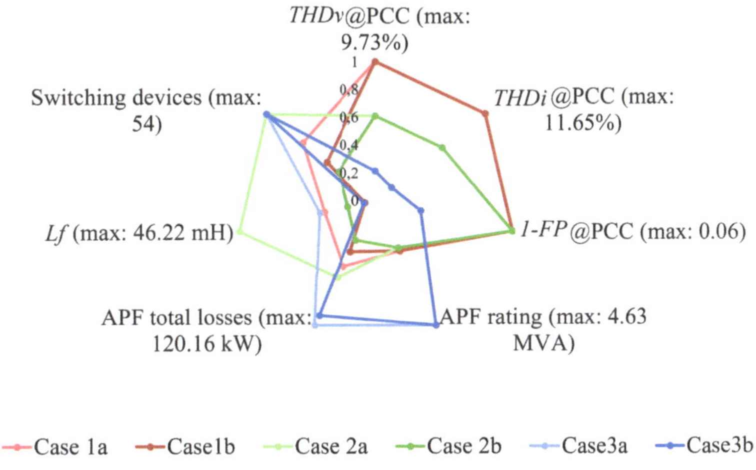

Figure 12. Comparison of SAPFs solutions, using pre-selection tool.

Table 7. Plot areas of each case in Figure 12.

Case 1a	Case 1b	Case 2a	Case 2b	Case 3a	Case 3b
95.41%	83.12%	100%	66.09%	78.73%	70.90%

It can be seen, for example, that Case 2.b (SiC-based SAPF connected to MV2 and only compensating the local load M1) offered the best compromise between the parameters, with intermediate PQ performance, but low total losses, semiconductor count and APF rating. As a comparison, the corresponding Si-solution (Case 2.a), despite providing equivalent PQ performance was penalized in terms of SAPF design (number of switches and filter size), and higher losses, therefore, should be disregarded. When the Si and SiC solutions were compared for each case, the latter options presented better or equal performance in all parameters.

5. Conclusions

The electrical power system of an offshore oil and gas installation is characterized by high energy consumption but with constrained physical space and weight. Most offshore platforms operate in isolated-mode, i.e., without connection to the onshore electrical system. The grid comprises generators, transformers, several loads, and different voltage levels, and must maintain the power quality within the requirements of the standards.

This paper analyzed the use of shunt active power filter (SAPF) in this scenario. Several aspects were addressed: semiconductor analysis, including the number of components and technology (Si and SiC), converter losses (conduction and switching), the design of passive components, and control strategies (resistive load synthesis (RLS) and sinusoidal current synthesis (SCS)). Overall, this paper has shown that SAPF can be a viable solution for isolated power grids, such as oil and gas platforms, where deteriorated power quality requires reactive and harmonic compensation. A SAPF pre-selection tool was designed based on several characteristics, e.g., SAPF connection point, losses, passive components, power quality, and semiconductor type. Moreover, the advantage of a SiC-based implementation of such SAPF has been presented and quantified. Finally, the RLS strategy is suggested for oil and gas applications, because it provides more damping to the system than the SCS strategy.

Author Contributions: Conceptualization, D.B. and E.T.; formal analysis, L.A.V., D.B. and E.T.; funding acquisition, E.T.; investigation, L.A.V., D.B. and E.T.; methodology, L.A.V., D.B. and E.T.; project administration, D.B. and E.T.; resources, D.B. and E.T.; software, L.A.V., D.B. and E.T.; supervision, D.B. and E.T.; validation, L.A.V., D.B. and E.T.; visualization, L.A.V., D.B. and E.T.; writing—original draft, L.A.V., D.B. and E.T.; writing—review and editing, L.A.V., D.B. and E.T. All authors have read and agreed to the published version of the manuscript.

Funding: This research was partly funded under the program PETROMAKS2 of the Research Council of Norway, within the project "Smart Platform" (grant number 308735).

Data Availability Statement: Not applicable.

Conflicts of Interest: The authors declare no conflict of interest.

References

1. Durham, R.A.; Brinner, T.R. Oilfield electric power distribution. In Proceedings of the 2014 IEEE Petroleum and Chemical Industry Technical Conference (PCIC), San Francisco, CA, USA, 8–10 September 2014; pp. 429–442.
2. Rajashekara, K.; Krishnamoorthy, H.S.; Naik, B.S. Electrification of subsea systems: requirements and challenges in power distribution and conversion. *CPSS Trans. Power Electr. Appl.* **2017**, *2*, 259–266. [CrossRef]
3. Sadik-Zada, E.R.; Loewenstein, W. Drivers of CO2-Emissions in Fossil Fuel Abundant Settings: (Pooled) Mean Group and Nonparametric Panel Analyses. *Energies* **2020**, *13*, 3956. [CrossRef]
4. Grainger, B.M.; Reed, G.F.; McDermott, T.E.; Mao, Z.; Kounev, V.; Tipper, D. Analysis of an offshore medium voltage DC microgrid environment—Part I: Power sharing controller design. In Proceedings of the 2014 IEEE PES T D Conference and Exposition, Chicago, IL, USA, 14–17 April 2014; pp. 1–5.
5. Schipman, K.; Delincé, F. The importance of good power quality. *ABB Power Qual. Prod.* **2010**, 1–20. Available online: https://electricalswitchboards.com.au/the-importance-of-good-power-quality/ (accessed on 2 February 2021).
6. 80005-1, I. ISO/IEC/IEEE 80005-1:2012 Utility Connections in Port—Part 1: High Voltage Shore Connection (HVSC) Systems—General Requirements. 2012. Available online: https://www.iso.org/standard/53588.html (accessed on 2 February 2021).
7. 45.1-2017—IEEE Recommended Practice for Electrical Installations on Shipboard–Design. Available online: https://www.wiley.com/en-us/Handbook+to+IEEE+Standard+45%3A+A+Guide+to+Electrical+Installations+on+Shipboard-p-9780738141015 (accessed on 2 February 2021).
8. IEC. IEC 61892-1—Mobile and Fixed Offshore Units. 2019. Available online: https://shop.bsigroup.com/ProductDetail?pid=000000000030350336 (accessed on 2 February 2021).
9. Kumar, D.; Zare, F. A Comprehensive Review of Maritime Microgrids: System Architectures, Energy Efficiency, Power Quality, and Regulations. *IEEE Access* **2019**, *7*, 67249–67277. [CrossRef]
10. Kale, M.; Özdemir, E. Harmonic and reactive power compensation with shunt active power filter under non-ideal mains voltage. *Electr. Power Syst. Res.* **2005**, *74*, 363–370. [CrossRef]
11. Shankar, V.A.; Kumar, N.S. Implementation of Shunt Active Filter for Harmonic Compensation in a 3 Phase 3 Wire Distribution Network. *Energy Procedia* **2017**, *117*, 172–179. [CrossRef]

12. Su, C.; Hong, C. Design of passive harmonic filters to enhance power quality and energy efficiency in ship power systems. In Proceedings of the 49th IEEE/IAS Industrial Commercial Power Systems Technical Conference, Stone Mountain, GA, USA, 30 April–3 May 2013; pp. 1–8.
13. Jaafari, K.A.A.; Poshtan, M.; Beig, A.R. Passive wide spectrum filter for variable speed drives in oil and gas industry. In Proceedings of the 11th International Conference on Electrical Power Quality and Utilisation, Lisbon, Portugal, 17–19 October 2011; pp. 1–6.
14. Verma, V.; Singh, B. Genetic-Algorithm-Based Design of Passive Filters for Offshore Applications. *IEEE Trans. Ind. Appl.* **2010**, *46*, 1295–1303. [CrossRef]
15. Dekka, A.R.; Beig, A.R.; Poshtan, M. Comparison of passive and active power filters in oil drilling rigs. In Proceedings of the 11th International Conference on Electrical Power Quality and Utilisation, Lisbon, Portugal, 17–19 October 2011; pp. 1–6.
16. Chen, J.; Yuan, X. Chain circuit active power filter for high voltage high power applications. In Proceedings of the 2014 17th International Conference on Electrical Machines and Systems (ICEMS), Hangzhou, China, 22–25 October 2014; pp. 2422–2425. [CrossRef]
17. Vitoi, L.A.; Brandao, D.I.; Tedeschi, E. Power quality enhancement by SiC Active Power Filters in Oil and Gas Platforms. In Proceedings of the 2019 IEEE Energy Conversion Congress and Exposition (ECCE), Baltimore, MD, USA, 29 September–3 October 2019; pp. 4299–4304.
18. Acosta-Cambranis, F.; Zaragoza, J.; Romeral, L.; Berbel, N. Comparative Analysis of SVM Techniques for a Five-Phase VSI Based on SiC Devices. *Energies* **2020**, *13*, 6581. [CrossRef]
19. Acharya, S.; She, X.; Todorovic, M.H.; Datta, R.; Mandrusiak, G. Thermal Performance Evaluation of a 1.7-kV, 450-A SiC-MOSFET Based Modular Three-Phase Power Block With Wide Fundamental Frequency Operations. *IEEE Trans. Ind. Appl.* **2019**, *55*, 1795–1806. [CrossRef]
20. Zhang, L.; Yuan, X.; Wu, X.; Shi, C.; Zhang, J.; Zhang, Y. Performance Evaluation of High-Power SiC MOSFET Modules in Comparison to Si IGBT Modules. *IEEE Trans. Power Electr.* **2019**, *34*, 1181–1196. [CrossRef]
21. Madhusoodhanan, S.; Mainali, K.; Tripathi, A.K.; Kadavelugu, A.; Patel, D.; Bhattacharya, S. Power Loss Analysis of Medium-Voltage Three-Phase Converters Using 15-kV/40-A SiC N-IGBT. *IEEE J. Emerg. Select. Top. Power Electr.* **2016**, *4*, 902–917. [CrossRef]
22. Hennig, T.; Mende, D.; Hofmann, L. Efficiency evaluation of offshore power systems with power electronics based on SiC technology. In Proceedings of the 2016 IEEE PES Asia-Pacific Power and Energy Engineering Conference (APPEEC), Xi'an, China, 25–28 October 2016; pp. 634–639.
23. Hu, B.; Lyu, X.; Xing, D.; Ma, D.; Brothers, J.; Na, R.; Wang, J. A Survey on Recent Advances of Medium Voltage Silicon Carbide Power Devices. In Proceedings of the 2018 IEEE Energy Conversion Congress and Exposition (ECCE), Portland, OR, USA, 23–27 September 2018; pp. 2420–2427.
24. Messiha, M.; Baraket, C.; Massoud, A.; Iqbal, A.; Soliman, R. Dynamic voltage restorer for voltage sag mitigation in oil gas industry. In Proceedings of the 2015 First Workshop on Smart Grid and Renewable Energy (SGRE), Doha, Qatar, 22–23 March 2015; pp. 1–6.
25. Madhusoodhanan, S.; Mainali, K.; Tripathi, A.; Vechalapu, K.; Bhattacharya, S. Medium voltage ($\geq$2.3 kV) high frequency three-phase two-level converter design and demonstration using 10 kV SiC MOSFETs for high speed motor drive applications. In Proceedings of the 2016 IEEE Applied Power Electronics Conference and Exposition (APEC), Long Beach, CA, USA, 20–24 March 2016; pp. 1497–1504.
26. Vechalapu, K.; Hazra, S.; Raheja, U.; Negi, A.; Bhattacharya, S. High-speed medium voltage (MV) drive applications enabled by series connection of 1.7 kV SiC MOSFET devices. In Proceedings of the 2017 IEEE Energy Conversion Congress and Exposition (ECCE), Cincinnati, OH, USA, 1–5 October 2017; pp. 808–815.
27. Alves, E.; Sanchez, S.; Brandao, D.; Tedeschi, E. Smart Load Management with Energy Storage for Power Quality Enhancement in Wind-Powered Oil and Gas Applications. *Energies* **2019**, *12*, 2985. [CrossRef]
28. Grassian, D.; Olsen, D. Detailed Energy Accounting of Electrical Submersible Pumping Systems. *Energies* **2020**, *13*, 302. [CrossRef]
29. Waterfield, T.; Germaine, B. The World Largest Injection Pumps. Sulzer Technical Review. 2003, Volume 4. Available online: https://www.researchgate.net/publication/291370966_The_World%27s_Largest_Injection_Pumps (accessed on 2 February 2021).
30. Devold, H. Oil and Gas Production Handbook. ABB ATPA Oil and Gas. 2006. Available online: https://www.amazon.com/Oil-Gas-Production-Handbook-Petrochemical/dp/132978345X (accessed on 2 February 2021).
31. Evans, I.C.; Solutions, H.; Uk, C.; Richards, M.J.; Corporation, A. The Price of Poor Power Quality. In Proceedings of the AADE National Technical Conference, Houston, TX, USA, 12–14 April 2011; p. 12.
32. Ma, D.; Chen, W.; Ruan, X. A Review of Voltage/Current Sharing Techniques for Series–Parallel-Connected Modular Power Conversion Systems. *IEEE Trans. Power Electr.* **2020**, *35*, 12383–12400. [CrossRef]
33. Wang, G.; Konstantinou, G.; Townsend, C.D.; Pou, J.; Vazquez, S.; Demetriades, G.D.; Agelidis, V.G. A Review of Power Electronics for Grid Connection of Utility-Scale Battery Energy Storage Systems. *IEEE Trans. Sustain. Energy* **2016**, *7*, 1778–1790. [CrossRef]
34. ABB. Voltage ratings of high power semiconductors—Application Note 5SYA 2051. Available online: https://docplayer.net/21285209-Voltage-ratings-of-high-power-semiconductors.html (accessed on 2 February 2021).

35. ABB. Applying IGBTs—Application Note 5SYA 2053-04 Applying. Available online: https://library.e.abb.com/public/ab11970 4d4797bc283257cd3002ac5e0/Applying%20IGBTs_5SYA%202053-04.pdf (accessed on 5 February 2021).
36. Duong, T.H.; Ortiz-Rodriguez, J.M.; Raju, R.N.; Hefner, A.R. Electro-thermal simulation of a 100 A, 10 kV half-bridge SiC MOSFET/JBS power module. In Proceedings of the 2008 IEEE Power Electronics Specialists Conference, Rhodes, Greece, 15–19 June 2008; pp. 1592–1597.
37. Reznik, A.; Simões, M.G.; Al-Durra, A.; Muyeen, S.M. *LCL* Filter Design and Performance Analysis for Grid-Interconnected Systems. *IEEE Trans. Ind. Appl.* **2014**, *50*, 1225–1232. [CrossRef]
38. Marafão, F.P.; Brandão, D.I.; Costabeber, A.; Paredes, H.K.M. Multi-task control strategy for grid-tied inverters based on conservative power theory. *IET Renew. Power Gener.* **2015**, *9*, 154–165. [CrossRef]
39. Nunez-Zuniga, T.E.; Pomilio, J.A. Shunt active power filter synthesizing resistive loads. *IEEE Trans. Power Electr.* **2002**, *17*, 273–278. [CrossRef]
40. ABB. *IGBT Module Datasheet-5SNA 0400J650100*; ABB: Zurich, Switzerland, 2016. Available online: https://library.e.abb.com/public/1eda3a30244c484fa0adc56d1d24af5c/5SNA%200400J650100_5SYA%201592-04%2005-2016.pdf (accessed on 5 February 2021).
41. Johannesson, D.; Nawaz, M.; Ilves, K. Assessment of 10 kV, 100 A Silicon Carbide mosfet Power Modules. *IEEE Trans. Power Electr.* **2018**, *33*, 5215–5225. [CrossRef]
42. ABNT. ABNT NBR IEC 61892-1 Unidades Marítimas Fixas e Móveis, 2016. Available online: https://www.normas.com.br/visualizar/abnt-nbr-nm/25069/abnt-nbriec61892-1-unidades-maritimas-fixas-e-moveis-instalacoes-eletricas-parte-1-requisitos-e-condicoes-gerais (accessed on 2 February 2021).
43. NEK. NEK IEC 61892-1—Mobile and Fixed Offshore Units. 2019. Available online: https://standard.no/no/Nettbutikk/produktkatalogen/Produktpresentasjon/?ProductID=1031986 (accessed on 2 February 2021).

energies

MDPI

Article

Cluster-Based Method to Determine Base Values for Short-Term Voltage Variation Indices

Paulo Vitor Grillo de Souza [1,2,*], José Maria de Carvalho Filho [1], Daniel Furtado Ferreira [2], Jacques Miranda Filho [3], Homero Krauss Ribeiro Filho [4] and Natanael Barbosa Pereira [5]

1 Institute of Electrical Systems and Energy, Federal University of Itajubá, Itajubá, MG 37500-903, Brazil; jmariacarvalho@gmail.com
2 Statistics Department, Federal University of Lavras, Lavras, MG 37200-000, Brazil; danielff@ufla.br
3 Electrotechnical Coordination, Federal Institute of Espírito Santo, Vitória, ES 29040-780, Brazil; jacques.filho@ifes.edu.br
4 Transmissora Aliança de Energia Elétrica S.A—TAESA, Brasília, DF 70385-080, Brazil; homerokrauss@yahoo.com.br
5 Energias de Portugal—EDP, São José dos Campos, SP 12210-010, Brazil; natanael.pereira@edpbr.com.br
* Correspondence: paulo.grillo@ufla.br; Tel.: +55-35-99105-4078

Abstract: This paper proposes a methodology for establishing base values for short-term voltage variation indices. The work is focused on determining which variables best describe the disturbance and based on that, establish clusters that allow a more adequate definition of base values for the indices. To test the proposed methodology, real data from 19 distribution systems belonging to a Brazilian electricity utility were used and consequently the index presented in the country standard was considered. This study presents a general methodology that can be applied to all distribution systems in Brazil and could serve as a guide for the regulatory agencies in other countries, to establish base values for their indices. Furthermore, the objective is to show through the results that, with the database used is possible to establish clusters of distribution systems related to the voltage sag and with these establish a base impact factor, distinct for each distribution system.

Keywords: power quality; voltage sag; clustering analysis; index

Citation: Souza, P.V.G.d.; Filho, J.M.d.C.; Ferreira, D.F.; Filho, J.M.; Filho, H.K.R.; Pereira, N.B. Cluster-Based Method to Determine Base Values for Short-Term Voltage Variation Indices. *Energies* **2021**, *14*, 149. https://doi.org/10.3390/en14010149

Received: 16 November 2020
Accepted: 26 December 2020
Published: 30 December 2020

1. Introduction

1.1. Relevance

Due to technological advances, always based on improving the productivity of industrial processes and providing well-being to all people, electro-electronic devices have had a great entry in the domestic sector but mostly at manufacturing sector. However, in general this electronics-based equipment has greater sensitivity to disturbances that affect the power quality, especially those related to short-term voltage variations. When there is a voltage sag in the electrical system, some industrial plant equipment may present malfunctions that could compromise the production process, or in extreme cases, it could cause a complete cessation of operations. Regardless of the type of interruption that occurs in the industrial process, there will always be losses due to lost productivity, loss of raw materials and the repair and replacement of damaged equipment [1].

The standard [2] presents methods for assessing the severity of individual voltage sag events (single-event characteristics) and identifies voltage sag indices to quantify the performance of multiple events in a specific location (single-site indices) or for the whole system (system indices), as an example the SARFI indices, voltage sag tables, voltage sag energy and voltage sag severity. References [3,4] do not present an index to assess voltage sags, and only suggest a way to account for voltage sags, using a table divided into residual voltage ranges and event duration ranges. Document [5] aims to standardize the approach in South Africa to the characterization of voltage sag performance, seven voltage sag categories have been established (Y, X1, X2, S, T, Z1, Z2), based on a combination of customer load compatibility and network protection characteristics. This standard also presents characteristic values for the number of sag events in each category obtained from a historical of the monitored sites. Currently [6] establishes an index called impact factor (IF in Equation (10)) to assess the severity of the incidence of short term voltage variations on substation buses and proposes a single reference value for this index of 1 p.u. One of the most controversial issues among the electricity sector agents was the reference value of 1 p.u. suggested for the IF index, as generally, large consumers of energy, have pointed out that the proposed value is soft and does not reflect the real needs of industrial consumers, because it allows many process stoppages to happen. Despite the numerous ways of assessing the voltage sags proposed in the standards [2–6], none of them establishes a compliance criterion, that is, they do not present limit values for their indices. Therefore, although voltage sags have a major negative economic impact for companies, electricity utilities are not penalized if the industrial consumer suffers process stoppages. For that reason, the question is how to properly establish limits for the IF index, since Brazil has a large territorial extension, it is one of the countries with the largest interconnected electrical system, has a great diversity of vegetation and climate. As distribution systems of different regions are prone to different levels of variables influencing the voltage sags occurrence, a credible way is to set distinct limits according to the characteristics of each distribution system. Regarding to the improvement of the standard, this work is proposing a methodology for the establishment of the base impact factor, considering distribution system clusters that have similar characteristics in relation to the variables that influence the occurrence of voltage sags, it is worth mentioning that the proposed methodology is generic and can be applied by regulatory agencies in other countries to establish base values for their indices.

1.2. State of Art

A survey of the main research databases in the field of electrical engineering, found articles that use cluster analysis to characterize the power quality phenomena. The following is a summary of each of these works. In [7] a method for the evaluation of the events of power quality considering different network operating conditions was proposed. The measured data may depend on the load changes, generation and different network configurations. For this reason, the author of the paper uses clustering techniques to divide acquired data into groups that reflect operating conditions. In work [8], a technique based on graphical cluster analysis was developed to be implemented in a smart power quality analyzer, to monitor electrical networks. In the presence of a fault, the equipment starts the measurement procedure and higher order statistics are calculated in the time domain to allow classification. The results showed the division into two groups of events (voltage sags and transients), with an accuracy of 80%. The paper [9] presents an algorithm that uses the k-means method to recognize and classify the voltage sags of measurement data from a large power grid in Shenzhen (China). The results showed that nearly all voltage sags disturbances can be classified into 11 clusters that probably represent the characteristics and causes of most events occurring in typical distribution systems. In [10], a method developed to determine the optimal number of groups to be formed in power quality measurement data is presented using a data mining algorithm based on the minimum message length (MML) technique. To test the proposed method, three different databases were used, and the test results confirmed the effectiveness of the proposed method, finding the optimal number of groups. A new approach to identify the severity profile of busbar voltage sags was introduced in [11], Voltage sags data caused by faults in all nodes of the system are separated into clusters using the k-means technique. By implementing the method, as a result, information is obtained from the buses that have the lowest occurrence of severe events, hence allowing the choice of installation of sensitive loads at such points of the system. In addition, knowing the most affected buses, the allocation of attenuation devices such as dynamic voltage restorers (DVRs) can be better evaluated.

It is presented in [12] a hybrid model for power quality analysis composed by a modification of the fuzzy min-max neural network (FMM) method added to a modification of the clustering tree (CT) technique. The results were compared with those obtained when applying other clustering algorithms, indicating a better accuracy of the proposed new method. A methodology for detecting and classifying power quality disturbances using a Stockwell transform was developed in [13]. The disturbances were generated by MatLab according to the standards established in the IEEE—1159. Several signal characteristics were extracted from the S-transform based multiresolution analysis. These characteristics are used to classify the disturbances by the fuzzy c-means clustering method. The effectiveness of the proposed algorithm was verified by satisfactory results from several case studies, showing an assertiveness of 99%. Reference [14] proposes a new method for reducing the training set size for the K–nearest neighbors (KNN) algorithm. The proposed method is based on an iterative process. Experimental results showed that the accuracy after sample reduction by recursive process had no difference compared to the original training set. However, the classification of a new signal became faster. For a signal from a real measuring device, the classification time has been reduced from 1.35 s to 0.09 s. The work [15] proposes a method to comprehensively evaluate the power quality based on the maximum tree (MT) algorithm for clustering by the fuzzy method. For the test, 4 indicators were selected: voltage deviation, frequency variation, voltage unbalance and harmonic. The results achieved in a practical case proved the viability of the method, which provides some scientifically based guidelines for the consumer to select the electricity utility and adjust the price paid for the energy according to the quality offered. The paper [16] proposes a methodology to locate the source of voltage sags, initially cluster analysis is used to divide data of voltage signals measured in different nodes into groups. Then, the set of decision rules is defined using the partial decision trees algorithm, which will confront the characteristics of each cluster and define which group the location of the disturbance source fits into. The IEEE 34-bus test feeder system was used to evaluate the methodology and the results showed a hit rate greater than 98%. The work [17] proposes and evaluates an alternative methodology to characterize and classify voltage sags. PCA and K-means clustering technique are applied to identify RMS voltage patterns and reduce the number of RMS voltage profiles representative of the events considered. Real data from 300 events collected at a wind farm in Spain were used to validate the methodology. The proposed methodology proved to be efficient to assess a large number of events. The paper [18] based on a statistical procedure that considers the correlation between the index and the number of equipment trips, proposes a methodology to determine different sensitivity regions and weighting factors from those established in [6]. Therefore, it proposes an improvement of the standard [6]. The research conducted in [19] shows a methodology for clustering distribution systems considering the variables related to voltage sags. The methodology is summarized in four processes: selection of the variables by their correlation with the frequency of voltage sags, implementation of the cluster analysis considering various methods for further investigation of the most appropriate, evaluation of the methods that generated the best clusters through analysis of variance between the response and the generated membership and finally robustness analysis made by including small noises in the input variable, observing which of the methods is more assertive in this condition. The results showed that Ward's method was the most appropriate to the considered database. In the paper [20] it is proposed to apply principal component analysis (PCA) to reduce 32 variable input data (with some level of redundancy) by seven principal components (PCs) which account for 97.9% of the information from the original variables, and from these PCs form clusters of substations, using the Ward's method, considering the Euclidean distance between the elements. The formed clusters allowed to classify the distribution systems in three categories regarding the number of occurrence of voltage sags (high, medium and low levels). Studies conducted in [21] show a novel methodology to increase discriminatory power in the estimation of voltage sag patterns using ellipsoidal functions. Ward's method was used to form clusters of substations with a similarity level to voltage sags, three distinct groups were found with small, medium and large amount of voltage sags. The work [21] is an evolution of that presented in [20]. The method showed results that are more precise, stable and reliable.

In articles [7–17], clustering techniques are used for purposes different from the objective of this paper, such as monitoring, identification and classification of events, location of the source and pattern recognition of voltage sags. These references were presented to identify the application of cluster analysis in the power quality area.

The paper [18] focuses on proposing different sensitivity regions and weighting factors from those established in [6]. While this paper, assuming that the regions of sensitivity and weighting factors established in [6] are adequate, using cluster analysis, proposes new values for the maximum frequency of occurrence of voltage sags and consequently a new base impact factor. Therefore, the works are distinct, although complementary.

Articles [19–21] test several methods of clustering, with the objective of evaluating which one is best suited to form groupings of distribution systems regarding the frequency of voltage sags. These works are the ones that are most related to this paper, but they are focused only on forming the groups, while this paper besides forming the groups, uses this information to establish a base value for the voltage sag index, distinct for each distribution system according to the performance of similar systems. Therefore, this paper complements the studies conducted in [19–21] with the aim of promoting improvements in [6]. None of the papers found use clustering techniques to determine the base values for short-term voltage variation indices, showing the innovation of the proposed methodology.

2. Theory

2.1. Multiple Regression Analysis

A regression model that contains more than one predictor is called a multiple regression model [22]. The purpose of multiple regression analysis is to use independent variables which values are known to predict the values of the dependent variable selected by the researcher. Typically, the dependent or response variable, y, may be related to k independent or predictor variables. The generic model of multiple linear regression with k variables is presented in Equation (1):

$$y = \beta_0 + \beta_1 x_1 + \beta_2 x_2 + \cdots + \beta_k x_k + \epsilon \quad (1)$$

Equation (1) describes a hyperplane in the k-dimensional space of the predictor variables. The parameters βj are called partial regression coefficients [22]. βj can be interpreted as the expected change in y due to the increase of one unit in x_j, with the other variables x_k, $k \neq j$ fixed. Suppose there are k predictor variables and n observations. This model is a system of n equations, which can be expressed in matrix notation by Equation (2):

$$y = X\beta + \epsilon \quad (2)$$

where $y = \begin{bmatrix} y_1 \\ y_2 \\ \vdots \\ y_n \end{bmatrix} X = \begin{bmatrix} 1 & x_{11} & \cdots & x_{1k} \\ 1 & x_{21} & \cdots & x_{2k} \\ \vdots & \vdots & & \vdots \\ 1 & x_{n1} & \cdots & x_{nk} \end{bmatrix} \beta = \begin{bmatrix} \beta_0 \\ \beta_2 \\ \vdots \\ \beta_k \end{bmatrix} \epsilon = \begin{bmatrix} \epsilon_1 \\ \epsilon_2 \\ \vdots \\ \epsilon_n \end{bmatrix}$.

The least-squares method can be used to estimate regression coefficients in the multiple regression model. Equation (3) gives the least squares estimate for β [23]:

$$\hat{\beta} = (X'X)^{-1} X'y \quad (3)$$

The adequacy of the model is evaluated through hypothesis tests related to its parameters. Therefore, the hypothesis test is given by Equation (4):

$$H_0 : \beta_i = 0 \quad H_1 : \beta_i \neq 0 \quad i = 0, 1, \ldots, k \quad (4)$$

If the p-value corresponding to the coefficient of a variable is inferior than or equal to a predetermined significance level α, H_0 is rejected and it is concluded that this coefficient is non-zero, i.e., the variable in question is a significant addition to the model. Otherwise, H_0 is not rejected and it is concluded that such variable has a non-significant effect. Another way of expressing the forecast accuracy level is with the coefficient of determination (R^2), as shown in Equation (5):

$$R^2 = \frac{SQ_{Reg}}{SQ_T} = \frac{\sum_{i=1}^{n}(\hat{y}_i - \bar{y})^2}{\sum_{i=1}^{n}(y_i - \bar{y})^2} \quad (5)$$

Thus R^2 is a global statistic to evaluate how much of the response variability of y is explained by the independent variables. In most surveys, there are a large number of independent variables available that can be chosen for inclusion in the regression equation. The step of selecting which variables will be part of the model is an important point in the model estimation process [23]. This research tested three sequential search methods to select variables called stepwise, forward and backward. Probably the most used variable selection technique is stepwise regression. A sequence of regression models is constructed iteratively, adding or removing variables at each stage. The criteria for removing or adding a variable at any stage are expressed in terms of a partial F test. To begin the process, the independent variable with the highest correlation coefficient with the dependent variable is chosen to generate a simple regression model. The next independent variables selected are based on their incremental contribution (partial correlation) to the regression equation. Each new independent variable introduced in the model is examined by the F test if the contribution of the variables that are already in the model remains significant, given the presence of the new variable. If not, the stepwise estimation allows variables already in the model to be eliminated. The procedure continues until all independent variables not yet present in the model have their inclusion evaluated and the reaction of the variables already present in the model is observed when these inclusions occur [23].

In the forward selection procedure, variables are added to the model one at a time, as long as their partial value of F exceeds a previously established limit. That is, this technique can be considered a variation of the regression stepwise.

The backward elimination algorithm begins with all k model predictors. Then the predictor with the lowest F statistic is removed if that F statistic is insignificant. Subsequently, the model with k–1 predictors is adjusted and the next predictor for potential elimination is found. The algorithm ends when no more predictors can be eliminated [22].

2.2. Cluster Analysis (Dynamic Method)

Cluster analysis is the set of multivariate techniques whose main purpose is to aggregate objects, items or individuals based on their characteristics [23]. The basic criteria used to group objects is their similarities. In this manner, objects belonging to the same cluster are similar to each other concerning the variables that were measured in them, and the elements of distinct clusters are dissimilar for these same variables [24].

To decide whether two database elements can be considered as similar or not, mathematical metrics are used. In this study, Euclidean distance was used as a measure of dissimilarity. Considering two elements X_l and X_k, $l \neq k$, the Euclidean distance between them is defined by Equation (6):

$$d(X_l, X_k) = \left[\sum_{i=1}^{p}(X_{il} - X_{ik})^2\right]^{\frac{1}{2}} \quad (6)$$

Clustering techniques are classified into two types: non-hierarchical and hierarchical, and these are again classified into agglomerative and divisive [24]. Although hierarchical and non-hierarchical methods have certain advantages, its application may not produce good results when analyzing the elements located at the borders between the different groups, as shown in Figure 1.

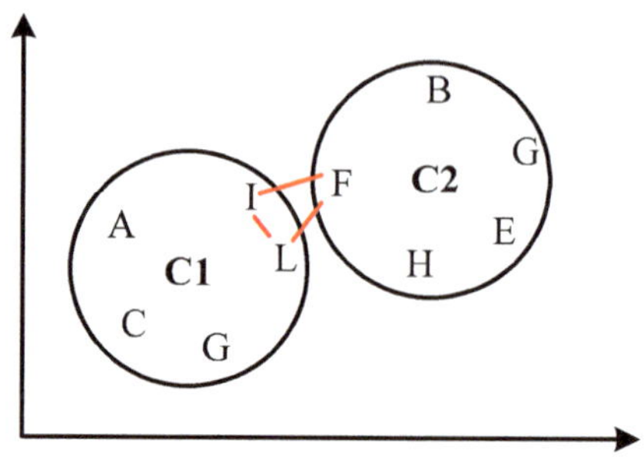

Figure 1. Similarity between border elements of different groups.

In Figure 1, it is noted that elements I and L belong to cluster 1 and the element F belongs to cluster 2. Therefore, such elements will be represented by the characteristics of their respective centroids. However, it is evident that the elements I, L and F are much more similar to each other than to their own centroids. To get around this problem [25] has created a new method, which works by establishing dynamic (changing) clusters from each element. In the dynamic method, for each element taken as reference, a grouping of elements that are most comparable to the so-called reference element will be formed.

In this method there is no formation of fixed clusters, as if there were distinct clusters for each element. This method is very appropriate when the sense of belonging to each cluster is extremely relevant. The algorithm for this technique consists of:

- Each element is adopted as the centroid of a group to be created;
- Once the centroid is defined, the distance of all elements to this centroid is determined;
- A cut-off criterion is established for the degree of similarity between the centroid and the elements;
- Each centroid is grouped with the most representative elements based on their similarities;
- The process is repeated for each of the elements.

The drawback of the dynamic method is that each sample element will generate a cluster. Consequently, for applications that have many elements, the algorithm must be implemented computationally.

2.3. Short-Term Voltage Variations and Index

Short-term voltage variations are defined as random events characterized by significant deviations in the voltage RMS value over a short period and are divided into voltage sags, swells and interruptions.

Voltage sags are the most frequent events among short-term voltage variation (STVV), having a much higher occurrence rate than voltage swells. The IEEE 1564 standard recommends that the handling of voltage sag and voltage swell events be done separately, due to the different effects they cause on equipment [2]. Therefore, this paper will prioritize the study of voltage sags. Although there are many studies and standards focused on voltage sags, there is no international consensus on which index best characterizes the disturbance. Standard [6] presents as parameters of an STVV the event amplitude (Equation (7)), the event duration (Equation (8)) and as an index of a bus or system the frequency of occurrence of events (Equation (9)):

$$V_e = \frac{V_{res}}{V_{ref}} \times 100 \tag{7}$$

where V_e is the event amplitude (in %), V_{res} is the residual voltage of the event (in Volts) and V_{ref} is the reference voltage (in Volts):

$$\Delta t_e = t_f - t_i \tag{8}$$

where Δt_e is the event duration, t_f is the event end time and t_i is the event start time:

$$f_e = n \tag{9}$$

where f_e is the frequency of events and n is the number of events recorded in the period.

Some standards such as [2–5] propose that event stratification be done in tables with certain amplitude and duration ranges.

Taking into consideration the particularities of the electrical system, the standard [6] establishes as shown in Table 1, nine sensitivity regions, to correlate the importance of each event with the sensitivity levels of different loads [18].

Table 1. Stratification based on sensitivity levels of various loads.

<table>
<tr><th rowspan="2">Magnitude (p.u.)</th><th colspan="7">Duration</th></tr>
<tr><th>[16.67 ms–100 ms]</th><th>(100 ms–300 ms]</th><th>(300 ms–600 ms]</th><th>(600 ms–1 s]</th><th>(1 s–3 s]</th><th>(3 s–1 min]</th><th>(1 min–3 min)</th></tr>
<tr><td>>1.15</td><td colspan="3" rowspan="2">REGION H</td><td colspan="4" rowspan="2">REGION I</td></tr>
<tr><td>(1.10–1.15]</td></tr>
<tr><td>(0.85–0.90]</td><td colspan="3" rowspan="2">REGION A</td><td colspan="4"></td></tr>
<tr><td>(0.80–0.85]</td><td colspan="4" rowspan="2">REGION G</td></tr>
<tr><td>(0.70–0.80]</td><td rowspan="2">REGION B</td><td colspan="2" rowspan="4">REGION D</td></tr>
<tr><td>(0.60–0.70]</td><td colspan="4" rowspan="7">REGION F</td></tr>
<tr><td>(0.50–0.60]</td><td rowspan="2">REGION C</td></tr>
<tr><td>(0.40–0.50]</td></tr>
<tr><td>(0.30–0.40]</td><td colspan="3" rowspan="4">REGION E</td></tr>
<tr><td>(0.20–0.30]</td></tr>
<tr><td>(0.10–0.20]</td></tr>
<tr><td><0.10</td></tr>
</table>

To describe the severity of the incidence of events in a single index, the Impact Factor (IF) index was established in [6], which has a 30 consecutive days calculation period, and is calculated by Equation (10):

$$IF = \frac{\sum_{i=A}^{I} (f_{ei} \times fp_i)}{IF_{base}} \tag{10}$$

where f_{ei} is the frequency of events over 30 consecutive days for each sensitivity region i, with i = A through I, fp_i is the weighting factor for each sensitivity region and IF_{base} is the base impact factor, calculated considering the weighting factors and the maximum frequency of occurrence for each sensitivity region. The maximum frequency of occurrence for each sensitivity region is presented in Table 2.

Table 2. Monthly maximum frequency of occurrence in the sensitivity regions [6].

Sensitivity Regions	Maximum Frequency of Occurrences
	1 kV < Vnominal < 69 kV
A	-
B	5
C	4
D	3
E	2
F	1
G	4
H	1
I	1

The weighting factors were stipulated by the regulatory agency in order to consider in the equation the sensitivity of the loads normally present in the industries, giving more weight to severe events, which have a high probability of causing equipment shutdowns and less weight for mild events, with a low probability of causing shutdowns. The weighting factor (fp) for each sensitivity region and also the base impact factor are shown in Table 3.

Table 3. Weighting factors and base impact factor [6].

Sensitivity Regions	Weighting Factor (fp)	Base Impact Factor (IF_{base})
		1 kV < Vnominal < 69 kV
A	0.00	
B	0.04	
C	0.07	
D	0.15	
E	0.25	2.13
F	0.36	
G	0.07	
H	0.02	
I	0.04	

The base impact factor currently adopted is the same for all distribution systems, not considering the levels of the variables that influence the occurrence of the event. The reference value set in [6] for the impact factor index for distribution systems is 1.0 p.u.

Therefore, the objective of this work is to define different base impact factors for each distribution system taking into account the performance of distribution systems that have similar characteristics with respect to the variables that influence the occurrence of voltage sags.

3. Material and Methods

3.1. Material

To make the proposed methodology applicable to all distribution systems with 1 kV < $V_{nominal}$< 69 kV in Brazil, starting from a larger database that is mandatorily sent by all electricity utilities to the regulatory agency were chosen by a specialist 9 attributes that include technical information of the distribution network that may be related to the occurrence of voltage sags. Besides the attributes, it is necessary the information that will serve as a goal to form clusters, which in the specific case of this research considered the frequency of occurrence of the phenomena. The average monthly frequency of voltage sag was obtained from measurements in 19 distribution systems belonging to a Brazilian electricity utility.

The complete database containing the values of the considered attributes and the frequency of voltage sags measured in each distribution system (DS) is shown in Table 4. The meaning of each abbreviation is listed in the Abbreviations section below.

Table 4. Database (attributes and frequency of voltage sags).

DS	NF	NRCU	D_DESC	PC_VRA	PC_TD_1F	PC_TD_R	AFL	FR	VA	FREQ
1	4	262	3.00	0.03	0.86	0.96	489.45	6.48	535.17	19
2	4	405	3.00	0.01	0.46	0.53	103.91	12.54	74.33	4
3	4	310	3.00	0.01	0.74	0.86	221.65	5.76	74.41	5
4	2	389	2.78	0.01	0.31	0.32	79.46	21.09	41.37	5
5	4	82	2.83	0.04	0.60	0.61	68.74	20.64	129.87	15
6	2	297	3.00	0.01	0.87	0.93	553.56	4.24	174.43	14
7	3	538	3.00	0.05	0.78	0.92	204.33	10.25	388.72	11
8	5	422	2.97	0.05	0.70	0.94	237.46	7.62	93.71	5
9	5	644	3.00	0.01	0.61	0.81	177.86	11.40	175.78	11
10	4	261	3.00	0.44	0.56	0.87	164.38	7.38	240.34	16
11	11	461	3.31	0.01	0.76	0.93	209.25	6.18	194.76	6
12	3	189	3.00	0.01	0.94	0.97	83.50	3.40	113.89	12
13	17	998	3.00	0.01	0.75	0.89	211.08	6.57	174.96	13
14	4	435	3.00	0.03	0.76	0.97	327.05	6.40	251.92	13
15	4	474	3.00	0.11	0.84	0.95	418.09	8.48	142.81	15
16	6	392	3.01	0.22	0.86	0.95	232.65	9.10	160.37	17
17	6	539	2.92	0.01	0.61	0.68	159.67	16.84	278.59	14
18	1	171	3.00	0.01	0.90	0.93	777.61	4.43	135.82	13
19	3	395	3.00	0.01	0.88	0.97	292.89	5.15	235.71	13

The number of feeders, is obtained by counting in the substation diagram, the number of rural consumer units provided by the electricity utility, the atmospheric discharge density was estimated from historical meteorological data, the percentage of remaining vegetation was established by processing satellite images, the percentage of single-phase transformers was obtained by the ratio of the number of single-phase transformers to the total number of transformers in the distribution system, the percentage of rural transformers was obtained by the ratio of the number of rural transformers to the total number of transformers in the distribution system, the average feeder length was obtained by the ratio of the total length of the distribution network to the number of feeders, the fault rate was obtained by averaging historical data, the vulnerability area refers to the substation bus and it was calculated considering failure impedance equal to zero. With the distribution system modeled in a simulation software, short-circuits are applied to all nodes in the network while the voltage on the substation bus is monitored, to check for voltage sag. All types of short circuit were considered and weighted by the typical probability of occurrence. The average monthly frequency of voltage sags was obtained through meters that were installed in the substations and measured during one year.

3.2. Methods

The proposed methodology can be summarized in the following steps:

- Variables selection through sequential search methods (explained in Section 2.1).
- Formation of distribution systems clusters through the dynamic method, using as input variables those selected in the previous step (explained in Section 2.2).
- Establishment of the base impact factor for each distribution system by averaging the frequency of occurrence found in similar distribution systems, this is the main point of the proposed methodology and will be exemplified in Section 4.3.

The flowchart in Figure 2, presents in more detail the process of the proposed methodology.

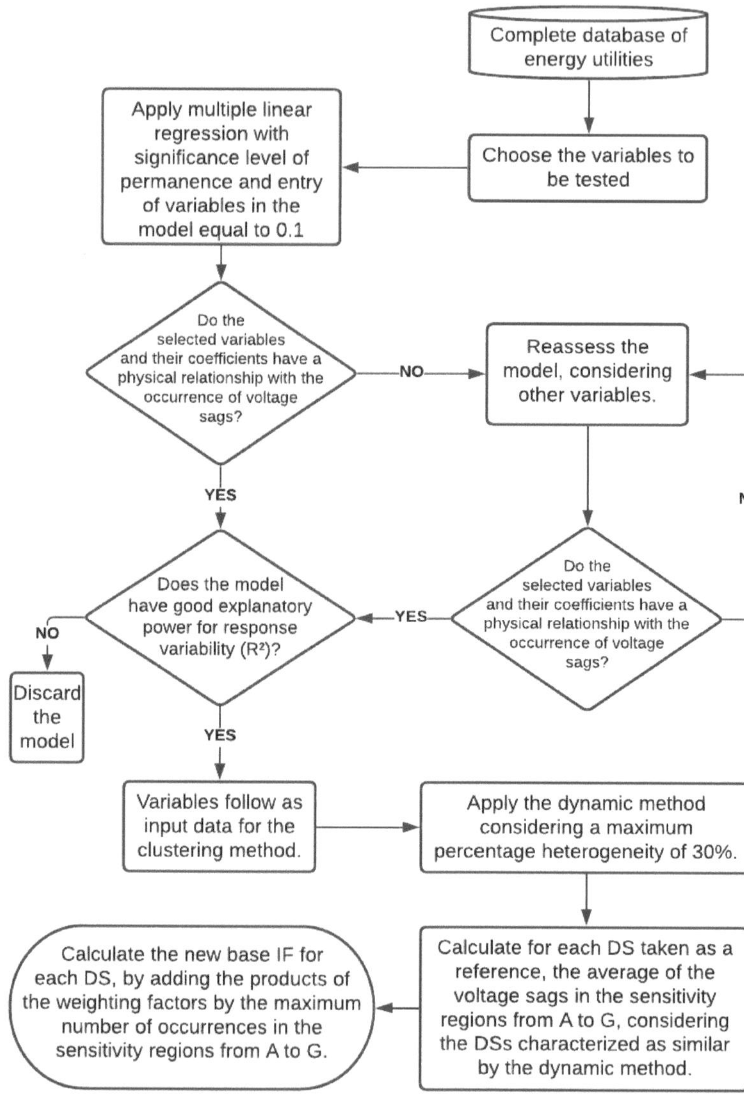

Figure 2. Flowchart of the proposed methodology.

4. Case Study and Results

4.1. Variable Selection

Given the number of variables available for analysis, and knowing the sensitivity that the clustering method has when considering a large number of input variables, a step in variable selection has been performed to define the smallest possible set that has a good capacity to explain the variability of the response. For this step, the stepwise regression, backward elimination, and forward selection procedures were tested. Considering a level of significance for entry and removal of variables in the model equal to 0.1 and applying the three regression techniques tested, the same model was obtained, whose main parameters (R^2, coefficients, regression equation) are shown in the Table 5.

Table 5. Regression analysis for the frequency of voltage sags.

Analysis of Variance					
Source	**DF**	**Adj SS**	**Adj MS**	***f*-Value**	***p*-Value**
Regression	4	271.26	67.816	9.98	0.000
PC_VRA	1	79.71	79.709	11.73	0.004
PC_TD_1F	1	106.12	106.123	15.61	0.001
FR	1	60.99	60.991	8.97	0.010
VA	1	36.81	36.810	5.42	0.035
Error	14	95.16	6.797		
Total	18	366.42			
Model Summary					
	R^2			R^2 adj	
	74.03%			66.61%	
Coefficients					
Term	Coef	SE Coef	*t*-Value	*p*-Value	VIF
Constant	−16.25	6.16	−2.64	0.020	
PC_VRA	20.88	6.10	3.24	0.004	1.10
PC_TD_1F	25.89	6.55	3.95	0.001	3.11
FR	0.589	0.197	3.00	0.010	2.81
VA	0.01321	0.00568	2.33	0.035	1.18
Regression Equation					
Freq = −16.25 + 20.88 PC_VRA + 25.89 PC_TD_1F + 0.589 FR + 0.01321 VA					

The generated model shows that all the selected variables presented P-Value below the 0.05 threshold, indicating to be significant in the model. Also, the VIF values are all less than 5, showing low multicollinearity between the selected variables. However, the parameter normally used to verify the adequacy of the model is R^2, the model adjusted for the number of occurrences of voltage sags, presented $R^2 = 74\%$ (satisfactory value), representing a model that although parsimonious (a small number of variables) still explains the variability of the response. Thus, in the subsequent steps of the methodology, the variables (PC_VRA—"percentage of remaining vegetation", PC_TD_1F—"percentage of single-phase transformers", FR—"fault rate" and VA—"vulnerability area") will be used. It is noteworthy that any model found by the statistical method should be appreciated by a specialist, to verify the selected variables and their coefficients, as to the physical meaning they have with the phenomenon under analysis. Making this critical analysis of the obtained model, it is valid to select the variable "percentage of remaining vegetation", since a short circuit source in the networks is the trees that can touch it. The variable percentage of single-phase transformers indirectly brings information on the percentage of rural networks, since single-phase transformers are commonly used in these. This way, the variable also has an explanation from electrical engineering, since rural networks are more exposed to the action of animals and tend to have less frequent maintenance compared to urban networks. The fault rate variable is also related to the occurrence of voltage sags, as some faults generate these events. The variable vulnerability area is strongly related to the occurrence of the phenomenon since it represents the area under which the occurrence of a fault will generate voltage sag. It is also noted that the coefficients linked to the variables present in the model are in agreement with the expected since these selected variables have a direct relation, i.e., an increase in the value of some predictor increases the value of the response.

4.2. Clustering Analysis

For the implementation of the dynamic method, it is necessary to create tables by increasingly sorting the distances between elements for each element taken as reference. For example, considering DS 8 as a reference, Table 6 shows the distances between elements.

Table 6. Distance and heterogeneity between elements (reference DS 8).

DS	Distance	Heterogeneity
8	**0.00**	**0.00%**
3	**0.60**	**9.96%**
13	**0.87**	**14.49%**
11	**1.04**	**17.32%**
15	**1.08**	**17.97%**
9	**1.19**	**19.80%**
14	**1.42**	**23.69%**
18	**1.46**	**24.32%**
6	**1.47**	**24.51%**
12	**1.73**	**28.78%**
19	**1.75**	**29.21%**
2	**1.80**	**29.96%**
16	1.98	33.00%
17	2.46	40.99%
5	2.57	42.91%
7	2.61	43.50%
4	3.53	58.87%
1	3.89	64.88%
10	3.96	65.94%

Percent heterogeneity is obtained by dividing the distance values by the maximum distance (denominator of Equation (11)). The maximum distance will be the distance between the reference DS and a hypothetical DS whose standardized attributes are three times the value of the reference DS attributes, in other words, a DS that is 3 standard deviations from the reference DS. Thus, the percentage heterogeneity formula is presented in Equation (11):

$$\text{Heterogeneity} = \frac{\text{Distance}}{\sqrt{k \cdot 3^2}} = \frac{\text{Distance}}{\sqrt{4 \cdot 3^2}} = \frac{\text{Distance}}{6} \tag{11}$$

where k is the number of attributes. From the analysis of Table 6, considering maximum percentage heterogeneity of 30%, DS 8 has 11 similar DSs.

4.3. Setting the Base Impact Factor

To establish the base impact factor, it is proposed to use the average of the values of the monthly average frequency of voltage sags in each sensitivity region in the DSs that most closely resemble the DS taken as reference. Starting with a determination of the maximum expected number of occurrences in each sensitivity region, with these values and using the weighting factors used by [2], a different IF_{base} is calculated for each distribution system. The differentiation of IF_{base} from each system allows the reference value set by [6] of 1 p.u. be maintained, but each DS will have a different goal according to the characteristics that most contribute to the occurrence of the phenomenon and according to the performance of systems that have similarities concerning these characteristics. Taking as an example the DS 8, Table 7 shows the average monthly frequency of voltage sags measured in these distribution systems stratified in sensitivity regions A to G.

Table 7. Frequency of voltage sags in the sensitivity regions.

DS	Frequency of Voltage Sags in the Sensitivity Regions						
	A	B	C	D	E	F	G
8	**3.08**	**0.00**	**0.00**	**0.50**	**0.25**	**0.50**	**0.58**
3	**3.25**	**0.08**	**0.00**	**0.50**	**0.00**	**0.17**	**1.25**
13	**6.67**	**0.67**	**0.33**	**1.92**	**1.75**	**0.42**	**1.42**
11	**2.08**	**0.75**	**0.17**	**0.83**	**0.67**	**0.92**	**0.83**
15	**8.55**	**1.36**	**0.09**	**2.73**	**0.18**	**0.18**	**1.45**
9	**3.58**	**0.17**	**0.08**	**1.33**	**0.33**	**1.50**	**1.25**
14	**7.60**	**0.40**	**0.00**	**1.10**	**0.30**	**0.70**	**2.70**
18	**6.13**	**0.75**	**0.13**	**4.25**	**0.38**	**0.63**	**0.88**
6	**7.08**	**0.67**	**0.33**	**3.58**	**0.33**	**1.00**	**0.92**
12	**6.00**	**0.75**	**0.42**	**2.25**	**0.67**	**1.58**	**0.67**
19	**8.89**	**0.33**	**0.22**	**1.33**	**1.11**	**0.56**	**1.00**
2	**1.83**	**0.50**	**0.25**	**0.92**	**0.08**	**0.25**	**0.25**
16	9.25	2.25	0.17	1.67	1.42	1.25	0.75
17	7.09	1.18	0.09	1.55	0.64	1.73	1.18
5	4.73	0.91	0.36	4.18	0.73	2.82	1.09
7	5.09	1.55	0.00	1.73	0.45	0.82	1.45
4	2.33	0.17	0.08	0.50	0.67	1.08	0.50
1	11.80	1.60	0.30	2.20	0.50	0.90	1.30
10	7.67	0.50	0.00	0.75	0.58	2.25	3.42

Considering the average of the data in bold type present in each column of Table 7, the maximum number of occurrences expected for each sensitivity region is obtained for DS 8. Table 8 shows the sensitivity regions A to G considered in the Impact Factor calculation, the weighting factor and the maximum number of occurrences relative to each sensitivity regions used by [6] and the calculated by the proposed procedure.

Table 8. Weighting factors and limits for voltage sag frequency at sensitivity regions.

Sensitivity Regions	Weighting Factor	Maximum Frequency of Occurrences (ANEEL)	Maximum Frequency of Occurrences (DS 8)
A	0.00	-	5.40
B	0.04	5	0.54
C	0.07	4	0.17
D	0.15	3	1.77
E	0.25	2	0.50
F	0.36	1	0.70
G	0.07	4	1.10
Base Impact Factor (IF_{base})		2.07	0.75

With the values of the maximum occurrences of DS 8, the new base impact factor for this system is calculated by summing the weighting factor products by the maximum occurrences, resulting in 0.75.

It is observed that the base impact factor found for the distribution system analyzed is lower than that established by [6], due to the lower maximum number of occurrences of voltage sags calculated for such a system. Using this calculation methodology, the IF_{base}(new method), IF(new method), IF_{base}(current) and IF(current) of the other DSs of this case study are shown in Table 9.

Table 9. IF_{base} and IF of distribution systems.

DS	IF_{base}(New)	IF(New)	IF_{base}(Current)	IF(Current)
1	0.89	1.07	2.07	**0.46**
2	0.53	**0.57**	2.07	**0.15**
3	0.79	**0.28**	2.07	**0.11**
4	-	-	2.07	**0.33**
5	1.56	1.26	2.07	**0.95**
6	0.84	1.30	2.07	**0.53**
7	0.81	1.02	2.07	**0.40**
8	0.75	**0.48**	2.07	**0.17**
9	0.68	1.37	2.07	**0.45**
10	-	-	2.07	**0.64**
11	0.79	**0.91**	2.07	**0.35**
12	0.78	1.51	2.07	**0.57**
13	0.79	1.29	2.07	**0.49**
14	0.80	**0.87**	2.07	**0.34**
15	0.84	**0.81**	2.07	**0.33**
16	0.95	1.28	2.07	**0.58**
17	1.34	**0.85**	2.07	**0.56**
18	0.78	1.35	2.07	**0.51**
19	0.79	**0.99**	2.07	**0.38**

The graph in Figure 3 shows the IF (new method) and the IF (current) compared to the reference value of 1 p.u.

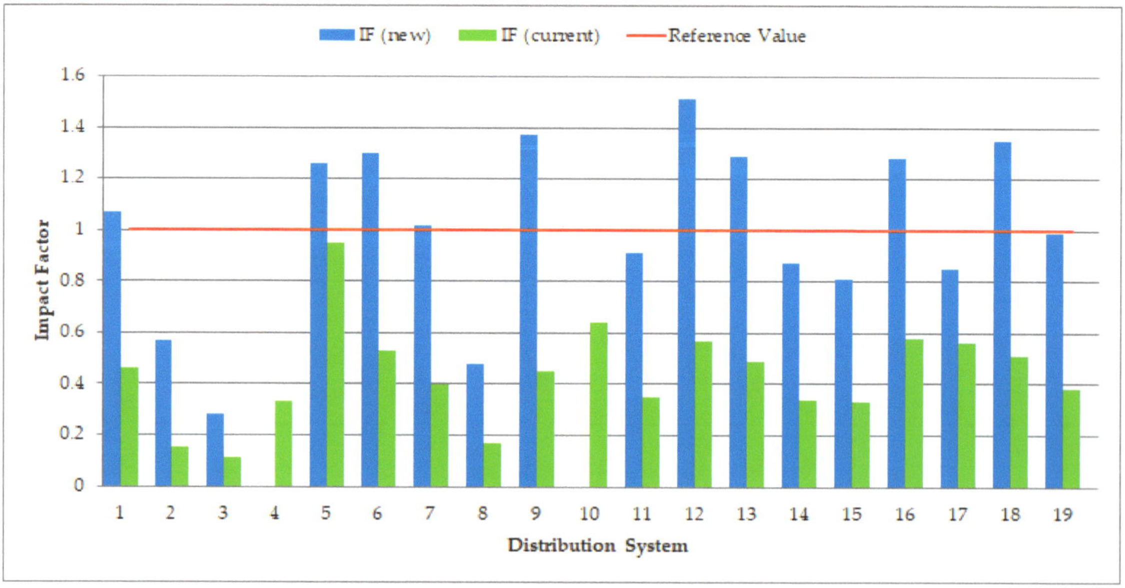

Figure 3. Comparison of the new and current IF with the reference value.

As shown in Figure 3, considering the current IF_{base} the index IF of all distribution systems are below the reference value of 1 p.u., showing that this IF_{base} is soft, because all distribution systems would be in accordance with the standard, not requiring actions by the electricity utility. Therefore, a hypothetical industrial consumer who is connected to any of these distribution systems and has a process sensitive to voltage sags characterized by regions D, E, F, G (Table 1), can suffer up to 13 process stoppages per month without the impact factor exceeding 1p.u. In many industrial sectors, this number of process stoppages would result in high financial losses.

On the other hand, with the calculation of the new IF_{base} considering the average of the voltage sags frequency of the cluster, about 53% of the DSs had an Impact Factor above the reference value of 1 p.u, if it was the methodology applied in the regulation, some distribution systems would need improvements, such as pruning the vegetation nearby the network, increasing the isolated compact network to adapt the index to the reference value. Therefore, for electricity utilities, the proposed methodology establishes hard values for the index, however it takes into account that similar distribution systems have to present similar performances and generates base impact factors that are aligned with the power quality demanded by industrial consumers.

5. Conclusions

Voltage sags cause major monetary losses to industrial consumers with sensitive loads. Hence, it is expected that in the future there will be changes in the standard for proposing limits and it is believed that the most appropriate procedure to be adopted should be the establishment of a distinct base impact factor for each DS according to the systems performance that it most resembles. In this context, this work is aligned with the aspirations of the electrical sector, presenting in a didactic way a methodology for the establishment of the base impact factor that is used in the calculation of the index that regulates voltage sags in Brazil.

The results showed that the proposed methodology was able to select the variables that are most related to the occurrence of voltage sags, to generate clusters of distribution systems in relation to these variables and to establish the base impact factor for each DS. The values found for the new base impact factors were lower than the current value, so it is tighter, if adopted it guarantees a better power quality for consumers.

The regulatory agency is able to implement the methodology for all distribution systems in Brazil, requesting the input data used from the electricity utilities. Other countries may adopt the proposed methodology to assign base values to their indices, even if the available variables are different, or if the chosen clustering technique is different, the suggested steps can be followed to find base values that take into account the performance of the similar systems with respect to the variables that influence the occurrence of voltage sags.

If the necessary data is available, in future research, the proposed methodology can be reevaluated considering a larger sample of distribution systems and other variables that may be relevant for the formation of clusters.

Author Contributions: Conceptualization, P.V.G.d.S.; Data curation, P.V.G.d.S. and H.K.R.F.; Formal analysis, P.V.G.d.S. and J.M.d.C.F.; Funding acquisition, N.B.P.; Investigation, P.V.G.d.S. and H.K.R.F.; Methodology, P.V.G.d.S. and J.M.d.C.F.; Project administration, J.M.d.C.F. and D.F.F.; Supervision, J.M.d.C.F. and D.F.F.; Validation, P.V.G.d.S. and D.F.F.; Writing—original draft, P.V.G.d.S.; Writing—review & editing, J.M.F. All authors have read and agreed to the published version of the manuscript.

Funding: This research was funded by Capes, CnPq and Fapemig agencies.

Institutional Review Board Statement: Not applicable.

Informed Consent Statement: Not applicable.

Data Availability Statement: The data presented in this study are contained in the article.

Acknowledgments: The authors would like to thank the Federal University of Itajubá and the Federal University of Lavras for the technological support and the EDP company for providing through a R&D project the data used in the case study.

Conflicts of Interest: The authors declare no conflict of interest. The funders had no role in the design of the study; in the collection, analyses, or interpretation of data; in the writing of the manuscript, or in the decision to publish the results.

Abbreviations

AFL	Average feeder length [km];
ANEEL	National Electricity Agency;
D_DESC	Atmospheric discharge density [lightning/km^2];
DS	Distribution system;
FR	Fault rate [faults/100 km/year];
FREQ	Frequency of occurrence of voltage sags;
IEEE	Institute of Electrical and Electronics Engineers;
IF	Impact Factor;
NF	Number of feeders;
NRCU	Number of rural consumer units;
PC_TD_1F	Percentage of single-phase transformers;
PC_TD_R	Percentage of rural transformers;
PC_VRA	Percentage of remaining vegetation;
PRODIST	Documents to standardize the technical activities related to the operation and performance of the electricity distribution systems in Brazil;
RMS	Root mean square;
STVV	Short-term voltage variation;
VA	Vulnerability area [km];

References

1. Bollen, M.H. *Understanding Power Quality Problems*; IEEE: New York, NY, USA, 1999.
2. IEEE P1564/D19. *IEEE Guide for Voltage Sag Indices*; IEEE: New York, NY, USA, 2013.
3. IEC 61000-2-8. *Electromagnetic Compatibility (EMC)—Part 2–8: Environment—Voltage Dips and Short Interruptions on Public Electric Power Supply Systems with Statistical Measurement Results*; International Electrotechnical Committee: Geneva, Switzerland, 2005.
4. IEC 61000-4-11. *Electromagnetic Compatibility (EMC)—Part 4–11: Testing and Measurement Techniques—Voltage dips, short interruptions and Voltage Variations Immunity Tests*; International Electrotechnical Committee: Geneva, Switzerland, 2004.
5. NRS 048-2. *Electricity Supply—Quality of Supply Part 2: Voltage Characteristics, Compatibility Levels, Limits and Assessment Methods*; Standards South Africa: Groenkloof, South Africa, 2003.
6. ANEEL—National Electricity Agency. PRODIST—Electricity Distribution Procedures in the National Electric System. 2017. Available online: https://www.aneel.gov.br/documents/656827/14866914/M%C3%B3dulo_8-Revis%C3%A3o_10/2f7cb862-e9d7-3295-729a-b619ac6baab9 (accessed on 4 March 2019). (In Portuguese)
7. Jasinski, M.; Sikorski, T.; Karpinski, J.; Zenger, M. Cluster analisis of long-term power quality data. In Proceedings of the 2016 Electric Power Networks (EPNet), Szklarska Poreba, Poland, 19–21 September 2016; pp. 1–6.
8. Florencias-Oliveros, O.; Agüera-Pérez, A.; González-de-la-Rosa, J.; Palomares-Salas, J.; Sierra-Fernández, J.; Montero, Á.J. Cluster analysis for Power Quality monitoring. In Proceedings of the 2017 11th IEEE International Conference on Compatibility, Power Electronics and Power Engineering (CPE-POWERENG), Cadiz, Spain, 4–6 April 2017; pp. 626–631.
9. Duan, R.; Wang, F.; Zhang, J.; Huang, R.; Zhang, X. Data mining & pattern recognition of voltage sag based on K-means clustering algorithm. In Proceedings of the 2015 IEEE Power & Energy Society General Meeting, Denver, CO, USA, 26–30 July 2015; pp. 1–5.
10. Asheibi, A.; Stirling, D.; Sutanto, D. Determination of the optimal number of clusters in harmonic data classification. In Proceedings of the 2008 13th International Conference on Harmonics and Quality of Power, Wollongong, NSW, Australia, 28 September–1 October 2008; pp. 1–6.
11. Ariyanto, N.; Anggoro, B.; Noegroho, R. New Probabilistic Approach for Identification Event Severity Index Due To Short Circuit Fault. In Proceedings of the IEEE International Conference on Electrical Engineering and Computer Science, Kuta, Indonesia, 24–25 November 2014; pp. 2–6.
12. Seera, M.; Lim, C.P.; Loo, C.K.; Singh, H. Power Quality Analysis Using a Hybrid Model of the Fuzzy Min–Max Neural Network and Clustering Tree. *IEEE Trans. Neural Netw. Learn. Syst.* **2016**, *27*, 2760–2767. [CrossRef] [PubMed]
13. Mahela, O.P.; Shaik, A.G. Recognition of power quality disturbances using S-transform and Fuzzy C-means clustering. In Proceedings of the 2016 International Conference on Cogeneration, Small Power Plants and District Energy (ICUE), Bangkok, Thailand, 14–16 September 2016; pp. 1–6.
14. Pan, D.; Zhao, Z.; Zhang, L.; Tang, C. Recursive clustering K-nearest neighbors algorithm and the application in the classification of power quality disturbances. In Proceedings of the 2017 IEEE Conference on Energy Internet and Energy System Integration (EI2), Beijing, China, 26–28 November 2017; pp. 1–5.
15. Duan, X.; Chen, K. Research on the application of maximal tree method based on fuzzy clustering for Power Quality Evaluation. In Proceedings of the 2014 China International Conference on Electricity Distribution (CICED), Shenzhen, China, 23–26 September 2014; pp. 1284–1287.

16. Filho, J.L.; Borges, F.A.D.S.; Rabelo, R.D.A.L.; Silva, I.S.; Junior, R.P.T.; Filho, A.O.D.C. Methods for voltage sag source location by Cluster Algorithm and Decision Rule Labeling with a Comparative Approach of K-means and DBSCAN Clustering Algorithms. In Proceedings of the 2020 5th International Conference on Smart and Sustainable Technologies (SpliTech), Split, Croatia, 23–26 September 2020; pp. 1–8. [CrossRef]
17. Garcia-Sanchez, T.; Lázaro, E.G.; Muljadi, E.; Kessler, M.; Molina-García, A. Statistical and Clustering Analysis for Disturbances: A Case Study of Voltage Dips in Wind Farms. *IEEE Trans. Power Deliv.* **2016**, *31*, 2530–2537. [CrossRef]
18. Costa, M.V.; Filho, J.M.C.; Leborgne, R.C.; Pereira, N.B. A novel methodology for determining the voltage sag Impact Factor. *Electr. Power Syst. Res.* **2019**, *174*, 105865. [CrossRef]
19. Souza, P.V.G. *Formação de Conjuntos de Sistemas de Distribuição quanto aos Afundamentos de Tensão*; Universidade Federal de Itajubá—UNIFEI: Itajubá, Brazil, 2016. (In Portuguese)
20. Filho, J.M.; De Carvalho Filho, J.M.; Paiva, A.P.; De Souza, P.V.G.; Tomasin, S. A PCA-based approach for substation clustering for voltage sag studies in the Brazilian new energy context. *Electr. Power Syst. Res.* **2016**, *136*, 31–42. [CrossRef]
21. De Almeida, F.A.; Filho, J.M.; Amorim, L.F.; Gomes, J.H.D.F.; De Paiva, A.P. Enhancement of discriminatory power by ellipsoidal functions for substation clustering in voltage sag studies. *Electr. Power Syst. Res.* **2020**, *185*, 106368. [CrossRef]
22. Montgomery, D.C.; Runger, G.C. *Applied Statistics and Probability for Engineers*, 3rd ed.; John Wiley & Sons, Inc.: New York, NY, USA, 2003.
23. Hair, J.F.; Black, W.C.; Babin, B.J.; Anderson, R.E. *Multivariate Data Analysis*, 7rd ed.; Pearson: London, UK, 2014.
24. Johnson, R.A.; Wichern, D.W. *Applied Multivariate Statistical Analysis*; Pearson Prentice Hall: Upper Saddle River, NJ, USA, 2007.
25. Tanure, J.; Tahan, C.; Lima, J.M. Establishing Quality Performance of Distribution Companies Based on Yardstick Regulation. *IEEE Trans. Power Syst.* **2006**, *21*, 1148–1153. [CrossRef]

energies

MDPI

Article

Considerations on Current and Voltage Unbalance of Nonlinear Loads in Residential and Educational Sectors

Gabriel Nicolae Popa *, Angela Iagăr and Corina Maria Diniș

Department of Electrical Engineering and Industrial Informatics, Politehnica University of Timișoara, 5 Revoluției Street, 331128 Hunedoara, Romania; angela.iagar@fih.upt.ro (A.I.); corina.dinis@fih.upt.ro (C.M.D.)
* Correspondence: gabriel.popa@fih.upt.ro; Tel.: +40-254207541

Abstract: Most often, electrical consumers in the residential and educational sectors are different from industrial electrical consumers. Whereas the vast majority of industrial electrical consumers are low-voltage, three-phase (with three or four wires), electrical consumers in the residential and educational sectors are low-voltage, single-phase. However, in practice, electrical consumers in the residential and educational sectors are in large numbers. Usually, current and voltage unbalances are lower in the industrial sector compared to the residential and educational sectors, where there are a large number of low-voltage, single-phase consumers that are connected/disconnected in an uncontrollable way and that need to be wired and balanced on each phase of power transformers from power substations. The purpose of this paper is to present the results of electrical balance and improve the power factor in the power substation from residential and educational sectors. The paper investigates the current and voltage unbalance of nonlinear con sumers in the residential and educational sectors. For this purpose, we performed measurements in the laboratory and the power substation to investigate the unbalance in the three-phase system. Laboratory measurements were made in the unbalanced operation of the single-phase electrical consumers connected at three-phase system. The measurements from power substation were carried out after the electrical consumers were uniformly spread among the three phases from the low-voltage power network, on two different days: a workday and a weekend day. The current and apparent power unbalance were reduced and the power factor was improved using the capacitive single-phase electric consumers (e.g., personal computers, which are in large numbers in such sectors) evenly across the phases.

Keywords: nonlinear consumers; power quality; unbalance; power factor

Citation: Popa, G.N.; Iagăr, A.; Diniș, C.M. Considerations on Current and Voltage Unbalance of Nonlinear Loads in Residential and Educational Sectors. *Energies* **2021**, *14*, 102. https://dx.doi.org/10.3390/en14010102

Received: 3 December 2020
Accepted: 21 December 2020
Published: 27 December 2020

1. Introduction

In the last decades, the widespread use of power electronics from equipment used in the home, as well as huge and costly industrial processes, have increased the awareness of power quality issues and concerns. The study of power quality issues has been a major effort of electric energy suppliers and industrial customers for many years. A perfect power supply that has a pure noise-free sinusoidal wave shape (for voltages and currents) that is always stable if voltage and frequency changes is difficult to obtain in practice. Most electric consumers impose disturbances on the systems that make deviations from ideal power supply. A large number of devices based on power electronics have been added to the home applications and industrial sector, which affects the power quality of the whole distribution system [1–6].

Power quality is important for electric power providers and customers and consists of transient and steady state electromagnetic disturbances in the electrical distribution system. Power quality contents include impulsive transients, interruptions, oscillatory transients, voltage sags, voltage swells, harmonic distortions of the voltage and current, and unbalance and flicker. A number of power quality indices are defined in the time domain (e.g., crest factors, RMS values, voltage sags, power factor) and others power quality indices are defined in the frequency domain (e.g., total harmonic distortion, THD, *K*-factor), depending on the purpose of the application [7–12].

The analysis of power quality has evolved into a substantially different form since the use of the microprocessors and digital signal processors. Today, the proliferation of power quality analyzers has forced a new definition of power quality to accommodate the needs of microelectronic equipment. No longer limited to power engineers, power quality now involves control, electronic, and manufacturing engineers [8,13–15].

Usually, in power plants, up to 92–93% of the produced energy is distributed to the electric consumers. The other 7–8% of the energy is dissipated in the transmission and distribution networks (including power transformers from the power substation) as losses. The special design of distribution transformers from the power substation can be used to decrease the losses (up to 21%) in transformer [16–20].

With the increase of nonlinear consumers in utility distribution systems, the voltage and current waveforms have become more distorted and the power quality has decreased. Utility engineers have to deal with analyzing and planning for the control of the power distortion. With the availability of these power quality analyzers, much more precise control of the processes has been developed, which makes the processes even more susceptible to the effects of power system disturbances. The power quality degradation may result in other adverse effects, such as overheating of the transformer, errors in measurement, voltage unbalances, and reductions in efficiency [10,21–23].

With repetitive operation, single-phase nonlinear loads such as the classical fluorescent lamp with capacitor, compact fluorescent lamp, sodium-vapor lamp, air conditioning system, laptop—normal operation, desktop PC with LCD display, and laser printer can cause current and, sometimes, voltage fluctuations at the point of common coupling (PCC, e.g., at the power substation). In the distribution system, mitigation devices for the fluctuation of voltages and currents can be used with the distribution static compensator [24–26]. For example, the current from fluorescent lamps with electromagnetic ballast has a nonsinusoidal shape because of two nonlinear components, the lamp tube and the iron-core ballast, and the effect increase when the saturation and hysteresis are present [27,28]. Typical appliances from the educational area are different devices like PCs with monitors, TV sets, electric motors, classical fluorescent lamps (with electromagnetic ballast), compact fluorescent lamps, and air-conditioning devices. The currents distortion from these devices depends on the design, operating mode, and voltage level [2,29].

The evolution of power electronics in the last decades, especially of switching mode power supply (SMPS) with large electrolytic capacitors, has increased the number of single-phase nonlinear consumers. The currents of these nonlinear loads are strongly deformed. The PCs and the compact fluorescent lamps with SMPS are responsible for odd current harmonics. The unbalanced consumers connected to the three-phase supply cause power quality problems. SMPSs themselves can be affected by nonsinusoidal voltages. PCs and classical and compact fluorescent lamps are the most significant items, because a large number of these can be connected to a single phase. In residential and education areas, nonlinear loads are increasingly found in classrooms, offices, etc. [30–32].

The unbalance phenomena must be monitored and corrected. An electric device under unbalance supply will create an important unbalance current. The unbalance can affect the single-phase and three-phase loads and cause voltage sags on the electric network. In three-phase power supply, the unbalance of loads has an influence on the magnetization currents through the power transformer but does not cause important output voltage distortion [3,21,33–37].

The improvement of power factor will reduce the network losses and the energy consumption and save money. For power energy providers, the power factor improvement reduces network losses and increases the electrical capacity for productive and can also help to maintain the voltage at the desired level. The consumers with low power factors are penalized with tariff clauses. For a constant active power, if the power factor decreases, the required apparent power and the electrical system losses increase [15,38–42].

Nonlinear loads generate harmonic currents. The LC filter can be used for nonlinear loads with two purposes: first, to improve the power factor of nonlinear loads, and second, to filter the harmonic load currents. There are differences between the harmonics (current and/or voltage) from industrial plants and residential and educational buildings. The consumers from industrial plants are usually large-power and three-phase -supplied (e.g., large variable speed drives). The low-voltage consumers (e.g., lighting) constitute a small part of the total power of the industrial plants [2,4,43–45].

In residential and educational buildings, the number of single-phase, low-voltage consumers is higher than the number of three-phase, large-power consumers. The total load of the residential and educational buildings is about tens or hundreds of kW (usually up to 200 kW), distributed among the three-phase, low-voltage network (e.g., 400V/50 Hz) [8,19,28,31]. The knowledge of balancing these single-phase nonlinear consumers is important to understand the behavior of these loads in different operation modes in low-voltage, three-phase power supplies. Usually in this sector, consumers supplied by SMPS are in the same rooms so they can supply in groups.

In the paper, we show the unbalance and balance of electric consumers (with power factor improvements) using laboratory experiments (different types of electrical loads) and measurements made in the PCC of a residential and educational electrical power substation, on two different days (workday and weekend day). Electric consumers in the residential and educational sectors are nonlinear and low-power, but there are many and they are connected/disconnected in an uncontrollable way (it is extremely difficult to simulate this for a large number of electric consumers). If they (groups of consumers, or different rezistive-inductive (R-L) types) are distributed uniformly approximately by phases, it is possible to achieve an approximate balancing on currents and an improvement of the power factor. This can be highlighted by working during the workday (when there are R-L electrical consumers and many consumers powered by SPMS, e.g., PCs, laptops) and on weekend days (when there are fewer electric consumers powered by SPMS).

The paper is structured in seven sections. In the second section, a review of power quality related to electrical power distribution is presented. Then, in third section, an experimental study on the unbalance condition (made in 2018, before the pandemic crisis, on a semester activity) for consumers from residential and educational sectors is made. In fourth and fifth sections, we show a case study and measurements from the power substation of the residential and educational sectors. Finally, in the sixth and seventh sections, the paper presents discussions and conclusions about the balance of electric consumers in residential and educational sectors.

2. A Review of Power Quality Issues Related to Electrical Power Distribution

Widespread use of nonlinear and time-varying single-phase or three-phase loads increasingly affects the operation of distribution networks in residential, commercial, and industrial sectors. Consequently, single-phase loads are nonsinusoidal, and three-phase loads are nonsinusoidal and unbalanced.

This section is focused mainly on the unbalance and main power factors in the case of three-phase systems with unbalanced and distorted waveforms, this being the most general and widespread case in power delivery systems.

The unbalance includes unequal voltage magnitudes at the fundamental system frequency, fundamental phase angle deviation, and unequal levels of harmonic distortion between the phases [2]. A major cause of voltage unbalance is the asymmetry of the loads, if the loads are not uniformly spread among the three phases. Additional causes of power system voltage unbalance can be single-phase traction, railroad systems, asymmetrical winding impedances transformer, open wye and open delta transformer banks, asymmetrical transmission impedances possibly caused by incomplete transposition of transmission lines, and blown fuses on three-phase capacitor banks [20,22]. Furthermore, the unbalance of voltages represents the most common fault type in electrical networks, which can occur in the case of voltage sags and can cause double-frequency power oscillations [10].

Below, we present some negative effects of voltage unbalances. An electric machine under unbalanced voltage has unbalanced currents at the phases, and the temperature in different parts exceeds the nominal temperatures. For this reason, large and more expensive electric machines may be fitted with protection to detect extreme unbalance. If the supply unbalance increases beyond a fixed limit, the single-phasing protection will trip the machine [43,46].

The unbalanced load creates unbalanced current components that generate harmonic powers flowing backward from the loads to the network [43]. Some electronic equipment, such as computers, may experience problems if the voltage unbalance is more than 2% or 2.5% [46]. Another negative effect of unbalance is increased network losses.

Compensation of the load imbalance reduces the energy loss and is usually combined with reactive power compensation. For this purpose, PWM-based switching compensators (SCs), reactive compensators (RC), or hybrid devices can be used [20,25,32,36].

It is very important to correctly quantify the distortions caused by the nonlinear and unbalanced loads. There are two ways to define the unbalance factors: the European system and the system used by the IEEE. In European standards, the supply voltage unbalance is evaluated using the method of symmetrical components, only for fundamental components (first harmonic, EN 50160) [2].

Negative sequence unbalance represents the ratio between the magnitudes (RMS measured values) of negative (U_1^-) and positive sequence (U_1^+) components of voltage (first harmonic) and current (first harmonic, I_1^-, I_1^+) [47,48]:

$$k_U^- = \frac{U_1^-}{U_1^+} \cdot 100,\ k_I^- = \frac{I_1^-}{I_1^+} \cdot 100 \tag{1}$$

Also, it can be defined the zero sequence unbalance, as the ratio between the magnitudes (RMS measured values) of zero sequence (U_1^0) and positive sequence (U_1^+) components of voltage (first harmonic) and current (first harmonic, I_1^0, I_1^+) [2]:

$$k_U^0 = \frac{U_1^0}{U_1^+} \cdot 100\ [\%],\ k_I^0 = \frac{I_1^0}{I_1^+} \cdot 100 [\%] \tag{2}$$

In most practical systems, $k_U^0 < 4\%$. An approximate way to calculate the negative sequence unbalance (in %) is [2]:

$$k_U^- \approx \frac{S_L}{S_{sc}} \cdot 100\ [\%] \tag{3}$$

In the above relation, S_L represents the apparent power of the load and S_{SC} represents the short-circuit power of the supply circuit.

It can be defined a total voltage (or current) unbalance factor (in %) [43,47,49]:

$$k_U = k_U^- + k_U^0\ [\%], \quad k_I = k_I^- + k_I^0 [\%] \tag{4}$$

For low-voltage (LV) and medium-voltage (MV) systems, the EN 50160 standard specifies that, under normal operating conditions, during each period of one week, 95% of the 10 min mean RMS values of the negative phase sequence component (fundamental) of the supply voltage shall be within the range of 0% to 2% of the positive phase sequence component (fundamental). This European Standard provides a mediation of the measured quantities for 10 min. Only values for the negative sequence component are given, because this component is the relevant one for the possible interference of appliances connected to the system (EN 50160). In some areas where there are some users with largely single-phase or two-phase loads, unbalance up to about 3% may occur at the three-phase supply terminal. The IEC recommends that the maximum voltage unbalance of electrical LV and MV supply systems be limited to 2% (IEC 61000-2-2:2002+A2:2019, IEC 61000-2-12:2003).

At the load terminals, an unbalance factor of apparent power (in %) can be defined using [49]:

$$S_{unb} = k_U \cdot k_I \tag{5}$$

The IEEE system uses the voltage unbalance in percent, defined as the ratio between the maximum deviation from average and the average of three phase-to-phase voltages (line voltages, IEEE Std 112TM: 2004). The phase angle unbalance does not appear in this definition, because it is based only on magnitudes.

The power losses that occur in AC power systems depended on frequency, waveform distortion, and unbalance [7,12,23,50]. In the presence of phase displacement, unbalance and waveform distortion it is difficult, but very important, to correctly define the apparent power (*S*) and power factor (*PF*), which represent a measure of the system's power delivery capability [5,12,14].

In the last seven decades, there have been many ways of defining and measuring the apparent power and power factor, both in single- and three-phase systems. Even today, these concepts have not been defined in one general and universally accepted way [2,5–12].

A good power theory should explain and describe power-related phenomena in electrical systems for all possible situations and should be able to be used for filter design and reactive compensation to improve the power factor [5,8]. Among the numerous approaches to the power theory, the Current Physical Components (CPC) theory, developed by Czarnecki [8,15,23], the Conservative Power Theory (CPT), developed by Tenti [6] and the *p-q* instantaneous power theory, developed by Akagi [12], are the most distinguished.

One of the main differences of these approaches consists in the description of the power properties of the systems in the time domain, and in the frequency domain respectively. Another difference is the use of instantaneous values of power, and the averaged values (over the period *T*) respectively [5,6,8,12,15,23].

All these power theories have certain limitations. Thus, CPT is formulated in the time domain and does not provide fundamentals for compensator design [6]. The *p-q* instantaneous power theory is formulated in the time domain and is not valid in systems with nonsinusoidal and/or asymmetrical voltages [12]. In addition, there is a major disadvantage of this theory due to the fact that it is based on instantaneous values and the power properties of the systems cannot be identified instantaneously.

The CPC theory, proposed by Czarnecki, introduces some current components: active current, reactive current, scattered current, load generated current, and unbalanced currents (of the positive, negative, and the zero sequence). The reactive current and the scattered current are defined in the frequency domain. The reactive current is related to the phase shift of the voltage and current harmonics. The scattered current is related to the load conductance, which changes with harmonic frequency. The CPC theory can explain the phenomena in single- and three-phase systems, at a sinusoidal or nonsinusoidal supply voltage, and with linear or nonlinear loads, respectively in three-phase systems with asymmetrical voltage [8,15,23].

A new, more general, formulation of power theory, in the frequency domain and in the elementary vector space linear algebra, has been described by the authors of [5]. This power theory is valid for poly-phase systems, with nonsinusoidal waveforms and DC components, voltage and current unbalance, and unequal and/or frequency-dependent wire impedances.

Further, we present the IEEE Std 1459-2010 approach. IEEE Std. 1459–2010 demonstrates that fundamental positive sequence active power P_1^+ is the only useful power supplied to the load and unity power factor means minimum possible line losses for a given total active power transmitted (IEEE Std 1459–2010). Maximum efficiency in the electric network is reached when only fundamental positive sequence active current is demanded and the voltages at the PCC only contain the fundamental positive sequence voltage [9,11,16,32].

For three-phase nonsinusoidal and unbalanced systems, the most general case, the basic quantity is the three-phase instantaneous power. For four-wire systems, the instantaneous power has the following expression (IEEE Std 1459–2010):

$$p = u_a \cdot i_a + u_b \cdot i_b + u_c \cdot i_c \text{ (W)} \tag{6}$$

where u_a, u_b and u_c are instantaneous line-to-neutral voltages, and i_a, i_b and i_c are instantaneous line currents.

IEEE Std. 1459-2010 defines the effective apparent power assuming a virtual sinusoidal and balanced circuit which has exactly the same line power losses as the real circuit (nonsinusoidal and unbalanced). In this approach, the effective apparent power is considered the maximal active power which can be transmitted by the virtual system (with sinusoidal and balanced voltages and currents), with the same voltage impact and the same transmission losses.

Effective apparent power is given by (IEEE Std 1459-2010):

$$S_e = 3U_e \cdot I_e \text{ (VA)} \tag{7}$$

where U_e and I_e represent the effective line-to-neutral voltage and the equivalent current.

The equivalent current I_e is defined as the RMS value of the three-phase currents from fictitious circuit (balanced and sinusoidal), which yields the same losses as the currents from the real circuit.

The RMS current can be separated into two components—the fundamental I_{e1} and the nonfundamental I_{eH} [9,14,16]:

$$I_e = \sqrt{I_{e1}^2 + I_{eH}^2} \text{ (A)} \tag{8}$$

In the case of four-wire systems, I_{e1} and I_{eH} are given by the relations (IEEE Std 1459-2010):

$$I_{e1} = \sqrt{\frac{I_{a1}^2 + I_{b1}^2 + I_{c1}^2 + \rho_1 \cdot I_{n1}^2}{3}} \text{ (A)}, \quad \rho_1 = \frac{K_{sn1} \cdot r_{nDC}}{K_{s1} \cdot r_{DC}} \tag{9}$$

$$I_{eH} = \sqrt{\frac{\sum_{h \neq 1} k_h (I_{ah}^2 + I_{bh}^2 + I_{ch}^2) + \rho_h \cdot I_{nh}^2}{3}} \text{ (A)}, \; k_h = \frac{K_{sh}}{K_{s1}}, \; \rho_h = \frac{K_{snh} \cdot r_{nDC}}{K_{s1} \cdot r_{DC}} \tag{10}$$

K_{s1} and K_{sn1} are the skin and proximity effect coefficient of the supplying line conductor and the neutral current path at fundamental frequency (50 Hz); K_{sh} and K_{snh} are the skin and proximity effect coefficients of the supplying line conductor and the neutral current path, respectively, computed for the h harmonic order, or any frequency component present in the currents spectra; r_{DC} is the DC line resistance and r_{nDC} is the DC resistance of the neutral current path.

In most practical applications, the ratios ρ_1, ρ_h and k_h are not known, being functions of temperature, network topology, and loading. Therefore, it is recommended to use the values $\rho_1 = \rho_h = k_h = 1$, which leads to the following practical expressions (IEEE Std 1459-2010):

$$I_e = \sqrt{\frac{I_a^2 + I_b^2 + I_c^2 + I_n^2}{3}} \text{ (A)} \tag{11}$$

$$I_{e1} = \sqrt{\frac{I_{a1}^2 + I_{b1}^2 + I_{c1}^2 + I_{n1}^2}{3}} \text{ (A)} \tag{12}$$

$$I_{eH} = \sqrt{I_e^2 - I_{e1}^2} \text{ (A)} \tag{13}$$

The practical expressions for the effective voltage, in the case of four-wire systems, are (IEEE Std 1459-2010):

$$U_e = \sqrt{U_{e1}^2 + U_{eH}^2} \text{ (V)} \tag{14}$$

$$U_e = \sqrt{\frac{3(U_a^2 + U_b^2 + U_c^2) + U_{ab}^2 + U_{bc}^2 + U_{ca}^2}{18}} \text{ (V)} \quad (15)$$

$$U_{e1} = \sqrt{\frac{3(U_{a1}^2 + U_{b1}^2 + U_{c1}^2) + U_{ab1}^2 + U_{bc1}^2 + U_{ca1}^2}{18}} \text{ (V)} \quad (16)$$

$$U_{eH} = \sqrt{U_e^2 + U_{e1}^2} \text{ (V)} \quad (17)$$

In the above relations: I_a, I_b, I_c and I_n are the line currents and neutral current; U_a, U_b, U_c are the line-to-neutral voltages; U_{ab}, U_{bc}, U_{ca} are the line-to-line voltages.

Effective apparent power can be separated into the fundamental effective apparent power S_{e1} and the nonfundamental effective apparent power S_{eH} (IEEE Std 1459-2010):

$$S_e = \sqrt{S_{e1}^2 + S_{eH}^2} \text{ (VA)} \quad (18)$$

$$S_{e1} = 3U_{e1} \cdot I_{e1} \text{ (VA)} \quad (19)$$

The load unbalance can be evaluated using the fundamental unbalanced power (IEEE Std 1459-2010):

$$S_{u1} = \sqrt{S_{e1}^2 - (S_1^+)^2} \text{ (VA)} \quad (20)$$

where:

$$S_1^+ = \sqrt{(P_1^+)^2 + (Q_1^+)^2} \text{ (VA)} \quad (21)$$

$$P_1^+ = 3U_1^+ \cdot I_1^+ \cdot cos\theta_1^+ \text{ (W)} \quad (22)$$

$$Q_1^+ = 3U_1^+ \cdot I_1^+ \cdot sin\theta_1^+ \text{ (VAR)} \quad (23)$$

In Equations (20)–(23), S_1^+ is the fundamental positive sequence apparent power; P_1^+ is the fundamental positive sequence active power; Q_1^+ is the fundamental positive sequence reactive power; θ_1^+ is the phase angle between the positive sequence components (first harmonic) of current and voltage (rad).

According to IEEE Std. 1459-2010, the main power factors are (IEEE Std 1459-2010):

- fundamental positive sequence power factor (or displacement power factor, DPF):

$$DPF = PF_1^+ = \frac{P_1^+}{S_1^+} = \frac{P_1^+}{\sqrt{(P_1^+)^2 + (Q_1^+)^2}} \quad (24)$$

- power factor:

$$PF = \frac{P}{S_e} \quad (25)$$

where P represents the active power:

$$P = \frac{1}{kT}\int_{\tau}^{\tau+kT} p \cdot dt = P_a + P_b + P_c \text{ (W)} \quad (26)$$

In Equation (26) p is the instantaneous power. According to Equation (24), increasing the PF_1^+ to the unity can be done by compensating the fundamental positive sequence reactive power ($P_1^+ = S_1^+$). Equation (25) shows that, in three-phase systems with nonlinear unbalanced loads, the ideal situation (maximum utilization of the line, PF = 1) is obtained when P = S_e.

Total harmonic distortion (THD) is a measurement of the distortion of voltages or currents due to harmonics. THD of voltages (or currents) is defined as the ratio of the RMS voltage (or current) of all the harmonic frequencies (from the second harmonic on) over the RMS voltage (or current) of the fundamental frequency [7,14,20]:

$$VTHD_i = \frac{\sqrt{\sum_{h=2}^{h=h_{max}} (V_{hi})^2}}{V_{1i}} \cdot 100, \ [\%] \tag{27}$$

$$ITHD_i = \frac{\sqrt{\sum_{h=2}^{h=h_{max}} (I_{hi})^2}}{I_{1i}} \cdot 100, \ [\%] \tag{28}$$

In Equations (27) and (28), V represents the phase voltage, I represents the line current, i represents the phase (i = 1, 2, 3), and h represents the order of the harmonic (h_{max} = 40, according to EN 50160).

Crest factor (CF) is calculated from the peak amplitude of the waveform divided by the RMS value of the waveform (voltage or current). When voltage and current have sinusoidal waveforms, the crest factor is 1.41 ($\sqrt{2}$). If CF > 1.41, then the waveform is a sharp nonsinusoidal waveform, and CF < 1.41 indicates a flat nonsinusoidal waveform [7,14,20].

$$VCF_i = \frac{V_{peak\ i}}{V_{RMS\ i}} \tag{29}$$

$$ICF_i = \frac{I_{peak\ i}}{I_{RMS\ i\ i}} \tag{30}$$

In Equations (29) and (30) i represents the phase (i = 1, 2, 3).

K-factor is a weighting of the harmonic load currents according to their effects on transformer heating. A high value of K-factor means large heating effects due to harmonic currents. K-factor is defined as a ratio between the additional losses due to harmonics and the eddy current losses at 50 Hz [7,14,20]:

$$K - factor = \sum_{h=1}^{h=h_{max}} I_h^2 \cdot h^2 \tag{31}$$

where I_h represents the RMS current at harmonic h, in per unit of rated RMS load current.

The duration of maximum power usage (consumption), T_{max}, over a time period t can be calculated with [7,14]:

$$T_{max} = \frac{\int_0^t i\ dt}{I_{max}} \tag{32}$$

and the duration of maximum losses, τ_{max}, over a time period t can be calculated with:

$$\tau_{max} = \frac{\int_0^t i^2\ dt}{I_{max}} \tag{33}$$

where I_{max} represents the current corresponding to the maximum power.

3. Laboratory Experimental Study in Unbalance Condition of Electric Loads Used in Residential and Educational Buildings

In this section, we present the experimental results from most often used consumers from the residential and educational sectors in an unbalanced operation. In order to analyze the unbalance of voltages and currents for different nonlinear electrical consumers in three-phase systems on each phase and different voltage amplitudes, laboratory measurements were performed.

The laboratory tests were made with nonlinear electric consumers. During the experiments, the voltage source had low distortion (THD < 1.5%).

To investigate the voltage unbalance and the current unbalance in three-phase power systems, an experimental test (Figure 1) was carried out using a PC (maximum 400 W) on phase *b* (*Load 1*), and two classical fluorescent lamps (2 × 20 W) on other phases *a* (*Load 2*) and *c* (*Load 3*). The lamps did not have capacitors to the improve power factor in the experiments.

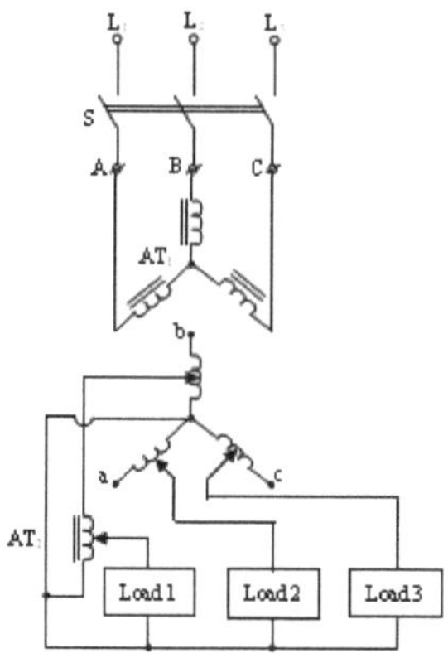

Figure 1. Laboratory experimental setup.

The electrical measurements (currents and voltages) were made on *Load 1*, *Load 2*, and *Load 3* using a power quality analyzer CA 8334B to measure the current were used current probe *MN 93A* (maximum 5A). A sampling frequency of 12.8 kHz was imposed in order to avoid aliasing and leakage errors. The measurements were made in the three-phase secondary windings of power autotransformer AT_1 (10 kW, Y/y connections).

In the following, we present four sets of experiments. During the experiment sets (set 1–4), the power quality analyzer CA 8334B was directly connected to *Load 1*, *Load 2*, and *Load 3*. In the experiments, we used two classic fluorescent lamps (with inductive ballast 2 × 20 W), identical in design but different at times of operation. In particular, we used an old lamp (10 years of operation) and a new one. Using the power quality analyzer, we measured the crest factor for the voltages and currents, the voltage and current unbalance for three-phase systems, the Fresnel diagrams, the power factor (PF), and the displacement power factor (DPF).

In the first set of experiments, the phase voltages were changed identically between 150 V and 230 V on the single-phase consumers: *Load 1* involved a PC with LCD display; *Load 2* involved an old classic fluorescent lamp 2 × 20 W; *Load 3* involved a new classic fluorescent lamp 2 × 20 W. In this set of experiments, AT_2 was missing from the experimental setup (*Load 1* was connected directly on winding *b* from Figure 1).

Figures 2 and 3 present the crest factor of phase voltages and the crest factor of currents depending on the supply voltages.

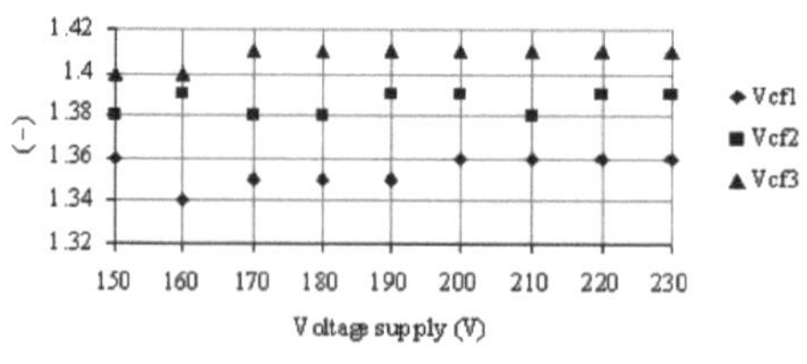

Figure 2. Crest factors V_{cf1}, V_{cf2}, and V_{cf3} for the phase voltages *a*, *b*, and *c* depending on the supply voltages (set 1).

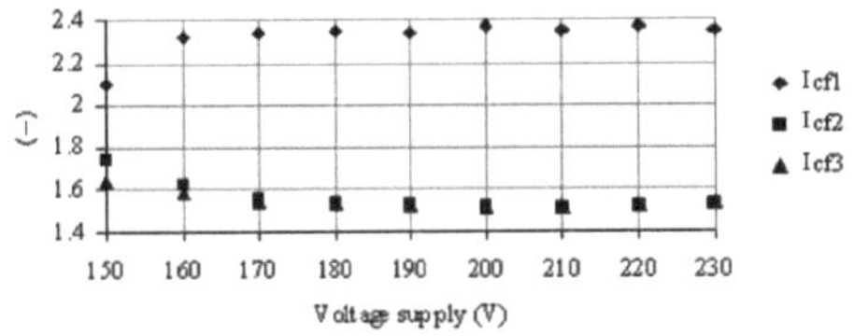

Figure 3. Crest factor I_{cf1}, I_{cf2}, and I_{cf3} for currents on phases *a*, *b*, and *c* depending on the supply voltages (set 1).

The crest factor of voltages and currents for the three consumers change with the supply voltage and the type of consumer are presented in Figures 2 and 3, respectively. With increasing voltage, the crest factor of voltages increased, but it was different even for the same type of nonlinear consumers (old and new classic fluorescent lamps). Deviation from sinusoidal waveform was evidenced by the crest factor values. For the PC, current crest factor had high values (Figure 3), whereas for fluorescent lamps, crest factor was closer to the value of sinusoidal waveforms (1.41).

The unbalance factor for voltages V_{unb} and currents I_{unb} are presented in Figures 4 and 5, respectively, depending on the supply voltages. It was found that V_{unb} was low, below 1.3% (the low-power, nonlinear consumers do not affect the voltages unbalance). For currents, I_{unb} was very high, reaching 58% for low-supply voltage (e.g., 150 V). Even at nominal voltage of the nonlinear voltage consumers, I_{unb} had high values (45%).

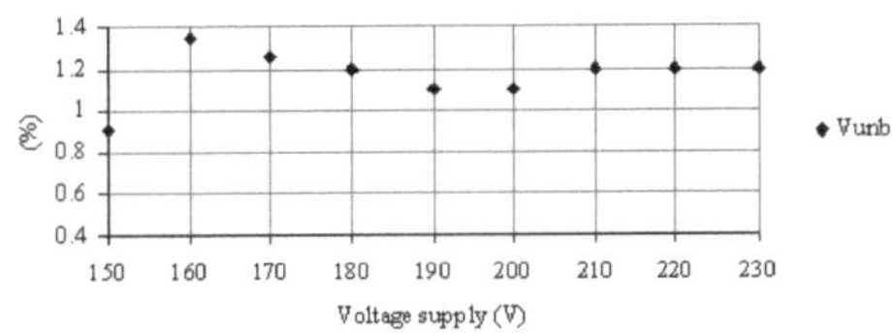

Figure 4. Unbalance factor for voltages V_{unb} depending on the supply voltages (set 1).

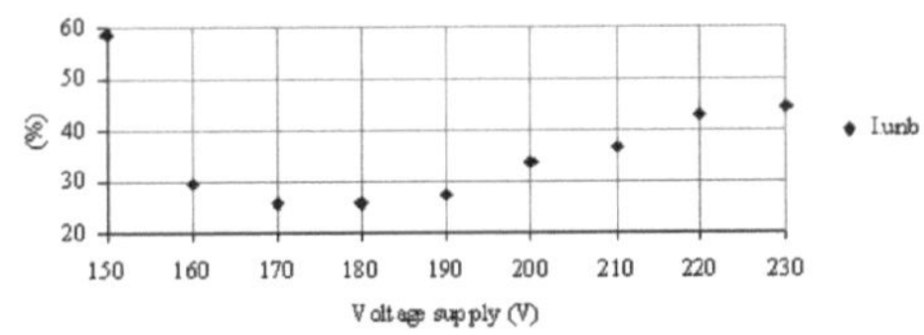

Figure 5. Unbalance factor for currents I_{unb} depending on the supply voltages (set 1).

The following figures show the Fresnel diagrams of voltages and currents for 190 V (Figure 6) and 230 V (rated voltage, Figure 7).

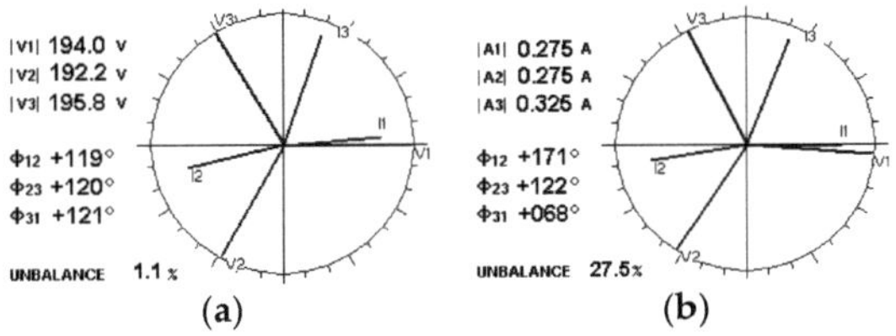

Figure 6. Fresnel diagrams for fundamental voltages and currents for 190 V phase voltage (set 1). (**a**) is voltages. (**b**) is currents.

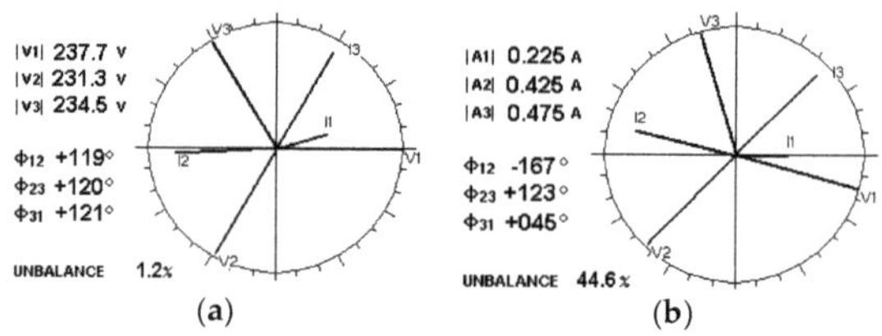

Figure 7. Fresnel diagrams for fundamental voltages and currents for 230 V (set 1). (**a**) is voltages. (**b**) is currents.

From Figures 6 and 7, we found a strong unbalance of currents characterized by different *RMS* values on the three different phases (a, b, and c) and different phase shifts (are not 120°). The Fresnel diagrams indicated the capacitive reactive type of the PC and inductive reactive type of fluorescent lamps.

Power factors PF_1, PF_2, and PF_3 (Figure 8) and displacement power factors DPF_1, DPF_2, DPF_3 (Figure 9) for the three phases decreased with increasing supply voltage. The values of PFs were under 0.8. In the case of the PC, PF values were between 0.8 and 0.69. We noticed a big difference between PF and DPF for the PC consumer ($DPF \cong 1$), indicating a large deviation of current from sinusoidal waveform.

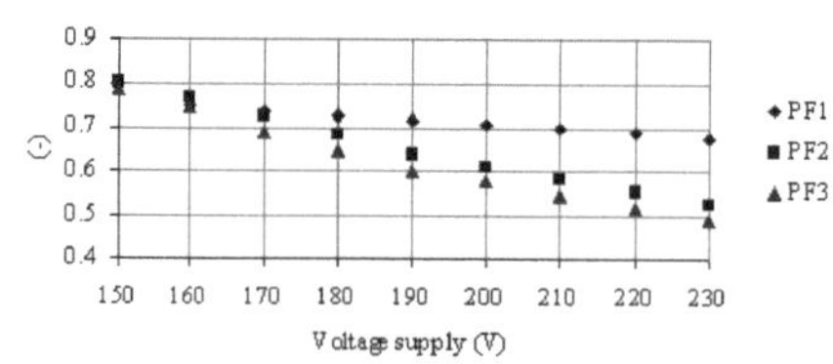

Figure 8. Power factors PF_1, PF_2, and PF_3 on each phase (a, b, c) depending on the supply voltages (set 1).

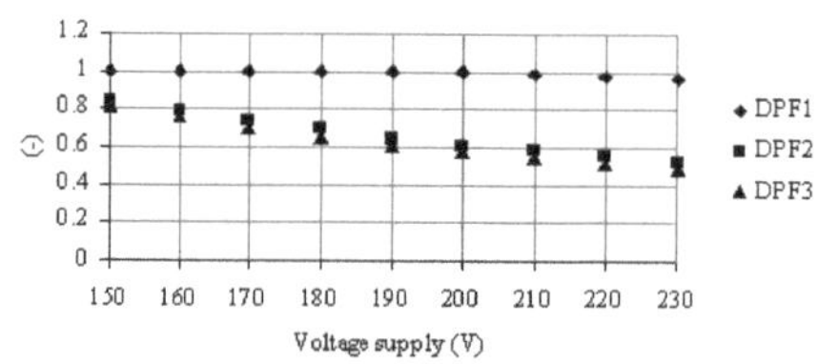

Figure 9. Displacement power factors (for fundamental components) DPF_1, DPF_2, and DPF_3 on each phase (a, b, c) depending on the supply voltages (set 1).

As the fluorescent lamps were not used capacitors to improve power factor, DPF values had an evolution between 0.8 and 0.5, and $PF \cong DPF$, indicating a closer sinusoidal waveform of current.

Set 2 of measurements is presented below. Phase *b* changed between 150 V and 230 V (Figure 1) using autotransformer AT_2, and the voltage on the other two phases (*a* and *c*) remained constant (at rated value, 230 V). The consumers were distributed identically with set 1 measurements: The PC with LCD display in phase *b*, and each fluorescent lamp with inductive ballast 2 × 20 W in phases *a* and *c* (Figure 4).

Figure 10 shows the evolution of V_{unb} and I_{unb} with increasing of supply voltage to the rated value for the set 2 experiments.

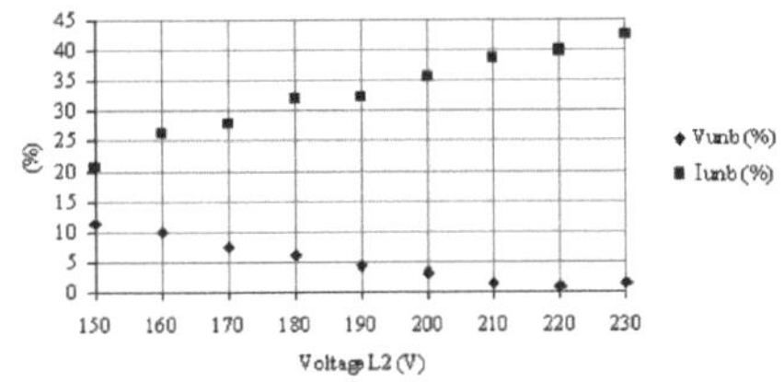

Figure 10. The unbalance factor for voltages V_{unb} and currents I_{unb} depending on the supply voltage of phase *b* (set 2).

We found a more disadvantageous operation than set 1 of experiments, because for voltages between 150 V and 190 V, the unbalance factor for voltage V_{unb} was above 5% (at low voltages). I_{unb} values were comparable to those of the previous set (set 1). If the voltage on *Load 1* was less than the other phases, V_{unb} increased and I_{unb} decreased. When the three phase voltages were equal, it obtained the same values as the set 1 of measurements (Figures 7 and 8).

The average values of power factor PF and DPF of the three-phase voltages were lower when using low voltage values (Figure 11). DPF varied between 0.79 and 0.69, and PF was modified between 0.67 and 0.59. The big difference between PF and DPF indicates deviation from the sinusoidal waveforms of the input current for the three consumers.

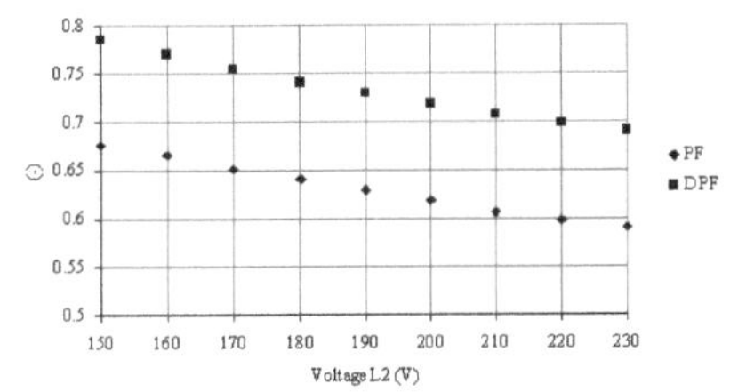

Figure 11. The average value PF and the average value DPF at the three phases depending on the supply voltage at phase *b* (set 2).

In set 3 of measurements, the phase voltage *a* (Figure 1) was modified between 150 V and 230 V and the voltage on the other two phases remained constant (at rated value). Phase *b* was connected to the PC with LCD display and other two phases were connected to the two fluorescent lamps 2 × 20 W (with inductive ballast).

Figure 12 shows the evolution of V_{unb} and I_{unb} when the phase voltage *a* (with PC consumer) was modified. The evolution of V_{unb} was similar to that of the set 2 experiments.

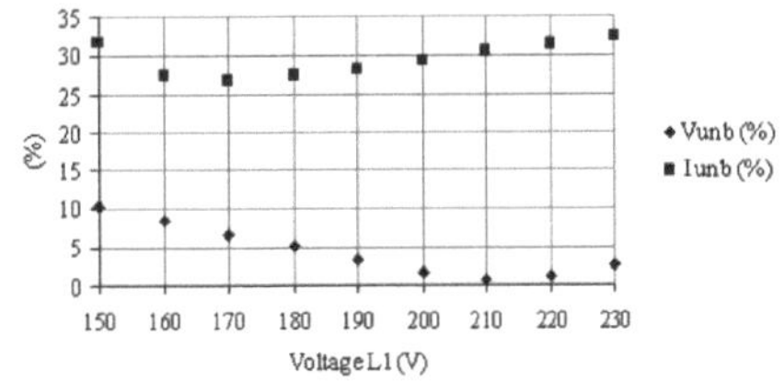

Figure 12. The unbalance factor for voltages V_{unb} and currents I_{unb} depending on supply voltage on phase *a* (set 3).

The average values of the PF and DPF at the three phases are shown in Figure 13 DPF had an almost constant value ($DPF \cong 0.695$) with increasing supply voltage, and PF decreased at minimum values at the rated voltage (230 V). The average values of the PF and DPF were slightly smaller than the set 2 of measurements.

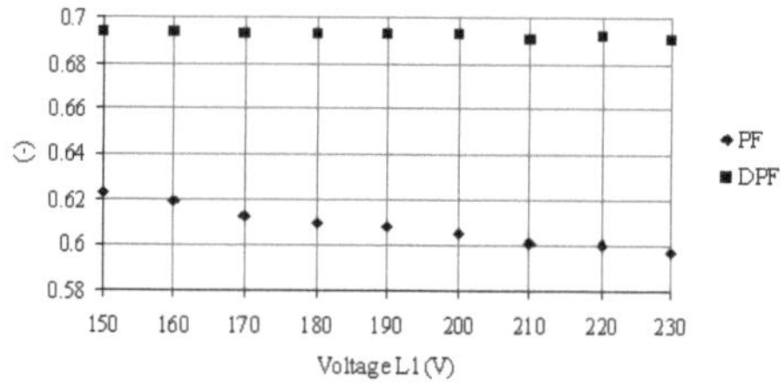

Figure 13. The average value PF and the average value DPF at the three phases depending on the supply voltage at phase *a* (set 3).

The goal of the set 4 measurements was to study neutral current for three identical nonlinear consumers. We used three compact fluorescent lamps (CFLs using SMPS) with the power of 18 W, with a star connection (each lamp was connected between a phase and neutral conductor).

In the series with each fluorescent lamp, we connected a breaker in order to disconnect/connect each lamp.

With power quality analyzer CA 8334B, we measured the voltage between L_1 and neutral conductor (0), and the current through neutral conductor with *MN 93 A* (max. 5 A) probe (Figures 14–16).

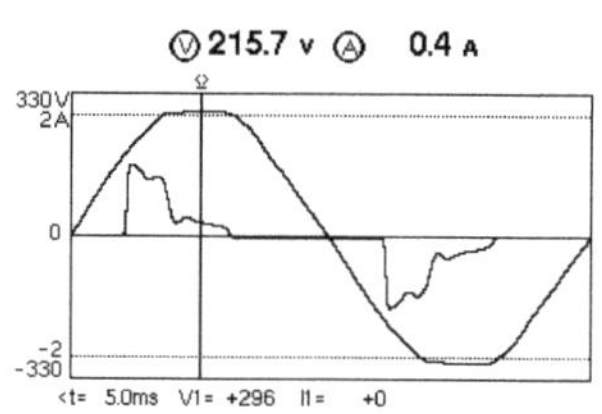

Figure 14. The phase voltage and the neutral current for a compact fluorescent lamp (set 4).

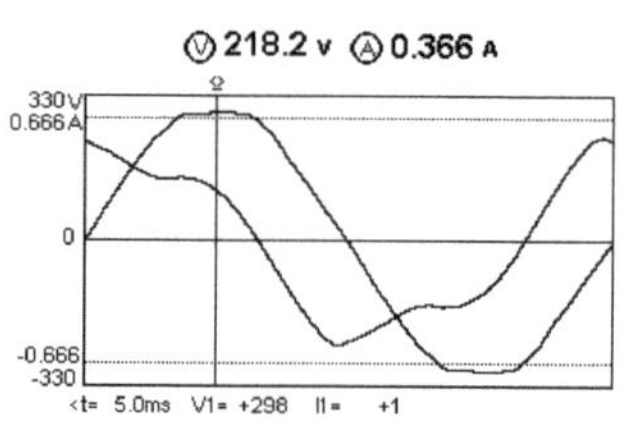

Figure 15. The phase voltage and the neutral current when two compact fluorescent lamps are supplied at different phase L_1 and L_2 (set 4).

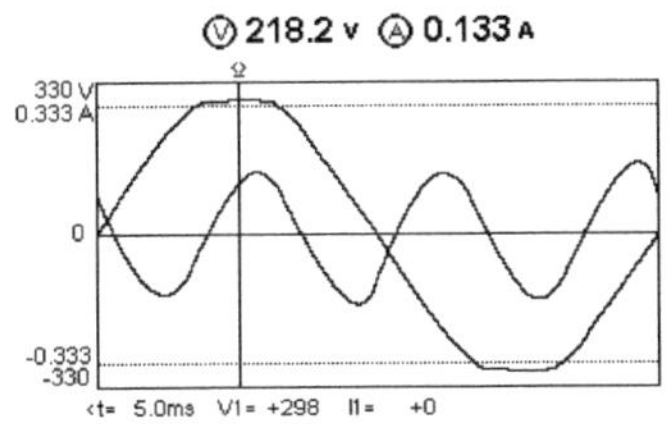

Figure 16. The phase voltage and the neutral current when all the three compact fluorescent lamps are supplied at different phases L_1, L_2, L_3 and neutral conductor (set 4).

The first measurement (Figure 14) was achieved when it was energized with only one CFL at one phase (L_1), with the other two CFLs being disconnected (from L_2 and L_3) by switches.

The neutral current (I_N) is the same as the value of the phase current through the fluorescent lamp. The current through the lamps was strongly deformed (with multiple current harmonics, the third order and multitude of three (9, 15, 21).

At the second measurement (Figure 15), two fluorescent lamps were connected at two different phase voltages (at phases L_1 and L_2), and the third fluorescent lamps was disconnected from the voltage (phase L_3).

From Figure 15, we found that the neutral current was less deformed and had higher values than in the first situation (current through one CFL, Figure 14).

Figure 16 measures the current through the neutral conductor when the three lamps were powered separately from each phase, L_1, L_2, and L_3, and the neutral conductor (0). Basically, the current through the neutral conductor was zero if the electrical consumers were the same and linear (extremely rare in practice). In the case of compact fluorescent lamps, the currents have harmonics of the third order and multiplies of three orders, and they add up algebraically through neutral conductor [34]. A current occurred through the neutral conductor with a frequency of 150 Hz (Figure 16). So, for nonlinear consumers, even if they are balanced on the three phases, there will be an important neutral current that cannot be reduced.

The set 1 experiments show that, to a relatively small extent, the crest factors for voltage depend on the amplitude of the supply voltage and the type of consumers (Figure 2). When using electrical consumers powered by SMPS, the crest factors were slightly lower (0.05) than the ideal case. It was found that R-L-type electrical consumers did not significantly reduce the crest factors values (the current was close to sinusoidal shape). Basically, the crest factor for the current decreased with the decreasing voltage (Figure 3). The crest factors for current depends on the type of electricity consumer: electricity consumers powered by SMPS had higher values compared with R-L consumers. The voltage unbalance was small (less than 1.4%) and depended on the supply voltage amplitude (Figure 4). In contrast, the current unbalance was higher (below 45%) and had higher values as the supply voltage increased (Figure 5). Electric consumers powered by SMPS had a capacitive character to which, with the decrease of the supply voltage, they had lower values. Using the R-L classic consumers, both the currents and the phase shifts between voltages and currents increased with the supply voltage amplitude (Figures 6 and 7). In both types of consumers, the PF decreased with increasing supply voltage: A smaller decrease (0.1) was found with consumers powered by SMPS compared to the larger decrease (0.25–0.3) found in the R-L consumer type (Figure 8). DPF (Figure 9) had higher values (1) for SMPS consumers compared to R-L consumers, which had lower values for DPF (0.55).

In set 2 experiments, when the amplitude of a phase decreased (Figure 10), the voltage unbalance changed a lot (toward 10%). The current unbalance increased with the supply voltage (values > 40%). The PF and DPF decreased with increasing supply voltage (Figure 11). In set 3 of experiments, when the amplitude of a phase decreased (Figure 12), the voltage unbalance changed (toward 10%), and the current unbalance increased with the supply voltage (values > 32%). The PF decreased and the DPF remained constant as the supply voltage increased (Figure 13).

If nonlinear single-phase and/or nonlinear three-phase consumers (with SMPS) are used in three-phase systems, which have harmonics of currents with rank-three and a multiple of three, the neutral current (I_N) zero cannot be diminished, because these current harmonics add up (in some situations, the neutral current can be comparable to the currents on the phases).

From the experimental measurements in the laboratory, it was found that the type of electrical consumers (powered by SMPS and R-L type) and the amplitude of the supply voltage (even a few volts) influenced the unbalance of voltage (to a small extent) the unbalance of current, and the PF and DPF values (to a large extent).

4. A Case Study about Residential and Educational Electrical Grid Distribution

The measurements (voltages and currents) from electric power station were made in the PCC of the LV power substation.

The residential and educational campus consisted of six buildings and comprised 12,800 m^2 of floor space.

Figure 17 shows a wire circuit diagram for typical residential and educational buildings electric consumers. The circuit consisted of two main power transformers T_1 (400 kVA, 6/0.4 kV, D/y 11) and T_2 (250 kVA, 6/0.4 kV, D/y 11) and two main low-voltage branches that supplied the consumers through two main breakers Q_1 (to supply the consumers from T_1) and Q_2 (to supply the consumers from T_2). A transversal circuit breaker Q_3 connected the two main branches. These branches were linear and had a lot of nonlinear consumers. Usually, the electrical consumers are connected to Q_1 and Q_3 circuit breakers from transformer T_1.

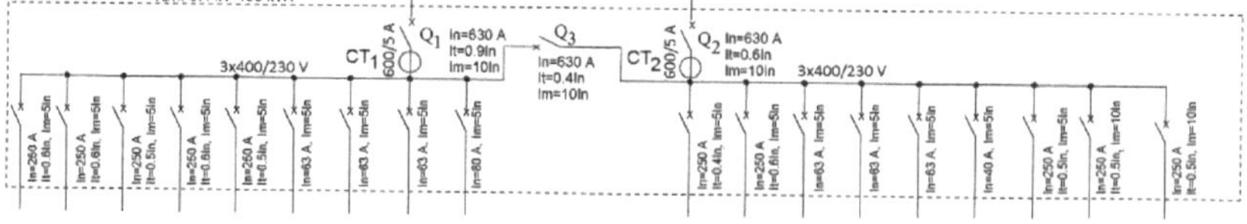

Figure 17. Wire electrical circuit diagram for residential and educational buildings.

The upper side of Figure 18a presents the results of the measurements of the power substation, the phase voltages, and currents from the network. Figure 18b–v shows the waveforms of voltages and currents from different nonlinear consumers used often in residential and educational buildings.

Initially, the electrical nonlinear consumers were distributed nonequally among the three phases.

The monitoring equipment (power quality analyzer CA 8334B) were connected in PCC after circuit breaker Q_1. The current was measured with *AmpFlex* probes (3000 A). The connection wired of the power quality analyzer used for the measurements was three currents (on each phase) and four wires for voltages (each phase and neutral conductor).

From Figure 18b–v, it was found that the current waveforms were different from one consumer to another. Some consumers had a current waveform closer to sinusoidal (e.g., classical fluorescent lamps, sodium-vapor lamps, refrigerator, high-speed angular grinder machines, three-phase induction motors (Figure 18)), and other consumers had a highly distorted current from sinusoidal waveform (e.g., compact fluorescent lamps, air-conditioning system, laptops, PCs, laser printers, single-phase static electric drive converters, induction furnace medium-frequency converters (Figure 18)).

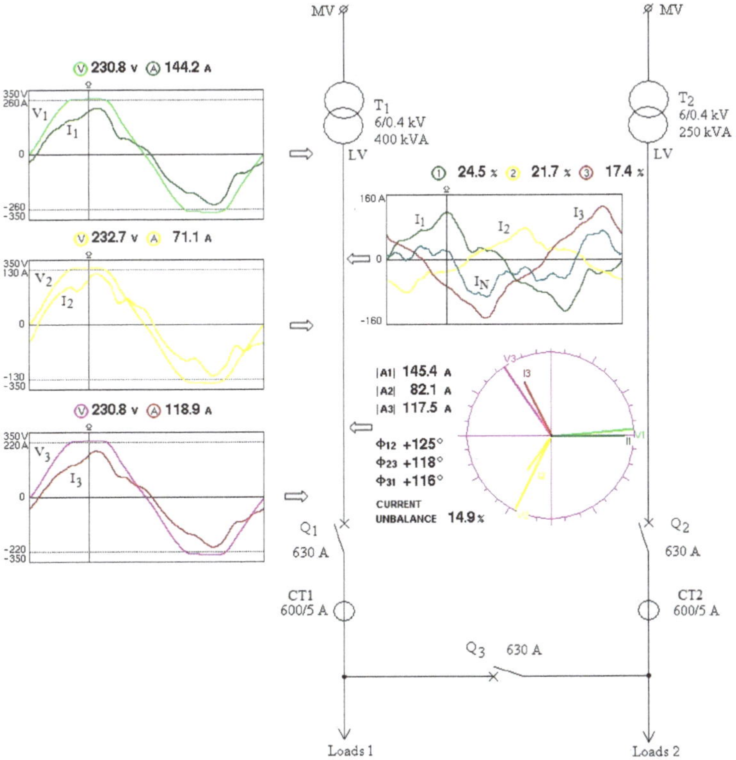

(a)

Figure 18. *Cont.*

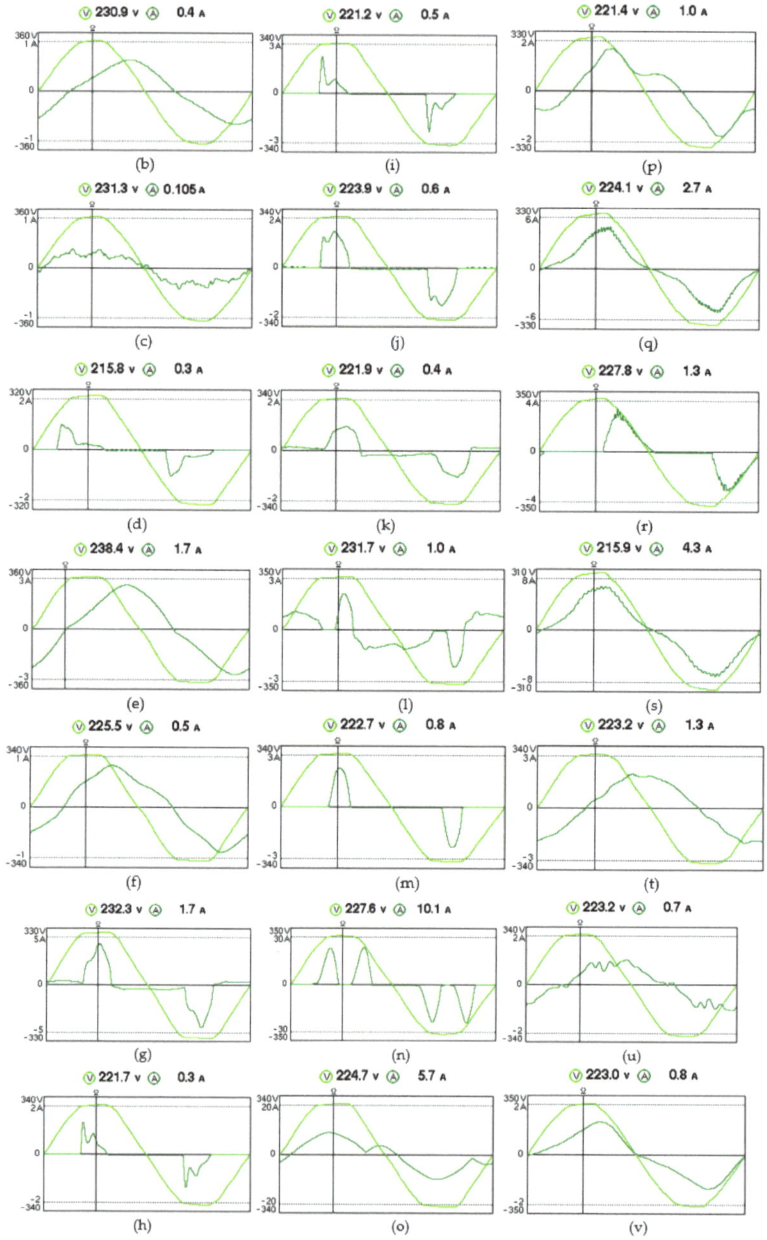

Figure 18. (**a**). Electrical measurements, voltages, and currents, in power substations. Electrical measurements, voltages, and currents, from different nonlinear consumers used in residential and educational buildings: (**b**) Classical fluorescent lamp; (**c**) Classical fluorescent lamp with capacitor; (**d**) Compact fluorescent Ip; (**e**) Sodium-vapor lamp; (**f**) Refrigerator; (**g**) Air conditioning system; (**h**) Standby laptop; (**i**) Laptop—normal operation; (**j**) Desktop PC with cathode tube ray monitor; (**k**) Desktop PC with LCD monitor; (**l**) Laser printer; (**m**) Electric drive single-phase using static converter; (**n**) Induction furnace medium frequency converter (measurement on one phase); (**o**) Microwave oven; (**p**) Portable grinder machine; (**q**) Portable finishing machine; (**r**) Portable high-speed milling machine; (**s**) Angular high-speed grinder machine; (**t**) Three-phase induction motor (measurement on one phase); (**u**) Three-phase induction motor with capacitor (measurement on one phase); (**v**) Single-phase induction motor using triac variator.

In residential and educational facilities, a lot of PCs (desktops and laptops) are usually working. The power sources of these electric consumers are SMPSs, with large electrolytic capacitors. These consumers are the capacitive reactive type, and using a large number of these consumers can improve the power factors for other inductive-reactive consumers (e.g., classical fluorescent lamps, electric motors, refrigerators, etc.).

The thermal images (obtained using thermal imagers FLIR 420 and Fluke Ti 25) of different parts of electrical installations in operation allowed the identification of thermal demands over the allowable limits and showed whether or not the electrical equipment were working correctly (Figure 19). Unbalanced loading phases can be observed through the thermal image in circuit brakers Q_1 and Q_3 (Figure 19).

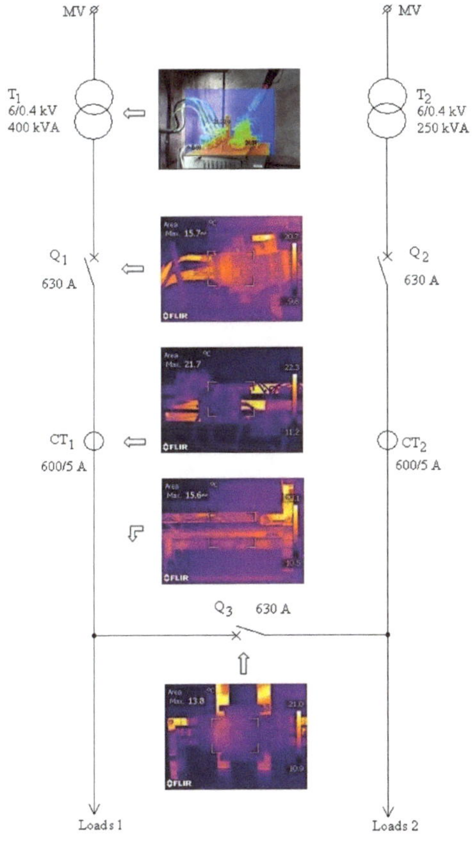

Figure 19. Thermal images in different sections of the power substation.

5. The Experimental Measurements in Power Substation after Uniform Distribution Balanced the Electric Consumers on Phases

The following measuring data were made in the power substation of the residential and educational sectors, after the electrical consumers (i.e., single-phase consumers from classrooms, offices, and libraries) were uniformly distributed and balanced among the three phases in the following situations: Transient measurements (Figures 20–22, during 80 ms), snapshot measurements (Figure 23), and recording measurements (Figures 24 and 25, during 24 h).

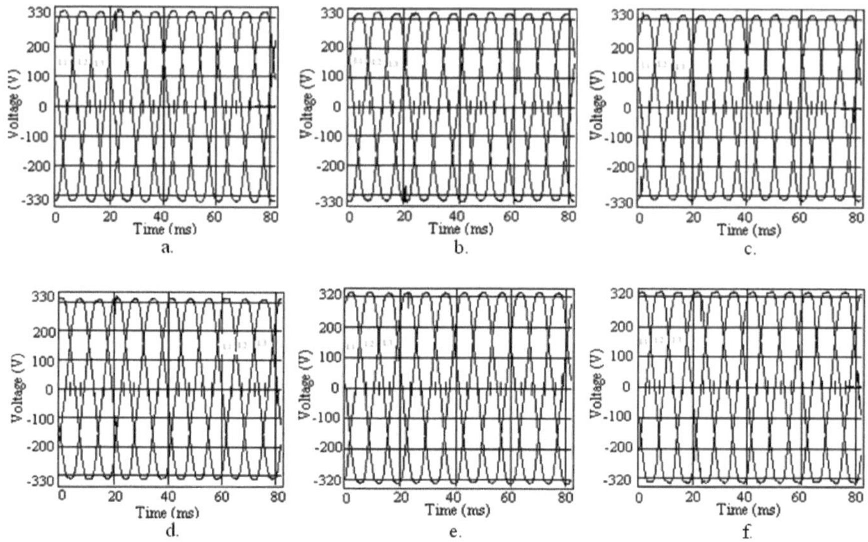

Figure 20. Fast voltage transients from the electrical power substation.

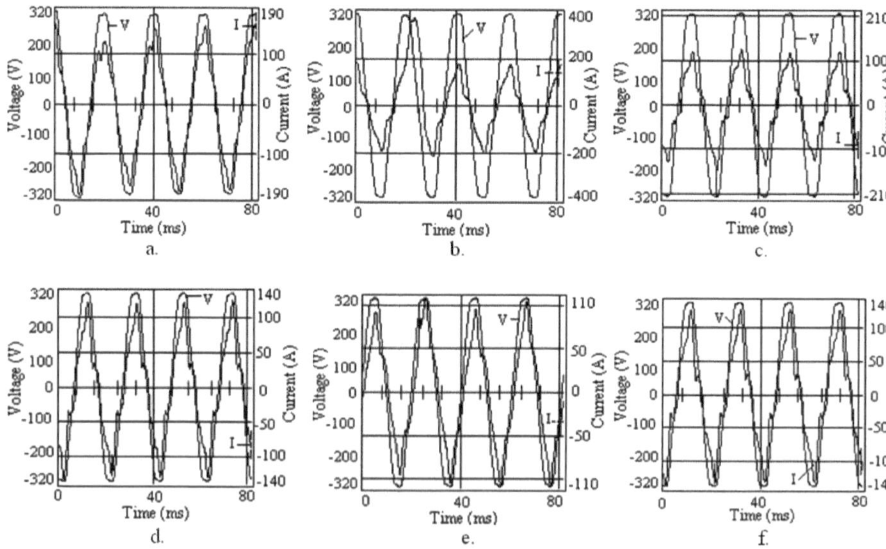

Figure 21. Fast current transients from the electrical power substation.

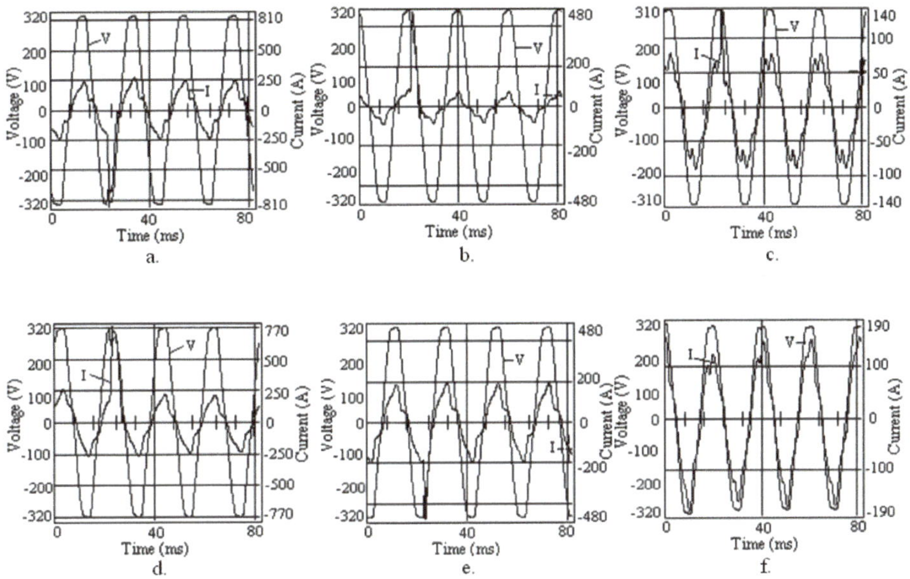

Figure 22. Current and voltage transients from the electrical power substation.

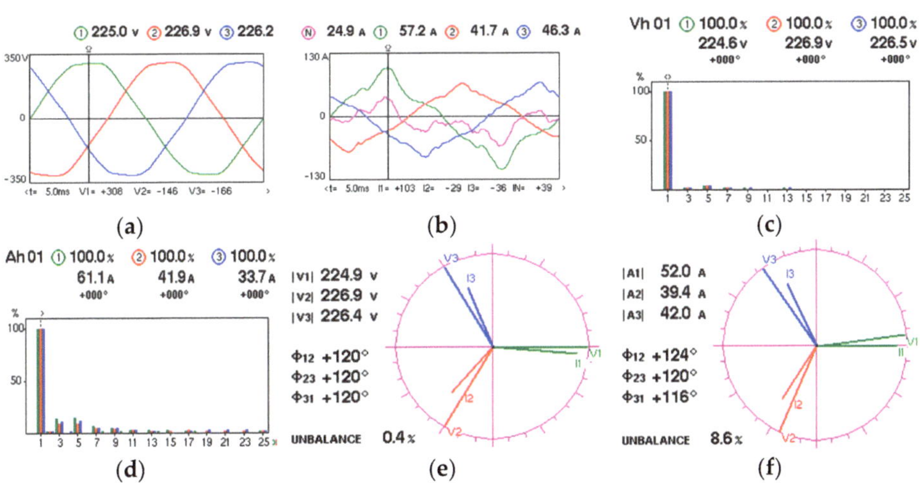

Figure 23. Instantaneous measurements of some electrical measurement from power substation of residential and educational area in a workday: (**a**) Three-phase voltages; (**b**) Three-phase currents; (**c**) Harmonic spectrum of phase voltages; (**d**) Harmonic spectrum of phase currents; (**e**) Fresnel diagram for phase voltages; (**f**) Fresnel diagram for phase currents.

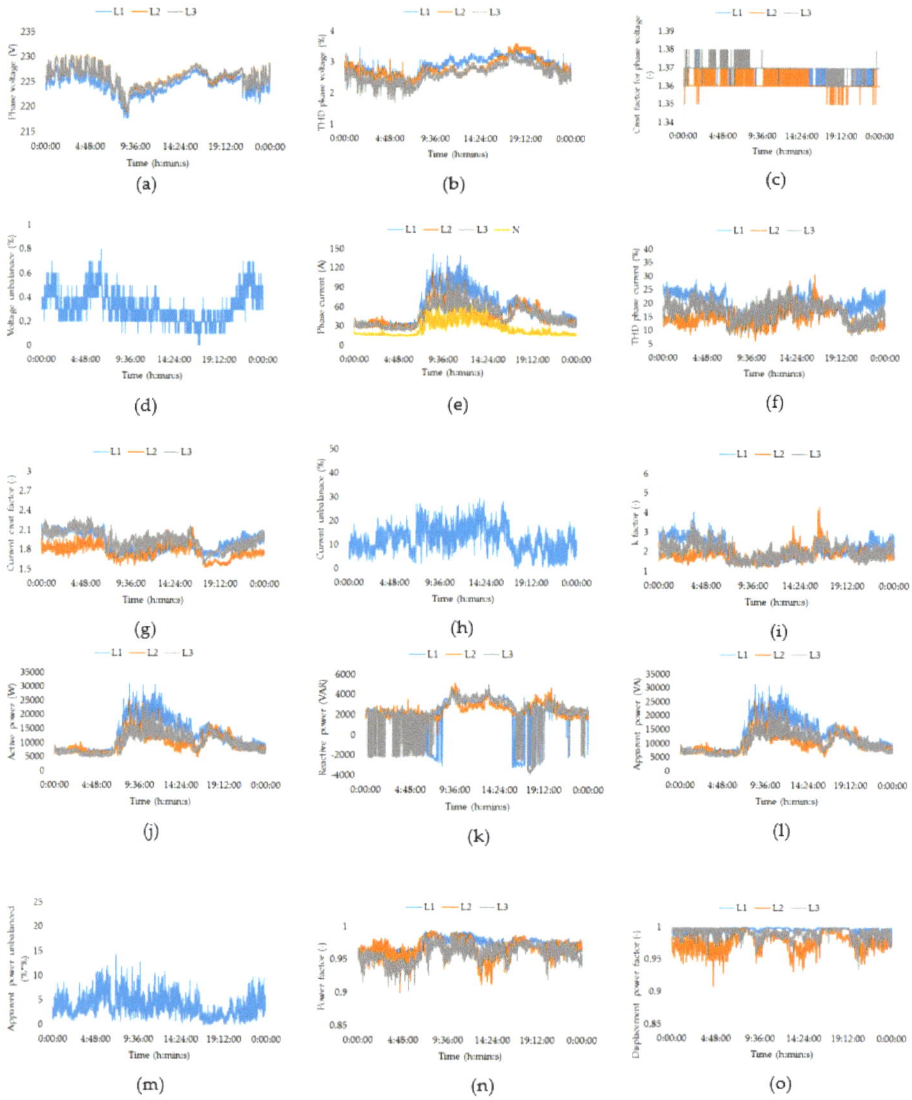

Figure 24. Recordings of some electrical measurement from power substation of residential and educational area in a workday: (**a**) Phase voltages; (**b**) THD for phase voltages; (**c**) Crest factor for voltages; (**d**) Voltage unbalance; (**e**) Phase currents; (**f**) THD for phase currents; (**g**) Crest factor for currents; (**h**) Current unbalance; (**i**) K-factor; (**j**) Active power; (**k**) Reactive power; (**l**) Apparent power; (**m**) Apparent power unbalance; (**n**) Power factor; (**o**) Displacement power factor (for fundamental harmonics).

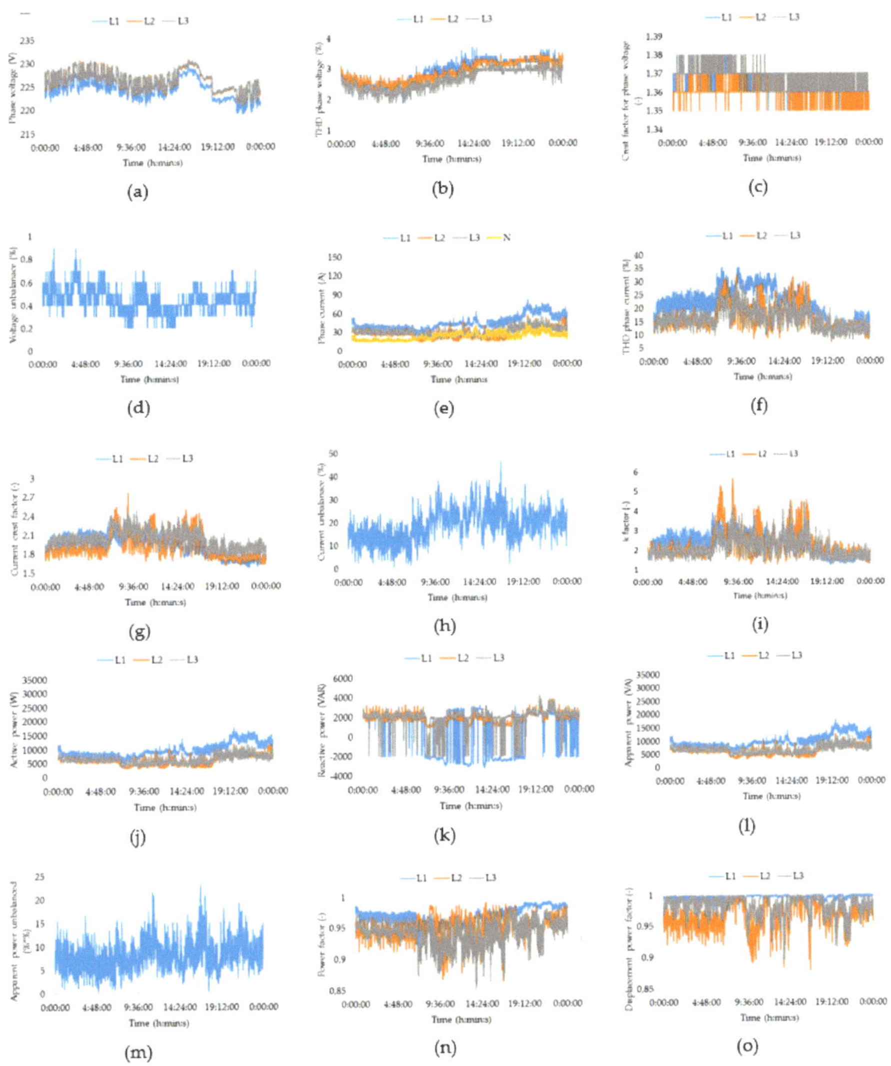

Figure 25. Recordings of some electrical measurement from power substation of residential and educational area on a weekend day: (**a**) Phase voltages; (**b**) THD for phase voltages; (**c**) Crest factor for voltages; (**d**) Voltage unbalance; (**e**) Phase currents; (**f**) THD for phase currents; (**g**) Crest factor for currents; (**h**) Current unbalance; (**i**) *K*-factor; (**j**) Active power; (**k**) Reactive power; (**l**) Apparent power; (**m**) Apparent power unbalance; (**n**) Power factor; (**o**) Displacement power factor (for fundamental harmonics).

During the operation of low-voltage electrical networks, transient events may occur that can change the waveforms of voltages and/or currents.

The measurement of transient waveforms from Figure 20 was made at the three voltage phases (L_1-0, L_2-0, L_3-0), and the transients from Figures 21 and 22 were measured only at one phase (L_1-0).

The different current and voltage switching disturbances are presented in Figures 20–22. The disturbances can be classified in fast and slow switching disturbances. Some impulses may be imposed on the current shape during the switching. Changing the current waveform without significant change of voltage supply (Figure 21) may be due to the start-up of electrical equipment that require time to reach nominal operating conditions (e.g., power transformer, electric motors, classical fluorescent lamps).

The waveforms from Figure 23 show instantaneous measurements of some electrical quantities from the power substation of the residential and educational area on a workday: three-phase voltages, three-phase currents, harmonic spectrum of phase voltages, harmonic spectrum of phase currents, the Fresnel diagram for phase voltages, and the Fresnel diagram for phase currents. The phase shifts between the voltages and currents, at each phase, were small, showing a weak inductive character and a high power factor.

The deforming voltages and currents waveforms from PCC are an effect of all of the nonlinear consumers (especially single-phase) connected in different places in the low-voltage network.

Further, we present the recordings of some electrical measurements from the power substation of the residential and educational area on a workday (Figure 24) and weekend day (Figure 25): phase voltages, THD for phase voltages, crest factor for voltages, voltage unbalance, phase currents, THD for phase currents, crest factor for currents, current unbalance, *K*-factor, active power, reactive power, apparent power, apparent power unbalance, power factor, and displacement power factor (for fundamental harmonics). For recording measurements (Figures 24 and 25) using the power quality analyzer, the sampling period (that is, an integration period in which we computed the average value) was 5 s.

Below, for the analyzed workday and weekend day, we present the main values (minimum, average, maximum, minimum/maximum) of active powers (Table 1), the main values (minimum, average, maximum, minimum/maximum) of reactive powers (absolute values, Table 2), the main values (minimum, average, maximum, maximum/average) of currents (Table 3), the average RMS effective current (Ie) and relative errors (ε_r) (Table 4), the main values (minimum, average, maximum, standard deviation) of voltage unbalance (V_{unb}), current unbalance (I_{unb}) and unbalance factor of apparent power (S_{unb}) (Table 5), the main values (minimum, average, maximum, standard deviation) of power factors (PF, Table 6) and displacement power factors (DPF, Table 7), respectively, the duration of maximum power usage (consumption), (T_{max}), and the duration of maximum losses (τ_{max}, for t = 24 h, Table 8).

Table 1. The main values of active powers.

	Workday			Weekend Day		
	L_1	L_2	L_3	L_1	L_2	L_3
P_{min} (kW)	6.193	5.147	4.6	4.95	3.028	3.422
P_{avg} (kW)	12.972	10.116	10.224	9.506	6.058	6.282
P_{max} (kW)	31.319	24.946	24.482	17.736	11.836	12.011
k_{uP} (-) P_{min}/P_{max}	0.198	0.206	0.188	0.279	0.255	0.285

Table 2. The main values of reactive powers (absolute values).

	Workday			Weekend Day		
	L_1	L_2	L_3	L_1	L_2	L_3
Q_{min} (kVAR)	0.001	0.101	0.001	0.001	0.478	0.001
Q_{avg} (kVAR)	2.059	2.53	2.063	0.634	1.927	1.849
Q_{max} (kVAR)	4.442	5.192	5.087	3.304	4.15	4.03
k_{uQ} (-) Q_{min}/Q_{max}	0	0	0	0	0.115	0

Table 3. The main values of currents.

	Workday				Weekend Day			
	L_1	L_2	L_3	N	L_1	L_2	L_3	N
I_{min} (A)	28.9	23.7	22.5	15.1	24.1	14.2	17.2	14.8
I_{avg} (A)	59.21	46.26	46.964	26.5	43.65	28.138	29.433	23.317
I_{max} (A)	143.1	113.3	111.1	63.2	81	56.1	56.1	43.9
K_F (-) I_{max}/I_{avg}	2.416	2.449	2.365	2.384	1.855	1.993	1.906	1.882

Table 4. The average RMS effective current I_e and relative errors.

	Workday				Weekend Day			
	I_e = 53.872 A				I_e = 37.321 A			
	L_1	L_2	L_3	N	L_1	L_2	L_3	N
ε_r (%)	9.9	−14.128	−12.823	−50.8	16.955	−24.605	−21.135	−37.523

Table 5. The main values of V_{unb}, I_{unb}, and S_{unb}.

	Workday			Weekend Day		
	V_{unb} (%)	I_{unb} (%)	S_{unb} (%)	V_{unb} (%)	I_{unb} (%)	S_{unb} (%)
MIN	0	0.4	0	0.2	0.9	0.45
AVG	0.319	12.327	3.867	0.437	18.63	7.942
MAX	0.8	29.6	14.34	0.9	46.6	23.3
SDT	0.123	5.563	2.198	0.105	6.921	3.184

Table 6. The main values of power factors (PF) (-).

	Workday			Weekend Day		
	L_1	L_2	L_3	L_1	L_2	L_3
MIN	0.945	0.9	0.902	0.907	0.865	0.853
AVG	0.973	0.965	0.959	0.967	0.949	0.938
MAX	0.995	0.994	0.99	0.992	0.99	0.981
SDT	0.011	0.014	0.014	0.015	0.02	0.021

Table 7. The main values of displacement power factors (DPF) (-).

	Workday			Weekend Day		
	L_1	L_2	L_3	L_1	L_2	L_3
MIN	0.965	0.908	0.935	0.971	0.88	0.897
AVG	0.996	0.977	0.991	0.996	0.964	0.983
MAX	1	1	1	1	0.999	1
SDT	0.004	0.013	0.008	0.004	0.021	0.015

Table 8. The duration of maximum power usage T_{max} and the duration of maximum losses τ_{max} (t = 24 h).

	Workday			Weekend Day		
	L_1	L_2	L_3	L_1	L_2	L_3
T_{max} (h)	9.928	9.797	10.143	12.933	12,037	12.591
τ_{max} (h)	4.82	4.49	4.798	7.403	6.477	6.891

6. Discussions

The measurements from power substation of the residential and educational buildings revealed that the currents were distorted (the higher current harmonics orders were 5, 3, 7, 9) and, through the neutral conductor, there was a considerable current (approximately 50% of the phase currents, Figure 23).

Single-phase electrical consumers had a resistive-inductive (R-L) character, and the current unbalance was an order of magnitude larger than the voltage unbalance.

When analyzing the quantities measured over a time interval of 24 h, on a workday and weekend day (Figures 24 and 25, Tables 1–7), it was found that the voltages had approximately the same evolution, except for the time interval 8–14 and 16–20 of the workday, when the voltage losses were more important (4–5 V). The THD for voltages and the crest factor of voltages had the same evolution in the same field (Figures 24 and 25a–c). During the daylight, the voltage decreased (by a few volts) and the currents increased due to the large number of electric consumers in operation (especially single-phase). Consequently, there was a voltage sag on the grid. During the night, only the refrigerators, air-conditioning systems, and some electric lamps (sodium-vapor lamp) were connected to the network. It was found that the voltage unbalance was slightly higher on the weekend day (Figures 24 and 25d). In statistics, the standard deviation (SDT) is a measure of the amount of variation or dispersion of a set of values. From Table 5, it can be seen that the SDT for the current unbalance I_{unb} and apparent power unbalance S_{unb} had lower values for the workday vompared to the weekend day. The voltage unbalance was almost the same for both days. A low standard deviation indicates that the values tend to be close to the mean value, while a high standard deviation indicates that the values are spread out over a wider range. The same conclusions can be made from Tables 6 and 7, as STD for PF and DPF had smaller values for the workday.

There are important differences between the evolutions and values of the currents, on the analyzed days (workday and weekend day). On the workday, the currents were 3–4-times higher in the active intervals 8–14, 16–20 compared to the weekend day (Figures 24e and 25e). The THD for the currents and the crest factor were higher on the weekend day, compared to the workday (Figure 24f,g and Figure 25f,g).

The unbalance of the currents was 5–10% higher on the weekend day (when the PCs in the laboratories did not work) compared to the workday (Figures 24h and 25h).

On the workday, the *K*-factor for the current did not exceed 4. Instead, on the weekend day, the *K*-factor exceeded 4 for several hours (Figures 24i and 25i).

The active power on the workday in the interval of the active program (8–14, 16–20) was greater than 2–2.5-times the active power on the week-nd day (Figures 24l and 25l), and the evolution of the active power differed from a day to another.

There were also differences in the reactive powers where different evolutions were registered from one day to another, with inductive and capacitive character (approximately the same values) and permanent modifications (Figures 24k and 25k).

The evolution of the apparent power in time was similar to the evolutions of the active powers on the same day (Figures 24l and 25l).

The unbalance of the apparent power was almost double between the weekend day (when the PCs in the laboratories did not work) compared to the workday (Figures 24m and 25m).

The power factor was higher, on average, by 0.02–0.03 on the workday compared to the weekend day, and the evolutions of the power factor differed in time from one day to another (Figures 24m and 25m).

The fundamental power factor had higher values than the power factor, being higher on the workday (over 0.95) compared to the weekend day (over 0.9). The differences between the two parameters (PF and DPF) show the large harmonic component (especially of the current, Figures 24o and 25o).

The results of the measurements on the workday and weekend day are summarized, for the most important parameters, in Tables 1–7. The conclusions are similar to those presented above.

In Table 8, we show that the duration of maximum power usage T_{max} and the duration of maximum losses τ_{max} (t = 24 h) were higher (with 30–40%) on the weekend day than on the workday.

During the experiments shown in Figures 20–25, we did not use the power factor regulator with capacitor banks connected to the power substation. From Figures 24 and 25, it can be seen that the power factor was high without using capacitor banks to improve the power factor. The only capacitive consumers used in the experiments were the PC with SMPS power sources. Usually, the capacitor compensation to improve the power factor will increase the voltage and, especially, the currents harmonic components in the network.

The voltage and current unbalance in three-phase power systems occurs due to the unbalanced electrical consumer connection or the persistence of unrepaired faults (e.g., two-phase operation of three-phase consumers).

In addition to household single-phase electrical consumers, which are very numerous but small-power, the main large-power consumers from the industry that cause unbalances in power system are mainly single-phase induction furnaces, arc furnaces, railway traction, etc. Although all power systems are provided with balancing equipment, the equipment may be missing or malfunctioning due to changes in the parameters and operations for these electrical consumers.

Nowadays, power systems do not have additional measurement systems for supplementary circulation of electricity and associated losses conditioned by the presence of unbalance.

The current measuring equipment of powers and energies from power systems do not distinguish between balanced and unbalanced electrical consumers, because measured powers and energies are actually received by consumers.

In this way, the balanced consumers, which are connected to a network with unbalanced consumers, become active power unbalance consumers and record a higher power consumption than actually useful, reducing their overall performance. Also, the losses in supply conductors (own technological consumption) increase. Although these losses are caused by unbalanced consumers, the increases are supported by the power systems.

The unbalanced operation of electrical consumers has following effects [28,31,32,36]:

- The production of additional technological consumptions in both consumer and power system networks;
- A reverse rotating field occurs in the electrical machines;
- Unbalanced operation negatively influences the electrical energy measurement using three-phase classical induction meters;
- Asymmetrical (unbalanced) operation has negative influences on electric motors, because besides vibration and braking, due the reverse rotating field. It also reduces their lifetime due to additional heat produced by this operating condition. For example, if the voltage of an electric motor has a degree of unbalance of 4% instead of 2%, the service life can be halved.

International standards (e.g., EN-50160 and the IEC 1000-3-x series) give limits for the unbalance ratio, defined by (2) of <2% for low voltage and medium voltage systems, and <1% for high voltage, measured as 10 min intervals, with an instantaneous maximum of 4%. The IEC standards recommend that, in general, single-phase consumers should not be connected to three-phase, low-voltage circuits supplying equipment sensitive to phase-voltage unbalance. Instead, a separate circuit should be used.

7. Conclusions

We analyzed the single-phase electric consumers in the residential and educational sectors. In the residential and educational areas, there are a lot of nonlinear single-phase electrical consumers that deform the current waveform more or less (less the voltages waveforms) depending on its nonlinear characteristics.

The unbalancing of nonlinear consumers leads to additional losses in the LV and MV electrical network, negatively affecting the power transformer from the power substation and the measurement of the electric energy of the three-phase electric consumers (e.g., variable speed drives) and single-phase electric consumers (e.g., PCs, laptops).

Today in the residential and educational sectors, there are a lot of small-power consumers with nonlinear characteristics. On the one hand, single-phase nonlinear electrical consumers have a resistive-inductive character (e.g., induction motors, classical fluorescent lamps), and on the other hand, single-phase nonlinear electrical consumers have a capacitive character (e.g., PCs, laptops, compact fluorescent lamps). Of course, in practice, it is impossible to achieve a perfect balance of single-phase consumer in the three phases.

By uniformly arranging (within the technical possibilities) electrical consumers among the three phases and using capacitive electrical consumers (with SMPS, e.g., PCs), balance was achieved. Also, we achieved the relative balance of the current and voltage, respectively, as well as the power factor improvement, without using fixed capacitor banks or a power factor controller with capacitors banks connected to the PCC of the power substations. Also, the type of electrical consumers (SMPS type and R-L type) and the amplitude of the supply voltage changed the unbalance of the voltage to a lesser extent, but the unbalance of current, PF, and DPF were strongly modified.

In order to improve the quality of electrical energy, as future research directions, three-phase interharmonic L-C passive filters (in different configurations), connected in the PCC on LV of the power substation, can be dimensioned and experimential in order to decrease the THD for currents on each phase. Online monitoring of the electrical parameters of the power substation can be achieved by the permanent measurement of voltages, currents, and phase shifts using power quality analyzers to set alarms to warn of exceeding limit values (e.g., unbalance current, power factors) of measured and calculated electrical parameters. An automation system can be developed that, depending on the unbalance of electrical consumers of each phase, introduces or removes larger groups of electrical consumers from one phase to another.

Author Contributions: "Introduction" G.N.P., and C.M.D.; "A Review of Power Quality Issues Related to Electrical Power Distribution" A.I.; "Laboratory Experimental Study in Unbalance Condition of Electric Loads Used In Residential and Educational Buildings", "A Case Study about Residential and Educational Electrical Grid Distribution" and "The Experimental Measurements in Power Substation after Uniform Distribution Balanced the Electric Consumers on Phases" G.N.P., A.I., and C.M.D.; "Discussions" and ""Conclusions"" G.N.P.; writing, review, editing, and supervision G.N.P., and A.I. All authors have read and agreed to the published version of the manuscript.

Funding: This research received no external funding.

Acknowledgments: The authors would like to thank Ioan Hodor, engineer, for interesting discussions about the topic of the paper.

Conflicts of Interest: The authors declare no conflict of interest.

References

1. Shin, Y.-J.; Powers, E.J.; Grady, M.; Arapostathis, A. Power quality indices for transient disturbances. *IEEE Trans. Power Deliv.* **2005**, *21*, 253–261. [CrossRef]
2. Copper Development Association. *Power Quality Application Guide*; Copper Development Association: Hemel Hempstead, UK, 2001; Available online: https://copperalliance.org.uk/knowledge-base/resource-library/power-quality-utilisation-guide/ (accessed on 11 December 2020).
3. Costa, L.F.D.O.; Filho, J.M.D.C. Electrical power quality and the challenges faced by power assemblies applications in petrochemical industry. *IEEE Trans. Ind. Appl.* **2016**, *52*, 4495–4502. [CrossRef]
4. Rodríguez, A.; Bueno, E.J.; Mayor, A.; Sanchez, F.J.R.; García-Cerrada, A. Voltage support provided by statcom in unbalanced power systems. *Energies* **2014**, *7*, 1003–1026. [CrossRef]
5. Kiranmai, S.A.; Laxmi, A.J.; Ai, Q. Hardware for classification of power quality problems in three phase system using Microcontroller. *Cogent Eng.* **2017**, *4*, 1–11. [CrossRef]
6. Balasubramaniam, P.M.; Prabha, S.U. Power quality issues, solutions and standards: A technology review. *J. Appl. Sci. Eng.* **2015**, *18*, 371–380.
7. Reid, W.E. Power quality issues-standards and guidelines. *IEEE Trans. Ind. Appl.* **1996**, *32*, 625–632. [CrossRef]

8. Lindig, S.; Louwen, A.; Moser, D.; Topic, M. Outdoor PV system monitoring-input data quality, data imputation and filtering approaches. *Energies* **2020**, *13*, 5099. [CrossRef]
9. Iagăr, A.; Popa, G.N.; Diniș, C.M. *Electric Power Quality—From Theory to Experiments*; Politehnica Publishing House: Timișoara, Romania, 2017. (In Romanian)
10. Tenti, P.; Mattavelli, P. A time-domain approach to power terms definitions under non-sinusoidal conditions. In Proceedings of the 6th International Workshop on Power Definitions and Measurement under Non-Sinusoidal Conditions, Milan, Italy, 13–15 October 2003.
11. Akagi, H.; Kanazawa, Y.; Nabae, A. Instantaneous reactive power compensators comprising switching devices without energy storage components. *IEEE Trans. Ind. Appl.* **1984**, *20*, 625–630. [CrossRef]
12. Mehebub, A.; Mandela, G. Power quality problems and solutions: An overview. *Int. J. Sci. Res.* **2014**, *3*, 1024–1030.
13. Czarnecki, L.S. Considerations on the reactive power in nonsinusoidal situations. *IEEE Trans. Instrum. Meas.* **1985**, *34*, 399–404. [CrossRef]
14. Malengret, M.; Gaunt, C.T. Active currents, power factor, and apparent power for practical power delivery systems. *IEEE Access* **2020**, *8*, 133095–133113. [CrossRef]
15. Ahmad, I.; Fandi, G.; Muller, Z.; Tlustý, J. Voltage quality and power factor improvement in smart grids using controlled DG units. *Energies* **2019**, *12*, 3433. [CrossRef]
16. Czarnecki, L.S. Orthogonal decomposition of the current in a three-phase non-linear asymmetrical circuit with nonsinusoidal voltage. *IEEE Trans. Instrum. Meas.* **1988**, *37*, 30–34. [CrossRef]
17. Olivares, J.C.; Liu, Y.; Canedo, J.; Escarela-Perez, R.; Driesen, J.; Moreno, P. Reducing losses in distribution transformers. *IEEE Trans. Power Deliv.* **2003**, *18*, 821–826. [CrossRef]
18. Seguí-Chilet, S.; Gimeno-Sales, F.; Orts, S.; Garcera, G.; Figueres, E.; Fillol, M.A.; Masot, R. Approach to unbalance power active compensation under linear load unbalances and fundamental voltage asymmetries. *Int. J. Electr. Power Energy Syst.* **2007**, *29*, 526–539. [CrossRef]
19. Tofoli, F.L.; Sanhueza, S.; De Oliveira, A. On the study of losses in cables and transformers in nonsinusoidal conditions. *IEEE Trans. Power Deliv.* **2006**, *21*, 971–978. [CrossRef]
20. Prakash, P.S.; Kalpana, R.; Singh, B.; Bhuvaneswari, G. Power quality improvement in utility interactive based ac–dc converter using harmonic current injection technique. *IEEE Trans. Ind. Appl.* **2018**, *54*, 5355–5366. [CrossRef]
21. Ozsoy, E.; Padmanaban, S.; Mihet-Popa, L.; Fedák, V.; Ahmad, F.; Akhtar, R.; Sabanovic, A. Control strategy for a grid-connected inverter under unbalanced network conditions—A disturbance observer-based decoupled current approach. *Energies* **2017**, *10*, 1067. [CrossRef]
22. Czarnecki, L.S.; Haley, P.M. Power properties of four-wire systems at nonsinusoidal supply voltage. *IEEE Trans. Power Deliv.* **2016**, *31*, 513–521. [CrossRef]
23. Mohammed, N.; Ciobotaru, M.; Town, G.E. Online parametric estimation of grid impedance under unbalanced grid conditions. *Energies* **2019**, *12*, 4752. [CrossRef]
24. Elnady, A.; Liu, Y.-F. A practical solution for the current and voltage fluctuation in power systems. *IEEE Trans. Power Deliv.* **2012**, *27*, 1339–1349. [CrossRef]
25. Pomilio, J.A.; Deckmann, S.M. Characterization and compensation of harmonics and reactive power of residential and commercial loads. *IEEE Trans. Power Deliv.* **2007**, *22*, 1049–1055. [CrossRef]
26. Serrano-Fontova, A.; Casals, P.; Bosch, R. Power quality disturbances assessment during unintentional islanding scenarios. A contribution to voltage sag studies. *Energies* **2019**, *12*, 3198. [CrossRef]
27. Chang, G. Characterizing harmonic currents generated by fluorescent lamps in harmonic domain. *IEEE Trans. Power Deliv.* **2003**, *18*, 1583–1585. [CrossRef]
28. Wang, Y.J.; O'Connel, R.M.; Brownfield, G. Modeling and prediction of distribution system voltage distorsion caused by nonlinear residential loads. *IEEE Trans. Power Deliv.* **2001**, *16*, 744–751. [CrossRef]
29. Jaiswal, G.C.; Ballal, M.S.; Tutakne, D.R.; Suryawanshi, H.M. Impact of power quality of the performane of distribution transformers. *IEEE Ind. Appl. Mag.* **2019**, *25*, 8–17. [CrossRef]
30. Moore, P.J.; Portugués, I. The influence of personal computer processing modes on line current harmonics. *IEEE Trans. Power Deliv.* **2003**, *18*, 1363–1368. [CrossRef]
31. Zheng, T.; Makram, E.B.; Girgis, A.A. Evaluating power system unbalance in the presence of harmonic distortion. *IEEE Trans. Power Deliv.* **2003**, *18*, 393–397. [CrossRef]
32. Hossain, E.; Tur, M.R.; Padmanaban, S.; Ay, S.; Khan, I. Analysis and mitigation of power quality issues in distributed generation systems using custom power devices. *IEEE Access* **2018**, *6*, 16816–16833. [CrossRef]
33. Masoum, M.A.S.; Moses, P.S.; Masoum, A.S. Derating of asymmetric three-phase transformers serving unbalanced nonlinear loads. *IEEE Trans. Power Deliv.* **2008**, *23*, 2033–2041. [CrossRef]
34. Czarnecki, L.S.; Haley, P.M. Unbalanced power in four-wire systems and its reactive compensation. *IEEE Trans. Power Deliv.* **2015**, *30*, 53–63. [CrossRef]
35. Montoya, F.G.; Baños, R.; Alcayde, A.; Montoya, M.G.; Manzano-Agugliaro, F. Power quality: Scientific collaboration networks and research trends. *Energies* **2018**, *11*, 2067. [CrossRef]
36. Casolino, G.M.; Losi, A. Load areas in radial unbalanced distribution systems. *Energies* **2019**, *12*, 3030. [CrossRef]

37. Coman, C.M.; Florescu, A.; Oancea, C.D. Improving the effciency and sustainability of power systems using distributed power factor correction methods. *Sustainability* **2020**, *12*, 3134. [CrossRef]
38. Aziz, M.A.; El-Zahab, E.A.; Zobaa, A.F. Power factor and your electrical utility bill in egypt. *IEEE Trans. Power Deliv.* **2003**, *18*, 1567–1568. [CrossRef]
39. Katsaprakakis, D.A.; Christakis, D.G.; Zervos, A.; Voutsinas, S. A power-quality measure. *IEEE Trans. Power Deliv.* **2008**, *23*, 553–561. [CrossRef]
40. König, W.; Löbbe, S.; Büttner, S.; Schneider, C. Establishing energy efficiency—Drivers for energy efficiency in german manufacturing small- and medium-sized enterprises. *Energies* **2020**, *13*, 5144. [CrossRef]
41. Pakere, I.; Lauka, D.; Blumberga, D. Does the balance exist between cost efficiency of different energy efficiency measures? DH systems case. *Energies* **2020**, *13*, 5151. [CrossRef]
42. Mendonça, H.; De Castro, R.M.; Martínez, S.; Montalbán, D. Voltage impact of a wave energy converter on an unbalanced distribution grid and corrective actions. *Sustainability* **2017**, *9*, 1844. [CrossRef]
43. Dell'Aquila, A.; Marinelli, M.; Monopoli, V.; Zanchetta, P. New power-quality assessment criteria for supply systems under unbalanced and nonsinusoidal conditions. *IEEE Trans. Power Deliv.* **2004**, *19*, 1284–1290. [CrossRef]
44. Abdelaziz, M.; AbouEl-Zahab, E.-D.; Ibrahim, A.; Zobaa, A.F. Practical considerations regarding power factor for nonlinear loads. *IEEE Trans. Power Deliv.* **2004**, *19*, 337–341. [CrossRef]
45. Emanuel, A.E. On the assessment of harmonic pollution [of power systems]. *IEEE Trans. Power Deliv.* **1995**, *10*, 1693–1698. [CrossRef]
46. Von Jouanne, A.; Banerjee, B. Assessment of voltage unbalance. *IEEE Trans. Power Deliv.* **2001**, *16*, 782–790. [CrossRef]
47. Emanuel, A.E. *Power Definitions and the Physical Mechanism of Power Flow*; Wiley-IEEE Press: Hoboken, NJ, USA, 2010.
48. Golovanov, N.; Postolache, P.; Toader, C. *Energy Efficiency and Power Quality*; AGIR Publishing House: Bucharest, Romania, 2007. (In Romanian)
49. Ionescu, T.G.; Pop, O. *Engineering of Power Electrical Distribution*; Technical Publishing House: Bucharest, Romania, 1998. (In Romanian)
50. Emanuel, A.E. Apparent power definitions for three-phase systems. *IEEE Trans. Power Deliv.* **1999**, *14*, 767–772. [CrossRef]

MDPI AG
Grosspeteranlage 5
4052 Basel
Switzerland
Tel.: +41 61 683 77 34

Energies Editorial Office
E-mail: energies@mdpi.com
www.mdpi.com/journal/energies

www.ingramcontent.com/pod-product-compliance
Lightning Source LLC
Chambersburg PA
CBHW041402210926
51850CB00013B/3159
9783725839124